161 027208 9
AF616170
Withdrawn

Chemistry for the Energy Future

INTERNATIONAL UNION OF PURE AND APPLIED CHEMISTRY

Chemistry for the Energy Future

A 'Chemistry for the 21st Century' monograph

EDITED BY

V.N. PARMON
Boreskov Institute of Catalysis
Siberian Branch of the Russian Academy of Sciences
Novosibirsk, Russia

H. TRIBUTSCH
Freie Universität Berlin and
Hahn-Meitner Institut
Dept. Solare Energetik
Berlin, Germany

A.V. BRIDGWATER
Bio-Energy Research Group
Chemical Engineering
Aston University
Birmingham, UK

D.O. HALL
Life Sciences Division
King's College London
UK

b

Blackwell
Science

Blackwell Science Ltd
Editorial Offices:
Osney Mead, Oxford OX2 0EL
25 John Street, London WC1N 2BL
23 Ainslie Place, Edinburgh EH3 6AJ
350 Main Street, Malden
MA 02148 5018, USA
54 University Street, Carlton
Victoria 3053, Australia
10, rue Casimir Delavigne
75006 Paris, France

Other Editorial Offices:
Blackwell Wissenschafts-Verlag GmbH
Kurfürstendamm 57
10707 Berlin, Germany

Blackwell Science KK
MG Kodenmacho Building
7–10 Kodenmacho Nihombashi
Chuo-ku, Tokyo 104, Japan

First published 1999

Set by Graphicraft Limited, Hong Kong
Printed and bound in Great Britain
by MPG Books Ltd, Bodmin, Cornwall

A catalogue record for this title is available from the British Library

ISBN 0-632-05269-4

Library of Congress
Cataloging-in-publication Data

Chemistry for the energy future : a chemistry for the 21st century monograph / edited by V.N. Parmon; H. Tributsch, A. Bridgwater, D. Hall.
p. cm.
Includes bibliographical references and index.
ISBN 0-632-05269-4
1. Energy development.
2. Chemistry.
I. Parmon, V. N. (Valentin Nikolaevich)
II. International Union of Pure and Applied Chemistry
III. Series
TJ163.2.C4829 1999
621.042—DC21 98-51140
CIP

DISTRIBUTORS

Marston Book Services Ltd
PO Box 269
Abingdon, Oxon OX14 4YN
(*Orders*: Tel: 01235 465500
Fax: 01235 465555)

USA
Blackwell Science, Inc.
Commerce Place
350 Main Street
Malden, MA 02148 5018
(*Orders*: Tel: 800 759 6102
781 388 8250
Fax: 781 388 8255)

Canada
Login Brothers Book Company
324 Saulteaux Crescent
Winnipeg, Manitoba R3J 3T2
(*Orders*: Tel: 204 837-2987)

Australia
Blackwell Science Pty Ltd
54 University Street
Carlton, Victoria 3053
(*Orders*: Tel: 3 9347 0300
Fax: 3 9347 5001)

For further information on Blackwell Science, visit our website:
www.blackwell-science.com

To the memory of Kirill Zamaraev,
the former IUPAC President,
who initiated the writing of this book but
unfortunately failed to see it published

Contents

Contributors

A.V. BRIDGWATER *Bio-Energy Research Group, Chemical Engineering, Aston University, Aston Triangle, Birmingham B4 7ET, UK*

D.O. HALL *Life Sciences Division, King's College London, Campden Hill Road, London W8 7AH, UK*

V.N. PARMON *Boreskov Institute of Catalysis, Siberian Branch of the Russian Academy of Sciences, Prospekt Akademika Lavrentieva 5, Novosibirsk 630090, Russia*

W. PLIETH *Director of the Institute of Physical Chemistry and Electrochemistry, Bergstraße 66b, 01062 Dresden, Germany*

D. RAHNER *Institute of Physical Chemistry and Electrochemistry, Bergstraße 66b, 01062 Dresden, Germany*

F. ROSILLO-CALLE *Life Sciences Division, King's College London, Campden Hill Road, London W8 7AH, UK*

H. TRIBUTSCH *Freie Universität Berlin and Hahn-Meitner Institut, Dept. Solare Energetik, 14109 Berlin, Germany*

Preface

The writing of this book was initiated by Professor Kirill Zamaraev, the President of IUPAC from 1993 to 1995. The main idea of this project was to demonstrate to the community of chemists how important chemical science will be in solving one of the most important problems of society, which is to provide inexhaustible and sustainable sources of energy. The book has been written by an international team comprising Professors V.N. Parmon (Russia), H. Tributsch (Germany), A.V. Bridgwater (UK) and D.O. Hall (UK), who represent different fields of chemistry and chemical technologies.

The book is not exhaustive and does not hope to encompass or discuss every possible contribution of chemistry and chemical technology to the energy problems of the future. However, the authors hope that the book is able to cover the most important directions of these contributions and that it will be of interest not only for chemists, but also for a wider spectrum of society who are concerned about the future of civilization.

The book was written as part of a sponsored IUPAC pool project entitled, 'Alternative scenarios for energy production in the future', and the authors express their cordial thanks to those in IUPAC for their continuous support, especially to the Executive Secretary of IUPAC, Dr M. Williams, who has just retired.

V.N. Parmon
H. Tributsch
A.V. Bridgwater
D.O. Hall

1 Introduction and Statement of the Problem

V.N. PARMON[1], A.V. BRIDGWATER[2], H. TRIBUTSCH[3] and D.O. HALL[4]

[1] *Boreskov Institute of Catalysis, Siberian Branch of the Russian Academy of Sciences, Prospekt Akademika Lavrentieva 5, Novosibirsk 630090, Russia*

[2] *Bio-Energy Research Group, Chemical Engineering, Aston University, Aston Triangle, Birmingham B4 7ET, UK*

[3] *Freie Universität Berlin and Hahn-Meitner Institut, Dept. Solare Energetik, 14109 Berlin, Germany*

[4] *Life Sciences Division, King's College London, Campden Hill Road, London W8 7AH, UK*

Two important phenomena seem to characterize the state of society on the eve of the 21st century:

1 A progressive exhaustion of economically and/or ecologically appropriate stocks of fossil hydrocarbons (firstly oil, and later natural gas and coal).

2 An increasing threat from the global 'greenhouse effect' and the resultant widespread demands to reduce pollution by CO_2 and other greenhouse gases.

These concerns were expressed in the 'Agenda for the XXI Century', adopted by the majority of countries represented at the *Earth Summit '92* in Rio de Janeiro, to elaborate strategic ways for sustainable development of the Earth's future. The concerns are a precursor to the inevitable reduction of the existing hydrocarbon basis for the four main technical pillars of modern civilization: heat, transport (particularly in relation to motor fuel), metallurgy (in relation to fuel and also reducing agents for ores) and the chemical industry (in manufacturing carbon-containing materials and ammonia).

Thus, significant changes are to be expected in the nature of the widely used and/or traditional energy carriers, as well as in the basic use of raw materials for many industries. Over a historic perspective this change will probably be considered as a jump in the evolution of society, similar to the well-known jumps in the evolution of life on our planet related to the stepwise progressive change of the main biochemical energy carriers operating inside living organisms.

The discussion on options for future energy technologies is influenced by short-term commercial, political and emotional arguments. However, independent of these, it may be agreed on rational grounds that the following conditions should be gradually fulfilled for energy generation: all energy systems must ultimately pay for all the costs (including social, political and environmental). This is at present not true either for the fossil energy cycle (generating pollution and the greenhouse effect) or for the nuclear energy cycle (which needs the reprocessing of waste and the insurance of nuclear power plants).

Until such a situation of realistic energy prices is achieved, renewable and non-polluting energies must be supported more than competing conventional fuels because of the growing world population and the limits of tolerance of the environment. Thus, as much renewable energy should be produced as possible and, for the same reasons, saving of energy in all fields of technology should have the highest priority.

What should be the answers to the most important strategic questions for the technical development of society? Should we follow the 'rigid' scenario and construct nuclear power plants to a high degree of safety (which may mean expensive underground power plants) for satisfying our energy needs or should we shift to 'softer' paths and orient much more towards renewable resources of energy and raw materials with biomass as a priority?

What are the appropriate criteria to make a correct choice?

For large-scale energy, a shift to energy resources that are not based on chemical fuels is possible: for example, there could be a predominance of nuclear (both fission and fusion) and solar energy. However, only solar energy is renewable because the generated entropy (heat) is radiated into the universe. Nuclear energy will lead to a gradual accumulation of nuclear waste and pollution, which already today has used more land than may ever be required for solar energy collection. Alternatively, the other three industrial demands (transport, metallurgy and chemical industry) unavoidably use chemical components for their energy basis and thus will involve chemistry for solving their problems.

In the case of metallurgy and transport, a reorientation towards hydrogen as both a reducing agent and a fuel is possible, although this may not necessarily be optimal for customer convenience. Water can serve as an inexhaustible raw material for hydrogen production.

The future basis of raw materials for large-scale chemistry of organic and other carbon-containing materials still remains unclear. Two developments are possible:

1 Reorientation to inexhaustible natural resources of inorganic carbon (in the form of either atmospheric CO_2 or fossil carbonates).

2 Intensification of the use of renewable resources of carbon-containing raw materials: the biomass of green plants as well as industrial and municipal solid wastes (MSW).

The production of hydrogen as well as the use of any of the inorganic resources of carbon inevitably require huge amounts of energy, which can be provided in the future only by nuclear (thermonuclear) or solar energy. A smaller consumption of energy is needed for processing biomass and carbon-containing MSW.

For this reason, one can expect that for sustainable development in the future an intensified use of plant biomass and MSW for the large-scale chemical industry and, probably, for the production of motor fuels and reducing agents for metallurgy would have to dominate. However, we also may expect that in the next century some chemical and catalytic technologies will be important not only for the chemical industry but also for other industries where today they are not so widely applied, e.g. in the production and storage of energy.

In this book, an overview is given of the possible unconventional areas of chemistry and chemical technologies applied to energy production and storage. For this reason, the main topics of the book concern the contribution of chemistry to the utilization of new and renewable resources of chemical energy carriers, and some new technologies for the combustion of fossil fuels. Also considered are the implementation of chemical technologies for nuclear energy, the production of electricity, the conversion and storage of solar energy and the conversion and storage of middle- and low-potential heat energy.

The goal of this book is not to go into details of the present state and perspectives of research and development (R&D) in all areas under discussion, but to discuss the main ideas that might guide those chemists who work at the borderline of chemistry and energetics or related areas.

Catalysis and catalytic technologies are known to play a major role in modern chemical technologies. Also, biological catalysts (enzymes) instigate the most complicated chemical reactions in nature. Therefore, it is no accident that a special emphasis on catalysis should be made in this book too.

On the eve of the 21st century we observe how catalytic technologies are becoming progressively more important in industries other than the chemical and petroleum areas. An obvious example is provided by catalytic technologies for the treatment of exhaust gases from motor vehicles and various industries, including power plants. In the future, new catalytic technologies

are expected to be developed that could influence still more serio
as energy production, metallurgy and biotechnology.

Chemical and especially catalytic technologies may be used
transformation of energy from one form to another is needed.
erally needs energy in the form of heat, electricity or mechanica
available primary energy sources, such as fossil fuels (oil, natural gas, coa
lar, wood), nuclear fuel, solar energy, low- and middle-potential subterranean hea
forms of mechanical energy from hydraulic, tidal or wind energy resources, typically doe
these consumer requirements. Therefore, an important problem is to convert the available primary energy into a suitable form, as well as to accumulate, store and transport the energy services to the site of consumption.

In this book, an overview is given of the possible applications of chemistry in the energy industry of the future. Mainly we discuss technologies that directly facilitate energy conversion processes or allow the utilization of still non-traditional primary energy carriers as well as the production of new energy carriers. We do not discuss various other chemical technologies such as those used for the processing and compounding of fuels or protection of the environment from the wastes of the energy industry; they are important for the energy industry but do not directly provide energy conversion.

As will be evident from the book, the challenge of society in the field of large-scale energy is primarily a challenge for chemists because the major role in the utilization of new sources of energy is played by chemistry.

It is clear that developing sustainable energy systems is an international challenge but the development and utilization of completely sustainable energy systems are more possible in the richer countries who can afford it, with the other countries continuing to use cheap and polluting energy.

A possible strategy towards a sustainable world energy system could involve restructuring of the funding of the United Nations, which could be financed by CO_2 taxes that are proportional to the non-renewable energy consumption of member countries. This would also provide the organization and economic basis for building up an international agency with the task of supporting research, development and implementation of appropriate sustainable energy technologies. The taxation of fuel for international air and ship traffic could be an alternative source of money for finding strategies towards a sustainable world energy system.

There is no doubt that in the future the present developing countries of the 'Third World' should achieve the same level of living standards, primarily in the consumption or availability of energy, as the developed countries have. But the developing world has little money for developing and implementing the newest technologies, which are more expensive than the non-sustainable conventional technologies. Thus, the 21st century will be an opportunity for developing countries to achieve such standards with the newest technologies. Note, too, that in developing countries there is much solar energy potential! Thus, the developed countries should take the lead in the development and adaptation of the more expensive new technologies of the future, and thus themselves bear most of the expenses for changing the technological background in order to transfer these technologies to developing countries.

Can we convince the people and governments of the developed countries to do this? It is possible that for our civilization there is no other choice! Provided that such a thrust for innovation is given, the chemists and the chemical industry are faced with many challenges, some of which are discussed in this book.

2 Strategies for Sustainable Energy Production: the Chemist's Approach

V.N. PARMON

Boreskov Institute of Catalysis, Siberian Branch of the Russian Academy of Sciences, Prospekt Akademika Lavrentieva 5, Novosibirsk 630090, Russia

1 Introduction

A major task of the scientific community for developing a consistent strategy for a sustainable world is to evaluate the status of the main resources for future large-scale energy. This chapter presents an approach for quantitatively estimating the situation for the world's people when the resources of hydrocarbon fossil raw materials are totally depleted and all the energy (including motor fuels) and chemicals need to be produced artificially from new types of resources [1,2]. This situation is rather hypothetical and almost improbable over the next century in view of the potential of geological prospecting, the discovery of new organic deposits and the large amounts (estimated energy equivalent of more than 10^{26} J [3]) of dispersed or difficult to access deposits of raw materials that are nowadays economically or ecologically difficult to exploit. However, the estimates given below may be of interest because they present the viewpoint of the chemist and the chemical industry on the potential of the Earth's resources.

The general scheme of these estimates is:

1 The average consumption of the main types of chemicals per capita is estimated.

2 The minimal energy resources required for the manufacture of these chemicals from expected *initial* raw materials is calculated.

3 The ability of the Earth to supply all its people with these chemicals is evaluated.

The estimates under discussion are based on several assumptions, some of which are objective and determined by the natural characteristics of the Earth and the physical–chemical characteristics of the chemical processes under consideration. Other rather important assumptions are mostly subjective and are based on the present speculations of mankind on its future life. These assumptions apply mainly to the average future consumption per capita of some chemicals and to the predicted total population of the Earth in this period. Below we shall consider this population to be stabilized at the level of 10 billion, which is the higher number given by present-day scenarios.

1.1 *Estimations of future average consumption (per capita) of energy and the main types of chemicals: the 'chemistry vector' of consumption*

A number of sufficiently reliable estimates of average energy consumption per capita and the structure of the primary energy carriers (the 'energy vector') are well known for the world in total and for different countries and groups of countries. This allows us, reasonably reliably, to predict world average energy consumption per capita and its structure in the future when living standards tend to approach the level considered optimal at the present moment for each inhabitant of the Earth. The level and structure of energy consumption in the USA or Western Europe are usually considered to be close to the optimal living standard. Later, we shall use the stabilized actual level of the average energy consumption per capita in the USA as a standard. This level, however, is thought by numerous experts to be excessive (in comparison to Western Europe) and may be

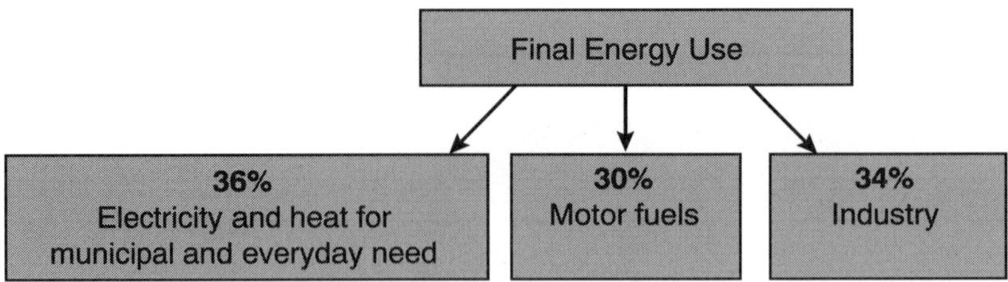

Figure 2.1. Final energy use in the USA. (From [3].) The main *carbon* and *hydrogen* components are present in the energy end-use of the following customers: transport (motor fuels), metallurgy (reducing agents) and the chemical industry (chemicals and construction materials).

considered an extreme limit of human needs. In the beginning of the 1990s, these characteristics were: total consumption of primary energy resources = *c.* 12 tons of oil equivalent (t.o.e.) per year per capita, including 12 000 kWh of electricity per capita [3]. At the end of the 1980s in the USA, 36% of the primary energy resources (see Fig. 2.1) and 34% of the electricity were provided for municipal and everyday needs.

Considerably less attention has been paid to estimating the existing and predicted '*chemistry vector*' of consumption per capita, i.e. the quantitative and structural distribution of the consumption of chemicals (at least of its main components). This should also be estimated for the 'chemistry vector' of energy consumption and the possible limitations for change in the future.

Later we shall consider those components of the 'chemistry vector' of consumption that are presently based on fossil organic raw materials. In future, these components should probably be produced from chemical materials of a different type. Clearly, these components are carbon and hydrogen, with the major part of the latter presently being produced via steam reforming of hydrocarbons.

1.2 *Chemical component of the current and predicted energy consumption*

In the beginning of the 1990s, the total annual production of the primary energy carriers on Earth was estimated at 4×10^{20} J (*c.* 10^{10} t.o.e.) with a population of 5.0 billion [4,5]. This means that each inhabitant consumed *c.* 80×10^{9} J (1.8 t.o.e.), with the composition of the 'input' energy vector shown in Fig. 2.2. For the industrially developed countries (1.2 billion people or 24% of the Earth's population), the total energy consumption was 2.6×10^{20} J (6.0×10^{9} t.o.e. or 66% of the total world energy consumption). It is seen that the average energy consumption per capita in the developed countries is 2.5-fold less than that of the USA. This supports the upper limit prediction based on data for the USA, especially since the assumed limiting world population of 10 billion is an upper estimate.

The composition of the 'input' energy in the distant future seems to include only nuclear (thermonuclear) power and the renewable energy resources: solar energy and its indirect forms such as hydro and wind energy.

A most important further estimate is the composition of the 'output' energy, i.e. the distribution of the final types of energy supplied to the consumer. Indeed, electricity and heat for municipal and everyday needs will not necessarily require chemical energy carriers in the future. At the same time, the use of chemical energy carriers in transport and other industries is preferable, if not inevitable. This is supported by the very high energy storage capacity of chemical energy carriers, as well as their simultaneous application as reducing agents, e.g. in metallurgy for the production

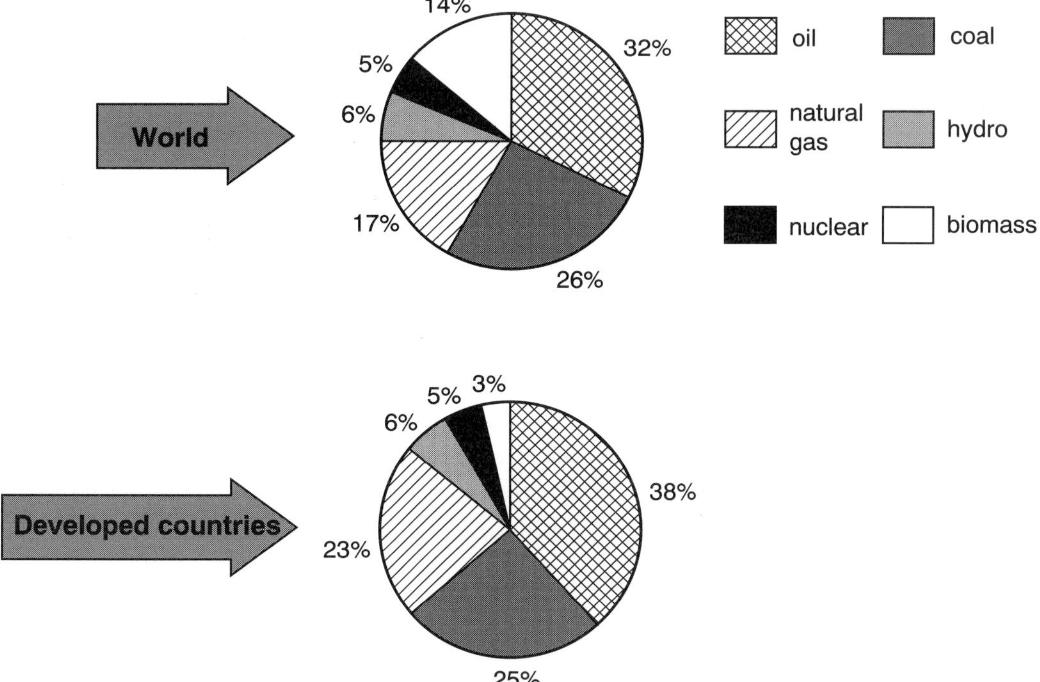

Figure 2.2. Components of the current input energy for consumption. (From [3].)

of metals from iron and nickel ores. In the chemical industry the main part of organic energy carriers is utilized for manufacturing the major chemicals.

1.3 *Chemical component of energy supply for transport (the motor fuels)*

At present in the USA, 30% of the primary energy resources are used as motor fuels in transport. The corresponding energy equivalent of the chemical component of the average energy consumption per capita may be estimated as a minimum of 3 t.o.e. per year. (For 2010, the optimal annual consumption standard for motor fuels in the former USSR is planned to attain 1 t.o.e. per capita [6].) If hydrocarbons are used as motor fuels, 3 t.o.e. corresponds to 214 kg-atoms of carbon accompanied by the same amount of kg-molecules of hydrogen (H_2). If one of the main ideas of the 'Hydrogen Energy Scenario' is realized and all the world's transport turns to H_2 as the motor fuel, 3 t.o.e. of a hydrocarbon fuel would be equivalent to 555 kg-molecules of H_2.

1.4 *Chemical component of metallurgy*

The chemical component of metallurgy may be estimated easily from the quantity of chemical reducing agents (H_2, C or CO) necessary for the existing and expected average production (per capita) of the main types of metals from their raw ores. We shall not consider scrap utilization in these estimates, even though scrap utilization decreases significantly the consumption of chemical reducing agents. We also shall not consider any change in composition of metal consumption, although it tends to be modified frequently.

Table 2.1 shows the average per capita consumption of the most common metals in the world and the USA in the early 1980s.

Table 2.1. The average per capita consumption of the most common metals in the world and in the USA in the early 1980s.

Metal	World consumption (kg per capita)	USA consumption (kg per capita)
Iron	165.0	272.0
Aluminum	7.0	19.2
Copper	3.71	8.62
Nickel	0.18	0.59

From [3].

Owing to modern technology, only iron and nickel production requires chemical reducing agents. Thus, if the average per capita world iron consumption is predicted to be 300 kg year^{-1}, the required amount of reducing equivalents can be estimated via the reaction:

$$Fe_2O_3 + 6e \rightarrow 2Fe + 3e_2O$$

where 'e' stands for the one-electron reducing equivalent. Thus 300 kg of iron corresponds to 5.4 kg-atoms of this element, which requires 16.2 kg-equivalents of one-electron reductants. The latter corresponds to *c.* 8 kg-molecules of H_2 or CO and to 4 kg-atoms of elementary carbon.

1.5 *Carbon and hydrogen components of the chemical industry*

The carbon component of the chemical industry can be estimated via the predicted rationalized consumption of monomers and the main types of organic semiproducts that are used for the production of the final chemicals and materials. All these monomers and semiproducts are mostly hydrocarbons. According to some expert estimates [1], for the next 20 years the consumption of such hydrocarbons may be the following (annual per capita): ethylene, 48 kg; propylene, 20 kg; benzene, 23 kg; xylenes, 11 kg; butadiene, 4 kg; isoprene, 4 kg; isobutylene, 1.5 kg — this makes a total of 112 kg of hydrocarbons and the carbon component is 8 kg-atoms. (In the 1980s in Western Europe the average annual consumption of ethylene per capita was *c.* 40 kg.) The hydrogen component corresponding to this amount of hydrocarbons is 8 kg-molecules of H_2.

The second major part of the monomers and semiproducts consists of alcohols, mainly methanol and ethanol. Their consumption per capita for the beginning of the 21st century was estimated as 20 kg of methanol and 2 kg of industrial ethanol. This amount of alcohol corresponds to *c.* 0.7 kg-atom of carbon and 2.1 kg-molecules of H_2.

An additional essential contribution to the consumption of hydrogen is provided by ammonia production, but its future production per capita seems unclear. However, according to the level of ammonia production attained in West Germany in the mid-1980s, we could estimate future production as 35 kg of ammonia per year. This corresponds to *c.* 3 kg-molecules of H_2.

2 Total predicted values of the carbon and hydrogen components of the chemistry vector

Comparing the above data, one can expect several variants in the predicted vectors of hydrogen and carbon consumption. Here we discuss only two of them.

2.1 *Variant A (maximal carbon consumption)*

This variant corresponds to the situation when hydrocarbons keep their major role as motor fuels in transport, whereas CO is still the main reducing agent in metallurgy. From the above data, the carbon and hydrogen components of the chemistry vector of consumption may be estimated as 230 kg-atoms of carbon and a minimum of 235 kg-molecules of H_2.

2.2 *Variant B (minimal carbon consumption)*

This variant corresponds to the situation when H_2 is both the dominating motor fuel and the reducing agent in metallurgy, with carbon being needed only for the production of valuable chemicals and construction materials. In this case, the carbon and hydrogen components of the chemistry vector of consumption may be estimated as 9 kg-atoms of carbon and 575 kg-molecules of H_2.

3 Raw materials for future production of H_2- and C-containing materials and energy consumption for this production: scenarios for 'fuels from water and atmospheric gases' and 'fuels from biomass'

3.1 *Hydrogen*

Water will most likely serve as the major raw material for H_2 production in the future.

The reaction of H_2 production from water:

$$H_2O \rightarrow H_2 + \tfrac{1}{2}O_2$$

corresponds to an increase of the Gibbs thermodynamic potential ΔG°_{298} = 237 kJ mol^{-1} and of enthalpy ΔH°_{298} = 286 kJ mol^{-1}. The minimum energy consumption for the production of 1 kg-molecule of H_2 cannot be lower than 237 MJ (65.8 kWh), even if the energy is used in the form of electricity. Modern electrolyzers provide *c.* 80% energy efficiency of water electrolysis. Thus, if electricity is obtained from nuclear (thermonuclear) reactors with 30% efficiency of heat conversion into electricity, a continuous 30 W heat power from a nuclear reactor will be needed for the production of 1 kg-molecule of H_2 per year. The same thermal power capacity will be required for H_2 production from water via purely thermal methods under the conditions of high-potential heat utilization.

If H_2 is obtained by means of solar energy (via either water electrolysis and photoelectrolysis or its photocatalytic cleavage), modern estimates (see Chapter 13) confirm the possibility of using solar energy for this purpose with an energy efficiency of up to 10%. (In our estimate we will not use the higher, more optimistic figures of 20–28% for this efficiency; see Chapter 13.) For regions in the middle latitudes with moderate solar radiation (*c.* 1000 kWh m^{-2}, which corresponds to an average annual power of 110 W m^{-2}), the production of 1 kg-molecule of H_2 per year will require a minimum area of 0.64 m^2. For deserts of low latitudes (Sahara, etc.), the required area may be reduced by a factor of 2–2.5.

In principle, H_2 may also be produced via processing the renewable resources of biomass (by its pyrolysis or steam reforming). However, this method of H_2 production requires preliminary manufacturing of carbon-containing raw materials and, thus, seems to be more wasteful of solar energy than by direct solar photolysis, although there may be other advantages from the biomass route.

3.2 *Carbon*

Fossil carbonates and renewable biomass resources supplied by atmospheric CO_2 seem to be the main raw materials for the production of carbon-containing materials in the future. Even in the far future, it would be rather unlikely that we could fix atmospheric or water-dissolved CO_2 directly in large quantities (except possibly by some biotechnological procedures) owing to the high energy cost of CO_2 concentration using these lean sources.

It is easy to estimate the minimal energy consumption for the production of carbon-containing materials from both of the above inorganic raw materials by assuming that the final products should be long hydrocarbons of formula $(CH_2)_n$. (Because only a small fraction of alcohols is needed in the overall amount of carbon-containing monomers and semiproducts, we restrict our crude estimations to hydrocarbons only and do not consider alcohols separately.)

If $(CH_2)_n$ molecules are obtained from fossil carbonates (e.g. limestone $CaCO_3$), we can follow the hypothetical reaction:

$$CaCO_3 + 3H_2 \rightarrow [CH_2] + 2H_2O + CaO \tag{1}$$

with the corresponding change of the Gibbs potential $\Delta G^{\circ}_{298} = +44$ kJ mol-1 and of enthalpy $\Delta H^{\circ}_{298} = -33$ kJ mol-1. (Here, and later, the estimates are given based on the assumption that *n*-heptane equals $(CH_2)_n$.)

For the fixation of free, but already concentrated, CO_2 via the reaction:

$$CO_2 + 3H_2 \rightarrow [CH_2] + 2H_2O \tag{2}$$

these values are both negative and equal to $\Delta G^{\circ}_{298} = -85.5$ kJ mol-1 and $\Delta H^{\circ}_{298} = -210.2$ kJ mol-1. Thus, reaction (2) can proceed spontaneously even at room temperature. However, as already mentioned, it is difficult and energy consuming to concentrate CO_2 from the atmosphere, where its content in 1995 was 355 ppm.

Thus, if we ignore the energy for the production of H_2, the minimum necessary net energy consumption for the production of 1 kg-atom of fixed carbon via reaction (1) is 44 MJ (12.2 kWh) in the form of electricity. At the present moment it is hard to predict the practically attainable energy efficiency of electricity utilization for implementation of reaction (1). However, due to the complexity of this reaction and the necessity to apply indirect routes for the chemical synthesis, this efficiency is rather unlikely to exceed 20%. Thus, a nuclear energy source with 23 W of heat power or a solar energy device of 0.24 m^2 area in the middle latitudes (at an overall solar energy conversion efficiency of 5%, i.e. lower than optimistic predictions) is required to produce 1 kg-atom of carbon in the form of hydrocarbons via reaction (1), accounting for the factors given in Section 3.1.

More precise and optimistic estimates are possible if hydrocarbons are considered to be produced from plant biomass. The woody or herbaceous biomass may be assumed to be a sufficiently homogeneous, large-scale organic raw material for the chemical industry. Moreover, in comparison to the sources of fossil organic materials, the biomass of plants is more environmentally favorable and more convenient for its processing, because the contents of sulfur and bound nitrogen in biomass are negligible. The approximate composition of the dry wood can be presented by the empirical formula $CH_{1.44}O_{0.65}$ [7]. Dry herbaceous biomass has a similar chemical composition. Thus the conversion of biomass into hydrocarbons should follow a process of hydrogenation via the reaction:

$$CH_{1.44}O_{0.65} + 1.21H_2 \rightarrow [CH_2] + 0.65H_2O \tag{3}$$

The expected thermodynamics of such hydrogenation may be estimated via hydrogenation of saccharose, which has a structure similar to that of the main component of wood, cellulose, and a similar chemical formula $(CH_{1.883}O_{0.917})_{12}$. The process should be:

$$CH_{1.883}O_{0.917} + 0.93H_2 \rightarrow [CH_2] + 0.917H_2O \quad (4)$$

with $\Delta G^\circ_{298} = -94.2$ kJ mol^{-1} and $\Delta H^\circ_{298} = -109.0$ kJ mol^{-1}.

There is no thermodynamically required energy consumption for reaction (4) to occur. Thus, the net energy consumption for reaction (3) will be small in comparison to that for reaction (1). For this reason, first of all it is necessary to estimate the expected consumption of solar (or other kinds of) energy required for cultivation and harvesting of biomass. According to recent data, for the middle latitudes of Western Europe, it should be possible to cultivate 'energy crops' with a shorter period of rotation and with an average yield of 12 tons or more dry biomass per hectare per year. The calorific value of 1 ton of such biomass is about 20 GJ. For some grass crops (e.g. sweet sorghum), the yield per hectare may be even higher and attain 16 tons of dry biomass [8]. In the tropics, the cultivation of sugar-cane may provide 35 tons of dry biomass per hectare [9].

The energy consumption for planting, cultivating (including fertilization) and harvesting energy crops with a yield of 16 tons of dry biomass per hectare per year does not exceed 8 GJ per ton of dry biomass [10]. Thus, the resulting net energy harvest, even in the middle latitudes, may be estimated at 80–160 GJ or 4–8 tons of dry biomass per hectare per year. This corresponds to an annual production of 1 kg-molecule of fixed carbon from a 15–20 m^2 area with the simultaneous support of an energy source with an average annual heat power of *c.* 18 W or from a 30–40 m^2 area without any additional supporting energy source.

4 'Hard' and 'soft' scenarios for the supply of hydrogen and carbon components of the chemistry vector

Returning to the evaluation of the hydrogen and carbon components of the chemistry vector, we can see that variants A and B (Section 2) should be corrected for the hydrogen component because reducing equivalents of H_2 are required for the reduction of new raw materials for the production of hydrocarbons.

1 *Variant A*. If hydrocarbons are obtained from carbonates or free CO_2 via reactions (1) and (2), the actual minimum average value of hydrogen consumption per capita should reach 695 kg-molecules of H_2 rather than 235. If hydrocarbons are obtained from biomass, the corrected value of the hydrogen component will be 511 kg-molecules of H_2.

2 *Variant B*. In the first case, the corrected value will be 593 kg-molecules of H_2 and not 575. In the second case, the correction will be less and the hydrogen component will reach 586 kg-molecules of H_2.

A comparison of the corrected chemistry vectors for variants A and B supports the apparent strategic superiority of variant B where all motor fuels in transport and chemical reducing agents in metallurgy are replaced by pure H_2, with chemically bound carbon being used only for the production of valuable chemicals or construction materials, etc. Moreover, the implementation of variant A, i.e. the production of large amounts of carbon-containing fuels from inorganic fossil raw materials like carbonates, does not solve the problem of pollution of the Earth's atmosphere by CO_2.

This choice allows us to estimate more precisely the required energy and/or territorial resources for satisfying the average per capita consumption. Two marginal scenarios should be considered:

'hard' and 'soft'. Under the 'hard' scenario all necessary energy resources are provided by centralized nuclear (thermonuclear) power plants, while under the 'soft' scenario solar energy is the sole energy source.

4.1 *'Hard' scenario*

If variant B is implemented to provide both H_2- and carbon-containing chemicals from inorganic fossil raw materials, the average per capita annual heat power:

$$593 \times 30 + 9 \times 23 \approx 18\ 000 \text{ W}$$

of a nuclear power plant is required to supply the hydrogen and carbon components. If other energy demands of an individual remain constant at the specified level of the flux of primary energy sources (see Section 1), the required per capita heat power of the nuclear power plant must be increased by 9.7 kW (excluding the already considered energy consumption for transport) and will attain about 27.7 kW. Land area requirements may be extrapolated from the fact that today, with a contribution to world energy production of 5%, more than 1.5% of the land is seriously contaminated with radioactivity.

If variant B is implemented via the production of carbon-containing materials from biomass, the direct energy consumption for the chemical vector will decrease slightly and reach:

$$586 \times 30 + 9 \times 18 \approx 17\ 700 \text{ W}$$

However, an area of 120–160 m^2 will also be required per capita to provide the carbon component of the chemistry vector. The latter data correspond to the situation of a 'moderate technocratic' methodology of biomass cultivation, when non-biotechnological sources of energy are also utilized (see Section 3.2).

For the predicted Earth population of 10^{10}, the total required heat power of nuclear reactors should reach *c.* 2.8×10^5 GW in both cases, which is equivalent to 9×10^{21} J of annual production of heat. Clearly, this level of production will be accompanied by a considerable exhaust of heat into the atmosphere and land loss due to radioactive contamination.

If the second case is implemented, an extra $(1.2–1.6) \times 10^6$ km^2 is required for biomass cultivation compared to no less than 2.8×10^5 km^2 if 1 km^2 is required for each gigawatt nuclear plant, including security zone and facilities for radioactivity disposal.

4.2 *'Soft' scenario*

When only solar energy is utilized for the implementation of variant B to provide both H_2- and carbon-containing products from inorganic fossil raw materials, an area of:

$$593 \times 0.64 + 9 \times 0.24 \approx 380 \text{ m}^2$$

in the middle latitudes will be required per capita to satisfy the chemistry vector, i.e. for 10^{10} people *c.* 3.8×10^6 km^2 (not necessarily suitable for agriculture) will be needed. If the other energy demands per capita (9.7 kW; Section 1) are also supplied only by solar energy with the same efficiency of *c.* 10%, a total area of 12.3×10^6 km^2 will be required in the middle latitudes for installing suitable solar energy converters. Note, however, that in fact there are significant possibilities for energy saving (through so-called solar architecture) and secondary use of land surfaces such as roofs, industrial land, etc.

In the long term, the average efficiencies of solar converters will increase beyond the assumed 10% because the thermodynamic limit for solar energy conversion (via cascade systems utilizing all wavelengths of solar radiation) is 68%. Already, today, solar cells with efficiencies of 20–35% are being tested in the laboratory and solar heat production efficiency is exceeding 50%. A 20% efficient solar energy utilization may therefore be expected to be reached within one century because electricity production efficiency has increased from 12–14% to nearly 40% from the beginning of this century to the present. This would reduce land consumption (including secondary land use on roofs, etc.) by solar energy systems to 6.2×10^6 km^2.

If with the moderately technocratic variant B the carbon-containing materials are obtained from biomass of specially cultivated crops, the supply of the chemistry vector per capita in the middle latitudes will require a territory of:

$$586 \times 0.64 + 9 \times (15\text{–}20) \approx 510\text{–}560 \text{ m}^2$$

including 130–180 m^2 of territory for highly productive biomass cultivation.

The total area required for covering the chemistry vector of mankind will reach *c.* $(5\text{–}6) \times 10^6$ km^2, whereas for covering all chemical and energy demands it will be close to 13.5×10^6 km^2 (compared to the world's land area of 1.3×10^8 km^2).

5 Conclusions on the strategy of energy production in the future world

Although the estimates given above are rather approximate, they allow us to make some interesting and important conclusions. Firstly, note that the present total annual consumption of primary energy resources is about 4.0×10^{20} J [4,5], and the annual amount of solar energy received by the Earth's surface, providing the heat balance of the planet, is 5.6×10^{24} J [11]. The total area of arable land is estimated at more than 10×10^6 km^2, the total area of forests (including low-productive boreal forests) is 40×10^6 km^2 [12] and relatively unused deserts and semideserts occupy $(46\text{–}49) \times 10^6$ km^2 [13]. The annual worldwide logging of wood exceeds 2.5×10^9 m^3, which corresponds in energy to *c.* 3×10^{19} J, and to the production of carbon-containing materials with a total carbon content of 4×10^{10} kg-atoms [5,10]. The annual anthropogenic emission of CO_2 into the atmosphere is *c.* 3.5×10^9 tons of carbon today, which is equivalent to 0.8×10^{11} kg-atoms of carbon.

The following conclusions can be drawn:

1 Geoclimatic conditions on the Earth allow us in principle to sustain, after all the fossil organic resources are exhausted, the existence of 10 billion people at a standard of living presently attained in the developed countries (with food supply problems solved independently). This possibility may be implemented under either a 'hard' scenario, when the main energy resources are provided by nuclear (thermonuclear) power plants, or a 'soft' scenario, when solar energy and its derivatives are the major energy source. No real threats, such as overheating of the Earth due to anthropogenic production of heat by thermonuclear power plants or lack of territory for installing the solar energy converters (situated in deserts not used now), seem to arise. However, construction of *c.* 2.8×10^5 nuclear breeders or similar will obviously pose serious security problems. Also, in the case of construction of such plants on land, local overheating is unavoidable.

2 In the future, the major fraction of energy is likely to be consumed for motor fuel production if the present structure of energy consumption continues. Unless significant progress in the development of energy-saving vehicles is achieved for economy in energy consumption and to decrease pollution, the inhabitants of a future Earth should be less individually mobile than the citizens of the USA at present.

3 Only H_2 seems to be the feasible major motor fuel of the distant future. This is mainly because the production of artificial carbon-containing fuels from inorganic raw materials is accompanied by an unjustifiably large energy consumption in comparison with that for H_2 production from water. The optimal annual H_2 production in the future should be close to 1 ton of H_2 per capita.
4 The fossil inorganic raw materials (carbonates) do not seem to be usable as the *main* source for the production of carbon-containing chemicals, even in the distant future. This conclusion will still be valid even if such chemicals are not utilized as motor fuels. With the postulated developments and with a dynamic balance in the output of carbon-containing materials and their later complete destruction, the emission of the so-generated technogeneous CO_2 into the atmosphere will remain at a dangerous level close to the existing one. Thus, the main production of carbon-containing materials is most likely to be based on the renewable resource of biomass, even if areas potentially also suitable for food production will be used for growing the 'energy crops'.
5 The above quite pessimistic situation with carbon-containing materials can be essentially changed only if an efficient *recovery* or *recycling* of the already produced carbon-containing materials is implemented. Note, however, that such a carbon recovery will affect only insignificantly the energy consumption of the chemical sector of the industry, because the major part of future energy will be used for H_2 as a motor fuel.

The above statements lead to some important conclusions concerning the forthcoming global transition period:
1 It is necessary to develop, as soon as possible, efficient methods for the cultivation of plant biomass and for processing the biomass and municipal solid waste into the traditional products of petrochemistry (hydrocarbons, firstly) in order to develop the large-scale technologies needed for the transition to renewable resources of carbon-containing raw materials. At present, the utilization of biomass for the economically supported production of hydrocarbon fuels might be a useful means for the development of such technologies. Later on, these technologies may be reoriented easily to the large-scale production of monomers, chemical semiproducts, etc.
2 Because the processing of plant biomass, secondary organic raw materials (e.g. municipal solid wastes) and coal requires large amounts of hydrogen, methane rather than water should serve as the major, most convenient and less costly source of hydrogen in the transition period. This results from the fact that the transition period will be characterized by enhanced mining and utilization of the natural gas or methane from the gas hydrates. The most useful method of methane utilization might be direct methanopyrolysis of the carbon-containing raw materials. In this process, methane is used as a chemical reducing agent and simultaneously as a carrier of the chemically bound hydrogen, thus increasing the yield of carbon-containing chemicals due to combining its own carbon atom. Moreover, in principle, these processes allow a cut in energy consumption, because they do not need an intermediate production of H_2 or syn-gas via energy-consuming steam or carbon dioxide reforming of methane, which is also accompanied by the release of CO_2. A promising means of H_2 production from natural gas without pollution of the atmosphere by CO_2 emission could be some of the new technologies discussed in Chapter 7. Although almost no attention has been paid to these new technologies previously, they might be developed rather rapidly in the case of real demand to reverse environmental problems.

6 References

1 Parmon VN. *Chem. Sustain. Dev.* 1993; **1**: 1.

2 Parmon VN. In Veziroglu TN, Klinter CJ, Baselt JP, Kreusa G (eds) *Hydrogen Energy Progress XI. Proceedings of the 11th World Hydrogen Energy Conference, Stuttgart, Germany, 23–28 June, 1996*, Vol. 1. Frankfurt: Schön and Kletzel, 1996; 51.

3 Skinner BJ. *Earth Resources*. New Jersey: Prentice-Hall, 1986.

4 Scurlock JMO, Hall DO. *Biomass* 1989; **21**: 75.

5 Hall DO. *Energy Policy* 1991; **19**: 711.

6 Makarov AA, Volfberg DB, Rudenko N. *Prognosis: Scenarios of the Development of the Energy Industry in the USSR*. Moscow: Mintopenergo, 1990.

7 Sonju OK. In Mitchell CP (ed.) *Bioenergy and the Greenhouse Effect. Proceedings of a Seminar at Stockholm, Sweden, April 15–16, 1991*. Stockholm: Narings-och teknikutvecklingsverket, 1991; 61.

8 Cherney JH, Johnson KD, Volenec JJ, Greene DK. *Energy Sources* 1991; **13**: 283.

9 Hall DO, Mynick HE, Williams RH. *Nature* 1991; **353**: 11.

10 Strauss CH, Grado SC. *Sol. Energy* 1992; **48**: 45.

11 Parmon VN, Zamaraev KI. In Serpone N, Pelizzetti E (eds) *Photocatalysis. Fundamentals and Application*. New York: Wiley, 1989; 565.

12 Bukshtykov AD, Groshev BI, Krylov GV. *The World of Nature. Forests*. Moscow: Mysle, 1981 (in Russian).

13 Babaev AG, Zonn IS, Drozdov NN, Freikin ZG. *The World of Nature. Deserts*. Moscow: Mysle, 1986 (in Russian).

3 Role of Chemistry in Improving Traditional Energetics via New Combustion Technologies and Chemical Heat Recuperation Techniques

V.N. PARMON

Boreskov Institute of Catalysis, Siberian Branch of the Russian Academy of Sciences, Prospekt Akademika Lavrentieva 5, Novosibirsk 630090, Russia

1 Introduction

Despite the projected expanding use of nuclear fission energy (frequently against public resistance) and the envisaged perspectives for the use of nuclear fusion energy, combustion of fossil fuels is still expected to provide the major source of energy during the next half-century. Mankind has successfully used flame combustion as the main source of energy since earliest times. However, the recent vast development of the energy industry has revealed certain important shortcomings of flame combustion technology. One of them is the formation of toxic compounds, such as CO, NO_x and SO_2. The presence of heteroatomic compounds in the fuel also contributes to the toxicity of flue gases. In addition, some other products of incomplete combustion with carcinogenic properties can be formed. Another shortcoming is the wasteful loss of a notable amount of chemical energy due to incomplete oxidation of the fuel to CO_2 and H_2O or the formation of certain undesirable by-products due to endothermic reactions at elevated temperatures (2000 K and higher). In these reactions, heat is lost and toxic products are formed.

Detoxification of flue gases from all these pollutants is rather expensive because of the necessity to process enormous gas fluxes that contain the whole blend of pollutants. Therefore, we face the challenge of designing new combustion technologies that will be efficient in energy production, safe for the environment and economical.

Another problem is how to increase the efficiency of transformation of chemical energy stored in a fuel into mechanical energy or electricity.

Currently, the main part of traditional energetics is based on the following energy transformations: combustion of fuel → heat → mechanical energy → electricity. The main limiting factor of the total efficiency of such an energy conversion is the Carnot cycle of heat-to-mechanical energy transformation, which depends on the highest (T_{max}) and lowest (T_{min}) temperatures of the cycle:

$$\{\text{Useful work 1}\} \leq \frac{T_{max} - T_{min}}{T_{max}} Q$$

where Q is the calorific value of the fuel equal to the change of enthalpy, ΔH, of the fuel mixture during the combustion: $\Delta H = -Q$.

For the common combustion processes, T_{max} does not exceed 1300 K and, actually, very often it is much less at the blades of turbines. At the same time, T_{min} is much higher than the temperature of the ambient. Thus, the Carnot factor does not exceed the value of 0.6–0.75. The overall efficiency of chemical-to-electrical energy transformation by conventional thermal power plants typically does not exceed the value of 0.3–0.4.

However, it is well known by chemists from lessons on chemical thermodynamics that, in fact, for the transformation of chemical energy to work there is another type of limitation:

$\{\text{Useful work 2}\} \leq -\Delta G$

where ΔG is the diminishing free Gibbs energy of the fuel mixture during its use. There is no Carnot factor for this transformation at all! Evidently, for many conventional fuels, 'useful work 2' is typically greater than 'useful work 1'.

Indeed, for processes of conventional combustion of a common organic fuel at temperature T:

$$Q = -\Delta H = -(\Delta G + T\Delta S)$$

When fuel is burned, ΔH and ΔG have negative values, while for most common types of fuel the entropy change ΔS is positive due to an increase in the number of moles of gases during the combustion (complete oxidation) process. For this reason, $Q = |\Delta H|$ should be greater than $|\Delta G|$. However, for the common energy-saturated fuels, Q remains usually still very close (only 10–15% greater) to the value of $|\Delta G|$ and thus the Carnot factor diminishes significantly the allowed 'useful work 1' in comparison with $|\Delta G|$.

This means that when burning chemical fuels in the traditional way we lose a huge amount of energy that could be transformed into work and electricity when following another path of energy transformation.

The chemist's best-known method to avoid this non-obvious loss of energy is the creation of so-called 'fuel cells' for direct transformation of chemical energy into electricity (see Chapter 8). However, one can suggest also some possible refinements in arranging the fuel combustion process as a whole and for overcoming the efficiency of the one-step Carnot cycle. These refinements are based on the chemical properties of some fuels and the process is usually known as 'chemical heat recuperation' (see Section 4 of this chapter) [1–3].

The main possible impact of chemistry in traditional energetics is thus either the application of catalytic technologies to aid the process of combustion or the implementation of principally new technologies based on ideas of 'chemical heat recuperation'.

2 Combustion via catalytic control

It is well known that chemical processes can be carried out in a controlled fashion when using the appropriate catalysts. Such catalysts permit a decrease in reaction temperature and can selectively accelerate desirable reactions and suppress undesirable side reactions. The combustion process is no exception.

The influence of catalysis on combustion can be traced from the works by H. Davy, E. Davy and J. Dobereiner as far back as in the 19th century. Humphry Davy discovered the catalyzing effect of platinum wire in the combustion of methane. Indeed, the studies on flameless catalytic combustion of hydrocarbons, alcohols and hydrogen over finely dispersed platinum were among the first scientific research in catalysis. The first practical applications — the miner's safety lamp by H. Davy and catalytic lighters by J. Dobereiner [4] — date from the same period. In the following years, this ecologically safe method failed to be applied to the energy industry because the problem of ecological safety was not, at that time, considered to be urgent for this industry.

Catalytic incineration differs principally from incineration in the common sense because the fuel is oxidized on the surface of solid catalysts at a low temperature and without the formation of a flame. For many types of catalysts, the process of complete oxidation (or 'heterogeneous combustion') of air–fuel mixtures consists of an interaction of fuel components with the 'surface' oxygen of the catalyst, with the successive regeneration of its reduced surface by oxygen from the gas phase.

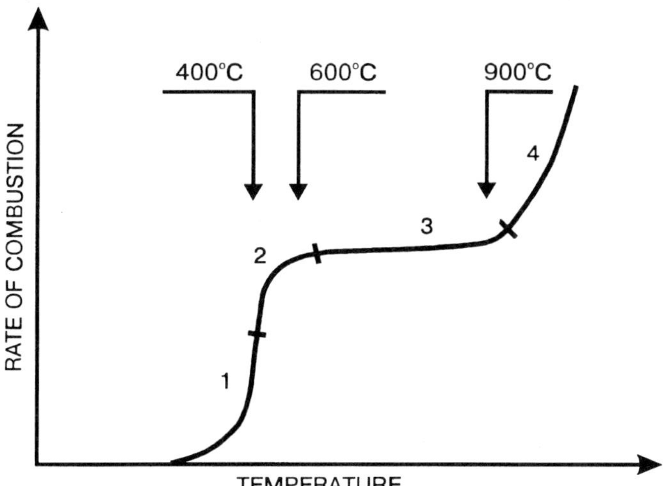

Figure 3.1. A typical temperature dependence of the reaction rate for catalytic combustion: 1, kinetic control; 2, internal diffusion control; 3, external diffusion control; 4, homogeneous combustion initiated over a catalyst. (According to [8].)

Until recently, applied research and development in the area of catalytic combustion were concentrated in the design of various low-power, low-temperature sources of heat used in household appliances. Nowadays, various catalytic burners, engine heaters, radiant burners and panels for room heating are available. The advantages of such appliances are compactness, safety and minimum harmful emissions. Within the last two decades, the aggravation of the ecological situation on a global scale has aroused a new wave of interest in catalytic combustion. Work is under way in many directions, including fundamental research, the development of thermostable and mechanically strong catalysts and supports and the creation of novel technological processes and devices for catalytic combustion (see review articles [1–3, 5–9]).

As temperature increases, four different modes of the process of catalytic combustion can be specified (see Fig. 3.1). At low temperatures, the rate of catalytic combustion is controlled by the rate of the heterogeneous reaction of catalytic oxidation: this corresponds to the mode of kinetic control (Fig. 3.1, zone 1). At higher temperatures, both the rate of the catalytic reaction and the rate of reagent diffusion within catalyst pores influence the overall rate of catalytic combustion: this corresponds to the mode of internal diffusion control (Fig. 3.1, zone 2). With further increasing temperature, the rate of catalytic combustion changes only slightly: this corresponds to the mode of external diffusion control, when the reaction rate is determined by the diffusion of reagents from the bulk of the reactor to the external surface of the catalyst (Fig. 3.1, zone 3). At still higher temperatures, homogeneous reactions that are initiated over the surface of the catalyst start to control the rate of catalytic combustion, and its rate again increases significantly with temperature (Fig. 3.1, zone 4).

Mode 1 is used mainly for low-temperature catalytic burning of small amounts of organic pollutants in industrial exhaust gases, but most low-power catalytic heaters and furnaces operate in modes 2 and 3. High-power stationary catalytic devices for fuel combustion that are currently developed operate in mode 4 at high temperatures (1200 K and over). The so-called catalytically supported combustion is also performed in this mode [8].

According to the mode of operation, the approaches to modeling catalytic reactors and selecting optimum catalysts will differ. In order to control low-temperature processes, one has to know the kinetics of heterogeneous oxidation reactions. In order to control reactors that operate in mode 4, the kinetic data on the generation and decay of radicals over the catalyst, as well as data on the kinetics of homogeneous gas-phase oxidation reactions, are needed. As far as low-temperature catalytic heat sources are concerned, the main demand of the catalyst is its high activity. As for the devices with catalytically supported combustion, the most important characteristic of the catalyst is its thermal stability.

2.1 *Catalysts for combustion*

Catalytic combustion of fuels is a particular case of catalytic reactions of complete oxidation, i.e. performed at a fuel-to-oxygen ratio close to the stoichiometric value and accompanied by high heat release; therefore, any catalysts active in the reactions of complete oxidation can be used, in principle, for catalytic combustion. However, efficient catalysts must also possess high thermal stability, mechanical strength and resistance to catalytic poisons that may be present in the fuel. The approaches to the choice of complete oxidation catalysts have been reviewed in [10,11]. Platinum, palladium and transition metal oxides can be used as catalysts, with mixed oxides, spinels and perovskites being more preferable to simple oxides. Porous granules and fibrous materials based on alumina or aluminosilicates are usually used as catalyst supports. In the devices operating at high rates of heat release, it is preferable to use monolithic honeycomb catalysts with a low pressure drop. Depending on the temperature at which the catalyst must work, the monolithic support may be fabricated of alumina modified by silica, zirconium ceramics, silicon carbide, silicon nitride, cordierite or mullite. Metal supports are also used. Structural–mechanical characteristics and thermal shock resistance of the monolithic ceramic supports are constantly being improved. Metal supports are very strong and they endure thermal shocks and mechanical loads, but their application is typically restricted within the temperature range of 1100–1300 K. Various modifications of alumina, including those with modifying additions of ZrO_2, CeO_2, Cs_2O or ThO_2, are used as high-temperature resistant support materials [7].

2.2 *Devices with a fixed bed of catalyst*

2.2.1 DEVICES FOR HOME USE

Today, numerous variations of low-power convective–diffusive catalytic heaters and furnaces of various shapes and applications are already widely accessible. In these devices, a fixed bed of porous or fibrous catalyst is typically used. The temperature of the operating bed is 900–1000 K and heat transfer to the environment occurs via radiation and convection.

Typically, the concentration of nitrogen oxides and carbon monoxide in effluents of such catalytic heaters does not exceed the allowable limits, but a considerable slip of the fuel through the catalyst bed (up to 5–20%) is sometimes observed, especially in the case of methane combustion. Another shortcoming of such heaters lies in the application of high-priced catalysts that are sensitive to catalytic poisons and become deactivated after long usage.

It has been pointed out that the process of combustion in the above devices is limited in rate by diffusional mixing of reagents in the catalyst bed due to the necessity to separate the fuel and air feeds. This, together with the fact that heat transfer occurs only by convection, considerably restricts the specific power loading of such heaters. Work is under way to improve gas-fired combustors of the convective–diffusive type. In some catalytic devices of this type, a preliminary mixing of the

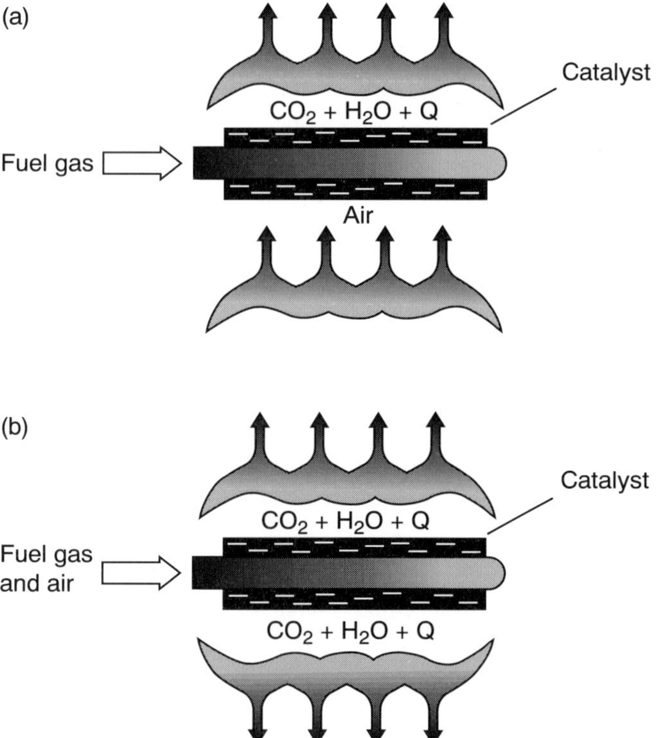

Figure 3.2. Catalytic heat-evolving units based on perforated tubular catalysts operating in either a convective–diffusive mode (a) or in a fuel mixture injection mode (b).

fuel and air is used. This reduces the slip of fuel and enhances the power of the device. For example, recently, a very simple and useful device was developed in Russia with catalytic heat-evolving elements on the base of perforated metal tubes surrounded by metal-support-based catalysts (Fig. 3.2). An advantage of such systems is that they can operate in a fuel mixture injection mode, which allows for better control of the combustion process and a diminishing heater size.

As an example of the implementation of advanced heating systems, catalytic heaters for tents and for cabins of cars and ships can be mentioned. They also have turned out to be very practical for short-term heating of living rooms during unusually cold days in houses without heating facilities in a mostly warm climate.

2.2.2 CATALYTIC GAS TURBINES

It may be attractive to use catalysts for intensification of fuel combustion in combustion chambers of gas turbines. Many laboratories and companies have undertaken such research and have carried out tests of pilot devices. At present, three directions of research aimed at the development of catalytic devices for turbines can be mentioned. These are devices for aircraft, cars and stationary installations [7].

Application of catalysts in combustion chambers of turbines permits combustion to be carried out at lower temperatures, and thus eliminates the necessity to dilute hot combustion gases with cold air in order to protect the turbine blades. Also, it becomes unnecessary to inject water into the combustion zone to reduce the production of thermal nitrogen oxides. Some schemes of catalyst implementation are based on a stepwise oxidation of fuels, including that after passing the first row of blades, etc., which improves both the completeness of the combustion and the efficiency of the turbine operation, as well as diminishing the formation of NO_x.

When using catalytic combustion for gas turbines, the emission of CO decreases by a factor of tens and that of the 'thermal' nitrogen oxides up to a factor of 100 compared to the flame combustion chamber [12–14]. However, the problem of reducing the emission of nitrogen oxides, which are formed from the nitrogen-containing compounds of the fuel, still seems to be unsolved following the tests of pilot plants described in [13–15]. As noted in [13], fuel-bound nitrogen almost completely converts into nitrogen oxides in a catalytic gas turbine combustor. However, it was possible to decrease considerably the amount of fuel NO_x by using a two-step combustion process in which a zone with fuel excess and thus with reductive properties is first formed, where fuel nitrogen is mainly converted to molecular nitrogen rather than to NO_x. In the second zone, the catalytic combustion of the residual unburned fuel takes place. However, the catalyst in the first bed tends to coke and lose its activity, which can lead to higher CO emissions.

2.2.3 BOILERS WITH CATALYTIC BURNERS

The use of catalytic combustion to produce thermal energy for water-heating or steam boilers is supposed to decrease pollutant emissions, primarily by way of NO_x reduction, and to improve the efficiency of the boilers. The design of water-tube and fire-tube boilers is in progress. In one of the modern projects, the fire-tube catalytic burner is fitted into the combustion chamber of a boiler. The burner is a tube made of thermally stable catalytically active fibers, closed at the end. The gas and air mixture is injected into the open end of the tube and diffuses through its walls to the outer surface, where complete combustion of fuel occurs. The burner's ignition is accompanied by a spark or by a flame at the outer surface. The attractive feature of this technology is that the fire-tube catalytic burner can be installed at the position of the flame burner without changing the design of a boiler. Commercial operation tests of 300 kW boilers with the fire-tube catalytic burners have been conducted already [7].

Another type of boiler that is being developed is the radiative catalyst/water-tube boiler. The catalyst is supported on the outer surface of tubes made out of a temperature-resistant material and located in the combustion chamber. The heat of the catalytic combustion process is transferred by radiation and convection to metal heat-exchange tubes that are located within the catalyst-carrying tubes or next to them [13].

2.2.4 DEVICES WITH MONOLITHIC CATALYSTS

Devices are also being developed for combustion of fuels in a fixed catalyst bed in the form of honeycomb monoliths (Fig. 3.3). The main advantage of a honeycomb monolithic catalyst bed is the low pressure drop. Because catalytic combustion in such structures is accompanied by a high adiabatic temperature rise, these monoliths should possess a number of unique properties. On the one hand, a minimum wall thickness is needed to obtain a low pressure drop. On the other hand, maximum thermal shock resistance and durability are required. Materials for monolith manufacturing are now mostly various expensive ceramics such as zirconia, silicon carbide, cordierite, mullite or special alloys and ferroalloy steels.

Platinum group metals that are typically used as the catalytically active component in devices with a fixed catalyst bed further raise the price of these devices. Platinum and palladium are also sensitive to catalytic poisons. Results of pilot plant tests with a fixed catalyst bed described in the

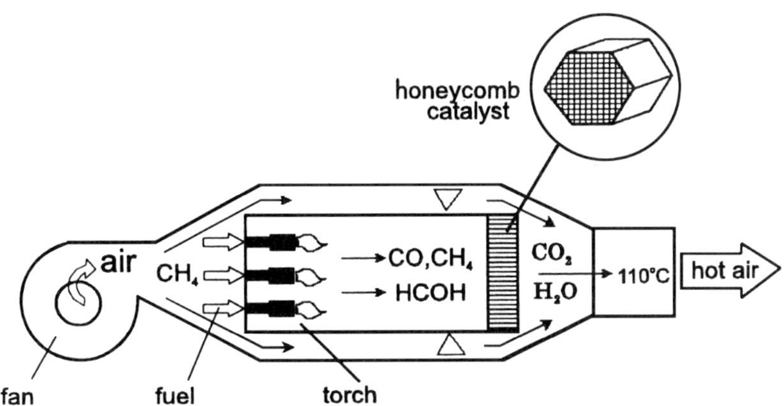

Figure 3.3. A device for two-step combustion using honeycomb monoliths to ensure the complete oxidation of fuel after the first step with its flame combustion.

literature suggest that it is possible to burn successfully only pure or gaseous fuels containing no catalytic poisons. Coal-derived liquid fuels and residual oil at present are unfortunately not suitable for combustion in such devices because they contain catalytic poisons. Also, the problem of reducing NO_x emission during the combustion of nitrogen-containing fuels on platinum group catalysts has not been solved satisfactorily as yet. Worthy of mention are some other difficulties related to the construction of more powerful devices, the production of honeycomb structures, the support of an active component and the stable performance of the combustion process. The cobalt or other transition metal perovskite-based honeycomb catalysts are now considered as the most promising for further development in this area.

Work on the improvement of catalytic combustion within a fixed catalyst bed as well as on the development of less expensive and more efficient catalysts is a strategic route for research and development in this area. However, alternative methods of catalytic combustion in a fluidized catalyst bed have also been proposed.

2.3 *Devices with a fluidized bed of catalyst*

A fruitful alternative to combustion in a fixed bed of catalyst is a process of fuel combustion in a catalytic heat generator (CHG) with a fluidized bed of granulated oxide catalyst (see Fig. 3.4) [1–3,9,10].

Complete fuel combustion is accomplished in the CHG at relatively low temperatures (800–1000 K). An advantage of the CHG as compared to the catalytic burner with a fixed catalyst bed is the possibility of combining the processes of heat release and heat removal within the same catalyst bed. Moreover, in the CHG, one can combine catalytic combustion with various technological processes consuming heat energy. The high heat transfer rates inherent in the fluidized bed, due to the heat movement with the solid heat carriers, permit the intensification of technological processes that consume heat energy.

Like any fluidized bed reactor, the CHG is a hollow apparatus filled with spherical granules of properly chosen catalyst; in the lower section of this apparatus an air-distributing plate is located. The fluidized bed is formed by an upward motion of the air–fuel mixture and combustion products through the layer of spherical catalyst particles of an appropriate size. The useful heat

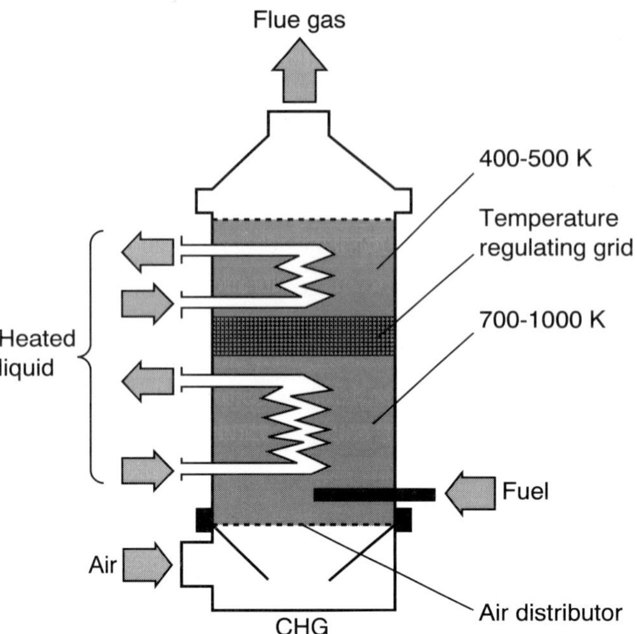

Figure 3.4. Scheme of a catalytic heat generator (CHG) with a fluidized catalyst bed (the darkened zones).

removal from the CHG can be improved either through the heat-exchange surface immersed in the fluidized catalyst bed or via direct contact of the hot catalyst granules with a substance that must be processed by heating. A principal peculiarity of the CHG is the presence of a horizontal section grid with definite sizes of cells. This grid serves as a non-isothermal barrier and divides the apparatus by height into two isothermal zones: a lower one with a temperature needed for complete fuel oxidation (800–1000 K) and an upper one with a temperature determined by conditions of heating of the working medium (Fig. 3.4). The presence of the non-isothermal grid and heat removal in the upper part of the apparatus cause the temperature of the flue gases to decrease to 420–450 K.

In the CHG, the catalyst is subjected to the influence of high temperatures, the chemical action of the stoichiometric air–fuel mixture (which can be destructive for oxides), the poisoning effect of sulfur, alkali metals, the aggressive or poisoning chemical elements that can be present in unpurified fuels (e.g. sulfurous fuel oils) and considerable mechanical shocks. For these reasons, certain special demands are imposed on the catalysts, namely: the ability to preserve sufficient activity at prolonged service (3000–4000 h) under conditions of minimum air excess and temperature up to 1000 K; endurance against mechanical attrition during operation in the fluidized bed, which should be less than 0.5% per 24 h; and stability to catalytic poisons during the burning of unpurified fuels and industrial wastes.

At present, oxide catalysts based on chromites of some transition metals as well as some iron- or manganese-containing oxides, etc. are employed in pilot plant and commercial CHGs. The catalysts are manufactured typically by supporting an active component over spherical granules of alumina with enhanced mechanical strength.

When burning fuels in a CHG, the most important ecological effect is the reduction of NO_x concentration. Indeed, a sufficiently low temperature of combustion allows almost complete elimination of 'thermal' NO_x (i.e. the NO_x formed upon oxidation of N_2 from the air at elevated temperatures), while application of selective oxide catalysts and an adjustment of combustion conditions markedly decrease the formation of fuel NO_x (i.e. the NO_x formed upon combustion of

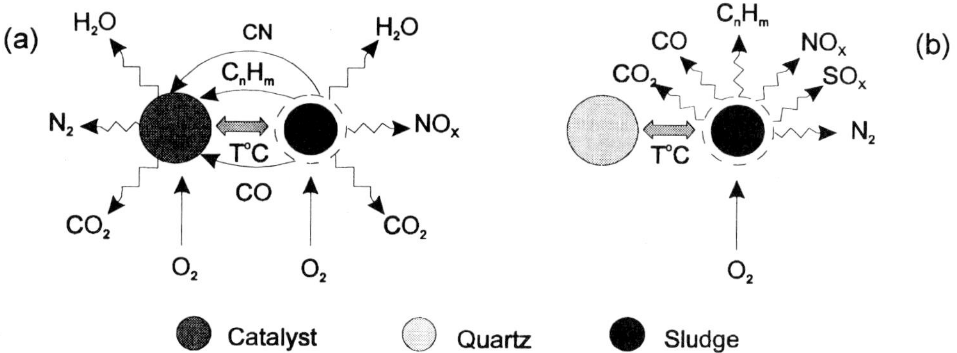

Figure 3.5. Combustion of a solid waste (e.g. sludge particles) on a heterogeneous catalyst (a) and on an inert material such as quartz (b).

nitrogen-containing admixtures in the fuel). When burning nitrogen-containing fuels in a CHG over the appropriate catalyst, the conversion of fuel-bound nitrogen to NO_x does not usually exceed 5–10%, compared to 40–100% with flame combustion. Thus, the NO_x concentration in flue gases (50–150 ppm) does not exceed the allowable limits for NO_x emission accepted in many countries of the world.

Nowadays, the following main directions of the efficient application of CHGs are identified [9]: heating and evaporation of liquids; drying of powdered materials, including coal; adsorption and contact drying of thermolabile chemicals, followed by adsorbent regeneration in a CHG; catalytic detoxification of highly concentrated exhaust gases and organic wastes with heat utilization; combustion of low-calorific fossil fuels such as wet biomass or wood wastes that are unfit for flame combustion; and processing of low-quality brown coals with simultaneous production of upgraded semicoke and/or syn-gas. In all these applications, the heat energy evolved in catalytic combustion is immediately utilized to accomplish an energy-consuming technological process.

In Russia, commercial CHG installations of *c.* 250 kW thermal power are under way and development is being performed to scale up these installations. Of top interest for the future is the possibility of using CHGs for the combustion of low-quality solid fuels, especially renewables such as dry and wet biomass, including sewage and other kinds of sludges, as well as municipal solid wastes, etc.

Catalytic combustion of solid fuels needs some comment here (see also [3,16,17]).

First of all, it is necessary to arrange optimum conditions for the catalyst to make the catalytic combustion of solid organic wastes more efficient. In many cases, the fluidized bed of dispersed catalyst allows a developed surface for contact of the catalyst with the solid waste. Moreover, in the fluidized bed, mass exchange is good at relatively low temperatures (700–1000 K). Solid organic particles react intensively with air, whereas the volatilized intermediates produced burn on the catalyst surface (see Fig. 3.5). Note that in flame or in the inert bed the volatile components and CO burn in the bulk.

Figure 3.6 shows typical data on conversion versus combustion temperature for some organic sediments burnt in a fluidized bed of catalyst and in an inert bed under identical conditions. The presence of the catalyst greatly increased the degree of combustion. Similar observations were obtained for the combustion of active sludge, micellar wastes from the production of drugs, city sewage and some agricultural wastes.

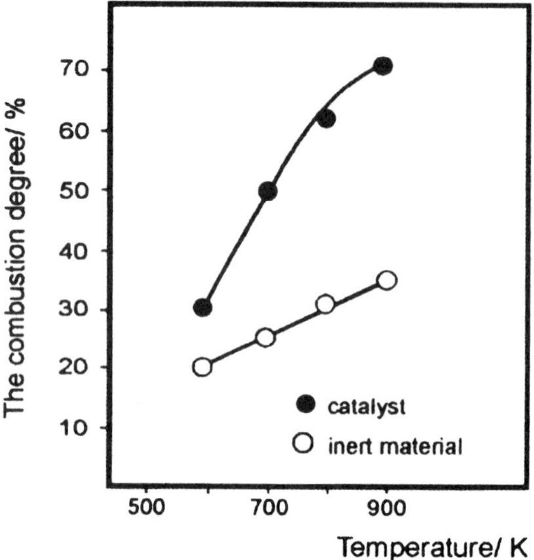

Figure 3.6. Comparison of the solid waste burn-off in fluidized beds of chromium-containing catalyst and of inert material (quartz) at different bed temperatures. (An example of the Baikal mill effluents, according to [16,17].)

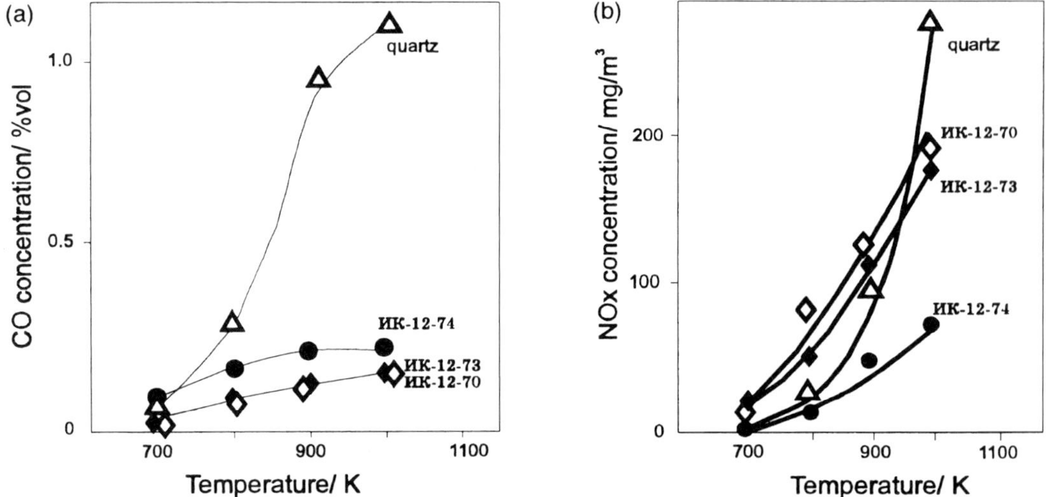

Figure 3.7. Change in CO (a) and NO_x (b) concentration in flue gases versus the temperature of the thermocatalytic treatment of an active sludge in fluidized beds of inert quartz and of different catalysts. (According to [16,17].)

Flue gases after the catalyst bed contain considerably smaller amounts of toxic products than after the inert bed (see the decrease in CO concentration in the active sludge combustion; Fig. 3.7a). Nitrogen entering the organic waste is mainly oxidized to molecular nitrogen. After the catalyst bed (Fig. 3.7b) the amount of nitrogen oxides does not exceed 150–200 mg m^{-3} (it is 300 mg m^{-3} or even higher after the quartz bed). No sulfur oxides were registered after the catalyst bed, while after the quartz bed at 1000 K the SO_x concentration was 200 mg m^{-3}. Sulfur compounds do not evolve into the gas phase owing to the low combustion temperature, but remain fixed in the mineral part of the waste in the form of sulfates [16,17].

The above technology of solid fuel combustion has numerous positive key features. The demands of the thermal and mechanical properties of the construction materials of the incinerator are not so severe; material wear due to erosion is reduced. Heat losses through the apparatus walls decrease, which allows the use of fuels with low calorific value and the autothermal regime to be achieved even with only 15–25% combustible organics in the fuel. The system operation also becomes simpler and the explosion danger decreases. The heat power loading of the furnace volume is very high and attains 10^8 kcal m^{-3} h^{-1} (10^5 kW m^{-3}); thus, the size and metal weight of the apparatus are reduced, heat losses with exhaust gases decrease; and there are no secondary endothermic reactions producing toxic compounds.

As the bed height or fluidization rate changes, the stay period of particles in the catalyst bed also changes. This allows the degree of combustion of the solid fuel to be controlled, decreases the evolution of volatile components and increases the carbonization of coke. The yield of solid product in the fluidized bed can attain 50–60%, depending on the sediment or solid fuel type. The texture of the resulting product allows its use as an adsorbent with a high specific surface area due to the low process temperatures. The products of the thermocatalytic treatment of solid wastes are competitive with activated carbons due to their adsorption properties.

Production of adsorbents via catalytic processing of solid industrial wastes allows an efficient combined production of technological heat with the utilization of sediments and the purification of industrial effluents. Large-scale pilot tests [16,17] have shown the possibility for implementing cyclic sediment treatment by using solid products for adsorption/coagulation purification of effluents. The excess of solid product is removed from the process and used as a commercial adsorbent.

To date, various bench and semi-industrial pilot plants based on these principles have been developed and tested by feeding with, for example, highly wet sludges from wastewater cleaning facilities and wastes from the pulp and paper industry [3,16,17]. The feasibility of using the same technology for ecologically pure gasification of biomass as well as for its drying has been shown.

Recently, a further modification of the fluidized bed catalytic reactor has been suggested by approaching an 'aerosol' rather than granular performance of the catalyst. The available data show a large increase (by several orders of magnitude) of the activity of the catalyst in aerosol mode, which no doubt will largely improve the economic parameters of the catalytic combustors.

2.4 *Outlook on new catalytic combustion and pollution control technologies*

Already, in the near future, ecologically safe catalytic combustion is expected to find wide industrial application in diverse technological processes that use the energy of organic fuels. Further development in this area is expected to be based on more detailed fundamental studies of the kinetics and mechanisms of heterogeneous and heterogeneous–homogeneous catalytic reactions of hydrocarbon oxidation and the development of new, efficient and cheap catalysts for combustion, as well as applied research into the development of new technologies for catalyst manufacture and new technological processes based on catalytic fuel combustion.

In view of the high efficiency and ecological advantages of catalytic combustion, it is expected to become widely used in energy production to meet local needs in various industries of the future.

One of the promising fields of application of catalytic combustion is space-heating devices using high-quality fuels: various heaters, catalytic panels, water-tube boilers, etc. Environmentally safe NO_x-free catalytic combustion is expected to find wide application for drying and treatment of agricultural products and for warming greenhouses with hot combustion gases free from pollutants. Another field of application is expected to be efficient energy-saving technologies on the basis of fluidized bed combustion for drying and thermal treatment of various powdered materials.

It seems expedient to look more closely at the perspectives for the use of catalytic combustion in commercial power plants because it provides a considerable reduction of pollutant emissions. This fact will be of much importance because of the growing consumption of low-grade fuels, which give rise to high pollutant concentrations in flue gases, with environmental emission standards becoming stricter. In recent years, heterogeneous catalysis has been applied for selective catalytic reduction (SCR) of NO_x in electric power plants (see Chapter 4). It is noteworthy that the amount of catalyst used in the SCR process is comparable to that needed for catalytic combustion. Hence, depending on such factors as the precise environmental situation, prices of catalysts and the progress in design and development of catalytic burning devices, the application of catalytic combustion in power plants might favorably compare to that of flame combustion together with the clean-up of flue gases.

3 New schemes of utilization of organic fuels in conventional thermal power energetics

In addition to improvement of the combustion processes with the help of catalysts under elaboration, there are also several approaches for increasing the efficiency of energy transformation in the chain: chemical fuel → heat → work. Many of these approaches to combustion improvement utilize particular chemical properties of carbon-containing fuels. Among these approaches one should mention precombustion of organic fuel for rotating the gas turbines.

One of the main ideas of the precombustion of fuel to increase the production of mechanical energy (work) or electricity is to utilize the possibility of stepwise combustion of organic fuels with intermediate formation of only partly oxidized fuel:

1 precombustion step: organic fuel + $O_2 \rightarrow CO, H_2O + Q_1$;

2 complete combustion step: $CO + \frac{1}{2}O_2 \rightarrow CO_2 + Q_2$.

An interesting peculiarity of CO oxidation is that it diminishes the entropy (by *c.* 86 J mol^{-1} K^{-1} at a not very high temperature) of the mixture due to a sufficient decrease of the mole numbers of the mixture. Entropic 'non-available' energy is thus liberated for the combustion process. The system, which, according to irreversible thermodynamics (linear range), approaches a minimum entropy production within the constraints of the system, may dissipate less energy.

This means that sometimes it is reasonable to convert into work, with the use of a gas turbine, the more mole-abundant CO-containing mixture after the precombustion step (which occurred in a deficiency of oxygen) and then add extra oxygen into this mixture to provide complete oxidation and utilize a new portion of hot gas for evolving the second portion of work. The amount of work produced during such a two-step combustion sometimes can be greater than after a single-step incineration. Simultaneously, the highest temperature of the mixture during such a two-step combustion can be lower than during one-step combustion, which sometimes is also helpful for both the turbine operation and the diminishing formation of nitrogen oxides.

It is of interest to note that there is also another way to increase the efficiency of gas turbines with the use of chemistry. It is well known that typically the technically available efficiency of gas turbines is limited by the maximum possible level of temperature allowed for the turbine's blades. Recently, some new approaches to cooling these blades during their operation were proposed; the approaches are based on covering the surface of the blades with catalysts that are able to provide endothermic catalytic cleavage or transformations of some non-oxidized molecules of the fuel. Consumption of the necessary amount of energy for such transformations on the surface of the blades will be followed by cooling of the blades.

4 Chemical processes and technologies for improving the production of mechanical energy: chemical heat recuperation

In the introduction to this chapter we have already mentioned that when using chemical fuels the Carnot cycle efficiency is, as a rule, much lower than the maximum work that can be produced by the same chemical fuels according to general thermodynamic laws. This follows from the fact that, for example, during fuel oxidation one may exploit not only evolved heat but, in principle, also all the changes in the Gibbs potential of the mixture under oxidation. Thus, the chemical machines for the production of work should be much more efficient than thermal engines.

This is well known from biology. Indeed, catalytic (enzymatic) processes form the basis for the metabolism of all living organisms, including mobile organisms [18]. This shows that catalysis can provide conversion of chemical to mechanical energy under rather mild conditions without transforming chemical energy into heat. Today, this possibility is not widely used in the energy industry, but perhaps it may become used in the future.

However, during the last decade, great interest has been paid to the application of catalysis for improving the efficiency of mechanical engines. One approach is based on improvement as well as stabilization or efficient control of fuel combustion in gas turbines or in piston engines.

Among the phenomena related to the conversion of chemical energy to mechanical energy with the help of catalysis, the provision of a catalytically active coating over gas turbine blades has been discussed. Perhaps this method could improve the efficiency of application of low-calorific fuels by their burning directly over the turbine blades. The possibility of introducing the catalyst inside the cylinder of internal combustion engines is also under consideration.

A second approach is based on the non-traditional chemist's idea of so-called '*chemical (catalytic) heat recuperation*' providing a significant (up to 10–15%) increase in the efficiency of a thermal engine, even over the value given by the conventional Carnot cycle of a one-step heat-to-mechanical energy transformation. This approach is based on the utilization of heat of exhaust gases for catalytic pretreatment of chemical fuels to enrich their calorific value [1–3]. By means of catalytic heat recuperation it is proposed to provide efficient utilization of middle- and even low-potential heat, which is now uselessly discharged into the atmosphere by internal combustion engines or gas turbines.

The main concept of such an approach is close to the principle of heat pumps and can be demonstrated by the example of a gas turbine with methanol as a fuel (Fig. 3.8). By the traditional mode of operation, the fuel is burned before the turbine inlet and the hot gas is directed to the blades in order to make them rotate; then, having served its purpose and being partially cooled but still reasonably hot, the gas is discharged into the atmosphere. To provide catalytic heat recuperation, it is suggested that before releasing this hot gas into the atmosphere it should be passed through a catalytic heat exchanger and its heat energy used for conversion of a certain amount of methanol by

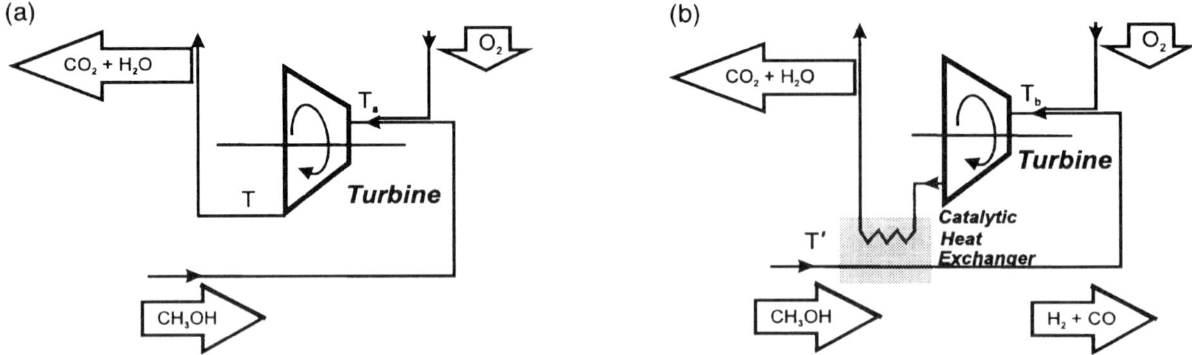

Figure 3.8. Schemes of gas turbine operation: (a) conventional; (b) catalytic heat recuperation. For (b), a hot steam–gas mixture of temperature T from the turbine is used to enrich the starting fuel (in this case methanol) with energy via accomplishment of an endothermic reaction in the catalytic heat exchanger; T_a and T_b are the temperatures of the steam–gas mixture just after burning the fuel and before feeding the mixture to the turbine blades ($T_a < T_b$); T and T' are the temperatures of the steam–gas mixture at the outlet of the devices ($T > T'$). Because the calorific value of the fuel (methanol) is increased by its pretreatment with hot exhaust gases of the turbine, it also increases the total energetic efficiency of the chemical-to-mechanical energy conversion.

means of an endothermic reaction to some energy-enriched products, e.g. syn-gas. For Fig. 3.8, this desirable reaction is [19,20]:

$$CH_3OH \rightarrow CO + 2H_2$$

which occurs easily on properly chosen catalysts at a temperature below 650–700 K.

Because the calorific value of the so-formed mixture of methanol and syn-gas is higher than that of pure methanol, its burning might, according to some estimates, provide a 10–20% increase in the efficiency of fuel utilization compared to the conventional method of simple methanol burning.

There are a number of projects of this type, some of which have already been tested on a pilot plant scale. Attention should be paid to the search for appropriate endothermic catalytic processes employing available fuels; the main difficulty here is the requirement to attain a sufficient endothermic conversion of the reaction mixture at a relatively low temperature of the exhaust gases (typically 500–600 K) that is much lower than the temperature level for operating similar processes for the nuclear and solar energy industries (see Chapter 12). Thus, new types of thermally sensitive fuels, as well as more active catalysts, should be designed for these endothermic processes.

An unexpected implementation of this idea is the use of 'catalytic heat recuperation' for cooling the overheated units of hypersonics with a simultaneous increase of the efficiency of the propulsion ability due to enrichment of fuel with energy when passing it through aircraft units to be cooled (see [3] and Fig. 3.9). Thus, in this situation, it is possible to take two coupled advantages of catalysis: make the aircraft's material demands easier and diminish the fuel consumption.

Recently, an interesting remark has been made about the possibility of efficient use of some 'monofuels' (i.e. fuels that do not need separation of their chemical components for storing the chemical energy) capable of experiencing reversible chemical transformations for direct driving of thermal engines (see Section 5 in Chapter 12 and [3,21]).

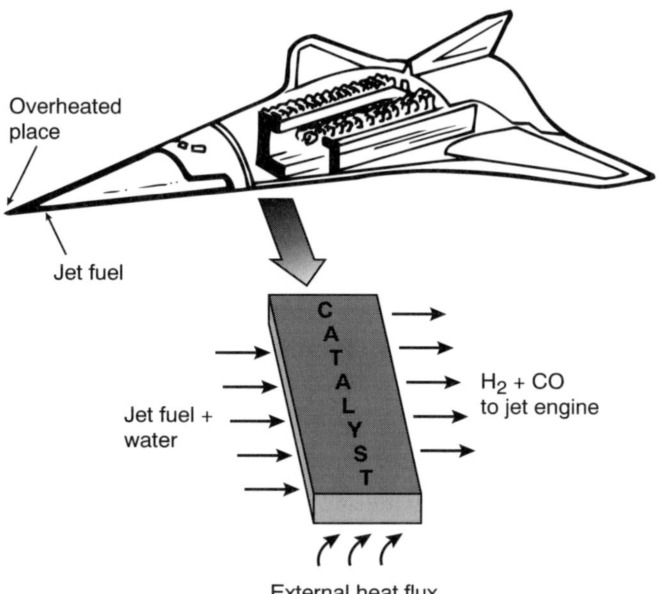

Figure 3.9. Thermocatalytic pretreatment of jet fuel conjugated with cooling of the aircraft in the Russian hypersonic (velocity *c.* 10 000 km h^{-1}) 'Neva' for the 21st century [3].

5 Improving traditional energetics via the creation of energochemical complexes and/or the energochemical industry

Another chemist's approach for improving the efficiency of utilization of available chemical fuel is the suggestion to erect so-called 'energochemical complexes' instead of conventional thermal power plants. The main idea of such complexes is the utilization of available fuels such as coal, biomass, etc. for *co-generation* of both mechanical energy (and/or electricity) and some valuable chemical products. Direct co-generation of electricity and valuable chemicals will be discussed in Chapter 8. Here we shall consider only energochemical complexes based on the conventional thermal power industry.

The most advanced projects of this type are based on the idea of coupling the production of heat (and then mechanical energy) with the large-scale synthesis of methanol. To do this, it is suggested that primary energy carriers (usually of moderate quality, such as coal or biomass) be used first of all for their total or partial pregasifying into syn-gas:

$$\text{Primary fuel} + O_2 \rightarrow CO, H_2 + Q_1$$

rather than making a complete incineration to carbon dioxide and water.

Heat Q_1 of hot syn-gas can be utilized in numerous ways, e.g. to rotate steam or gas turbines, etc., while the obtained syn-gas is used for providing the exothermic synthesis of methanol:

$$CO + 2H_2 \rightarrow CH_3OH + Q_2$$

The syn-gas mixture can also be used to power a fuel cell. Note that the heat Q_2 evolved during the methanol synthesis (which occurs at *c.* 500 K) can also be used for some needs of the energy industry.

There is a very important advantage of such technological schemes over simple complete incineration of primary fuel. Indeed, optimization of this scheme includes the jointly optimized production of not only energy for immediate use but also a more valuable chemical fuel such as methanol (see Chapter 6) that can be either stored at the location of the energochemical complex in order to be used at the same thermal power plant at some selected period or transported somewhere as a chemical or a fuel for sale. Optimization of this energochemical industry can be done in various ways, either from the point of view of total energetic efficiency or from the point of view of economic assessment, etc.

The existing data of such analysis show the clear potential of the mentioned energochemical approach for improving the production or consumption of energy in many situations. The most advantageous seems to be erection of the energochemical industry near deposits of cheap but low-quality coals whose transportation, without upgrading, is not profitable.

6 Conclusions

The above data show the existence of numerous ways for chemistry and chemical technologies to influence the future of the energy industry based on conventional carbon-containing combustibles. Some of the chemist's approaches here are based on already well-developed and approved chemical technologies such as catalytic combustion, and their application can be scaled up immediately. On the contrary, the other very interesting approaches are based on new non-traditional ideas to increase the production of work when using the same amount of fuel via improvements of the irreversible thermodynamic basis of the processes with the participation of chemically reactive systems. These ideas are still to be tested and approved experimentally.

Note that nearly all of the approaches discussed in this chapter are based on *catalytic* processes. This confirms once more the great importance of catalysis in all that is done in the joint area of chemistry and energetics.

7 References

1 Parmon VN, Kerzhentsev MA, Ismagilov ZR. In Thomas JM, Zamaraev KI (eds) *Perspectives in Catalysis. The IUPAC Chemistry for the 21st Century 'Monograph'*. Oxford: Blackwell Scientific Publications; IUPAC 1992; 337.

2 Zamaraev KI, Parmon VN, Ismagilov ZR. In Behrens D (ed.) *Strategies 2000. Proceedings of the Fourth World Congress of Chemical Engineering, Karlsruhe, Germany, 16–21 June, 1991*. Karlsruhe: Dechema, 1992; 49.

3 Parmon VN. *Catal. Today*. 1997; **35**: 153.

4 Collins PM. *Platinum Met. Rev*. 1986; **30**: 141.

5 Trimm DL. *Appl. Catal.* 1983; **7**: 249.

6 Prasad R, Kennedy LA, Ruuckenstein E. *Catal. Rev. Sci.-Eng.* 1984; **26**: 1.

7 Kesselring JP. In *Advanced Combustion Methods*. London: Academic Press, 1986; 238.

8 Pfefferle WC, Pfefferle LD. *Catal. Rev. Sci.-Eng.* 1987; **29**: 219.

9 Boreskov GK, Levitskii EA, Ismagilov ZR. *Zh. Khim. Obtsch. im. D.I.Mendeleeva* 1984; **29**: 379 (in Russian).

10 Boreskov GK. *Heterogeneous Catalysis*. Moscow: Nauka, 1986 (in Russian).

11 Boreskov GK. In Anderson JR, Boudart M (eds) *Catalytic Science and Technology*, Vol. 3. Berlin: Springer Verlag, 1982; 39.

12 Enga BE, Thompson DL. *Platinum Met. Rev.* 1979; **23**: 134.

13 Krill WV, Kesserling JP, Chu EK, Kendall RM. *Mech. Eng.* 1980; **102**: 28.

14 Tucci ER. *Hydrocarbon Process*. 1982; **61**: 159.

15 Dalla Betta RA. *Catal. Today*. 1997; **35**: 129.

16 Simonov AD, Yazykov NA, Jepifantseva EI, Wolf IV, Chernoberezhsky YuM. *Chem. Sustain. Dev.* 1993; **1**: 298.
17 Wolf IV, Chernoberezhsky YuM, Jepifantseva EI, Simonov AD. *Water Sci. Tech.* 1991; **24**: 357.
18 Gray CJ. *Enzyme-Catalyzed Reactions*. London: Van Nostrand Reinhold, 1991.
19 Spilrain EE. *Izv. Akad. Nauk. SSSR, Ser. Energ. Transp.* 1985; **6**: 115 (in Russian).
20 Tada A, Watarai Y, Takahashi K, Imuzu Y, Iton H. *Chem. Lett.* 1989; 543.
21 Prokop'ev SI, Aristov YuI, Parmon VN. *Int. J. Hydrogen Energy* 1997; **22**: 415.

4 Role of Chemistry in Improving Environmental Safety and Pollution Control of Traditional Sources of Energy

V.N. PARMON[1] and A.V. BRIDGWATER[2]

[1] *Boreskov Institute of Catalysis, Siberian Branch of the Russian Academy of Sciences, Prospekt Akademika Lavrentieva 5, Novosibirsk 630090, Russia*

[2] *Bio-Energy Research Group, Chemical Engineering, Aston University, Aston Triangle, Birmingham B4 7ET, UK*

1 Introduction

Traditional large- and medium-scale sources of energy are based mostly on the incineration of various kinds of available organic fuels, polluting the environment in a variety of ways.

Evidently, the pollution from traditional energies arises mostly from:

1 transfer of poisonous or harmful contaminants contained in the fuel at the start (e.g. sulfur, arsenics, heavy metals, etc.) to the ambient through the flue gases or wastewaters;

2 harmful by-products forming with conventional fuel use when it is incinerated in the presence of air (the most typical examples of this type are nitrogen oxides and oxygenates, as well as dioxins);

3 incomplete use of the fuel accompanied by harmful exhausts of, for example, soot, residual hydrocarbons, oxygenates, carbon monoxides, etc.;

4 pollution with CO_2;

5 heat pollution.

A large body of specialized and popular literature exists on the subject of environmental improvement of traditional sources of energy, and chemistry is well recognized in all spheres to be the main tool for protecting the environment from most kinds of pollution.

In this chapter, we will not discuss all the possible applications of chemistry and the respective achievements in this area, but will outline only the main directions and ideas of the possible impact of chemistry on the environmental safety of the traditional thermal energies. Very specific chemical and physicochemical problems of environmental safety are detailed elsewhere for traditional nuclear energies.

Mechanical treatment and cleaning of flue gases to get rid of dust and other particulate contaminants are not discussed. Here, we concentrate on only those contaminants needing chemical transformations for their removal. The disposal of solid and liquid wastes from large sources of energy and liquid wastes from other industries and transport is not discussed, nor is the possibility of using ash from coal-based thermal power plants as a rich feedstock for metallurgy and non-organic chemical industries and as an ingredient of cement or concrete.

There are three main chemist's approaches to solving the problem of cleaning the exhausts of thermal power plants: to clean the flue gases and wastewaters themselves; to clean and upgrade the initial fuel for the plants; or to arrange better combustion of the fuel.

2 Chemical tools for cleaning the flue gases

The easiest way to diminish the contents of sulfur, arsenics and heavy metals in the flue gases is to lower the temperature of the flame (or other type of fuel oxidation process), thus creating the conditions for these contaminants to remain bound with ash as their non-volatilized oxide or salt forms at temperatures below 900–1000 K. Evidently, this approach can be utilized only at the stage of catalytic oxidation of the fuel. A distinct success with this technique has been achieved when oxidizing fuel using a fluidized bed of catalyst (see Section 2.3, Chapter 3).

However, in the common thermal power plants, the temperature of combustion is too high to prevent emission of the mentioned oxides. For this reason, these are very common components of flue gases and their removal is needed. Chemical processes are used nowadays for the removal of volatile sulfur and arsenic oxides, the other toxic compounds being removed mostly by their absorption or mechanical filtering. Chemical transformations are also needed for preventing the emission of nitrogen oxides (N_2O, NO and NO_2).

The main chemical and technological problem in the chemical cleaning of virtually any flue gas from thermal power plants is usually the small concentrations of contaminants and the huge flows of the gases. A typical magnitude for these flows can be roughly estimated as 1000 m^3 MW^{-1}.

2.1 *Eliminating sulfuric contaminants*

The main contaminant of flue gases of thermal power plants is SO_2. Its emission in the atmosphere due to the operation of thermal power plants is huge and exceeds 100 million tons per year, the energy industry being among the largest contaminators of the atmosphere with this compound. Tremendous local concentrations of SO_2 in the atmosphere over some industrial territories are the main reason why acid rain has a serious influence on forests, harvesting, architectural monuments, etc. Emission of other toxic sulfuric compounds, such as H_2S, in the flue gases of common thermal power plants is not significant due to the typically clear oxidative conditions when the fuel is burned.

Indeed, the best way to prevent pollution with sulfur compounds is to preclean the initial fuel to eliminate sulfur contamination, but this appears to be very costly when using coal or non-synthetic liquid fuels produced from fossil oil. Thus, the elaboration of tools for eliminating sulfur compounds from the flue gases will remain a long-term problem for the large- and medium-scale sources of energy of the first half of the 21st century.

Sulfur dioxide as a gas is not very chemically reactive so it does not bind very easily, in contrast to the higher oxide, SO_3. For this reason, with conventional thermal power energies, the chemical reactions important for environmental protection are generally the various types of complete oxidation of SO_2 with oxygen in the air:

$$SO_2 + O_2 \rightarrow SO_3$$

and further conversion of SO_3 either to commercial sulfuric acid or its absorption by limestones, etc. to form sulfate salts. The main problem for SO_2 oxidation is the very low content of SO_2 in the very large quantities of flue gases. For this reason, the best way would be to apply SO_2 oxidation catalysts operating at the temperature of the power plant outlet gases. Unfortunately, such catalysts are still not available on a commercial scale. Thus, in order to exclude additional expensive heating of large flows of gases (to push the oxidation catalysts into operation) and thus reduce the consumption of energy for the cleaning process, some sophisticated catalytic techniques (e.g. 'reversing' the gas stream in order to improve recovery of the heat evolved at oxidation) or intensive physical methods (e.g. electron beam irradiation of the flue gas stream) for stimulating the SO_2 oxidation reaction could be used. Note, however, that modern 'pure' catalytic techniques of SO_2 oxidation can consume only a negligible amount of energy for their operation but need sophisticated equipment, and the utilization of electron beams demands a sufficient amount of electricity, sometimes reaching 5–10% (or even more) of the total electricity output of the plant.

For these reasons, an ordinary and widely applied way of removing sulfur oxides from smoke gases of thermal power plants is their simple absorption by wet or suspended calcia, calcium carbonate or limestone. When absorbing SO_2 and/or SO_3 from the flue gases with calcinated limestone, gypsum is created:

$$SO_2, SO_3 + CaO, CaCO_3 (+ O_2) \rightarrow CaSO_4\downarrow + CO_2\uparrow$$

Thus, each modern power plant equipped with such cleaning systems appears to be a huge generator of gypsum, which can be used further as a construction material.

Another chemist's approach to eliminating SO_2 is its absorption in scrubbers by specially buffered water or other solvent, with consecutive low-temperature 'wet' oxidation with air oxygen over homogeneous or heterogeneous catalysts. Arsenic oxides can also be removed by such scrubbing with consecutive wet catalytic oxidation.

Thus, a variety of chemical methods are known for avoiding pollution of the atmosphere by sulfuric compounds emitted by thermal power plants. The main problem in implementing these technologies is their construction or energy cost.

2.2 *Eliminating nitrogen oxides*

Nitrogen oxides such as NO and NO_2 are emitted in enormous amounts by thermal power plants and are generally denoted typically as NO_x. They are well known as being not only toxic but also ozone-depleting gases. The source of NO_x in flue gases is nitrogen either from the air or from some nitrogen-containing fuel components existing both in coal and oil. Formation of NO_x from air originates from the particular thermodynamic properties of molecular nitrogen creating nitrogen oxides at high temperatures (above 1200 K); thus, the creation of 'thermal' NO_x can be avoided only by lowering the temperature of combustion (see Chapter 3).

Nitrogen oxides are nevertheless formed and present in flue gases and the main method for their removal is selective reduction of NO_x to N_2, typically in the presence of residual O_2. The most inert oxide for selective reduction is NO. Widely applied now is the so-called DENOX catalytic technique where ammonia is used as a selective reducing agent for NO and NO_2 and is capable of providing reduction at low enough temperatures of the outlet flue gases even in the presence of residual oxygen:

$$NO_x + NH_3 (+ O_2) \rightarrow N_2 + H_2O$$

Active catalysts with a very low pressure drop were specially designed for the DENOX technique to reduce NO_x in large streams of flue gases. At the moment these catalysts are mostly honeycombs of V–W–Ti oxides. Nevertheless, new kinds of selective catalysts and respective techniques for operation in moderately hot smoke gases are under intensive elaboration (technologies, e.g. NO_x reduction with pulverized urine, are under test).

In spite of the wide practical application of NO_x reduction with ammonia, this method has numerous shortcomings. Indeed, ammonia itself is expensive and toxic, thus demanding very precise techniques for controlling its introduction into large streams of flue gases, etc. For this reason, selective reduction of NO_x with more available, cheaper and non-toxic reducing agents other than ammonia is now under intensive development. Available hydrocarbons or cheap oxygenates such as methanol and formaldehyde are generally considered suitable chemical reductants.

Experiments show that catalytic reduction of NO_x is easily accomplished with either light olefins or a propane–butane mixture. However, of most interest is the use of methane or natural gas as the reducing agent. Unfortunately, methane at the moderate temperatures of the smoke gases is hardly activated and suitable non-expensive catalysts have still not been found.

A possible way to overcome the inertness of methane and other light alkanes could be their pre-activation with the formation of more chemically reactive species serving as secondary reducing agents for NO_x at its low-temperature reduction. One of the possible approaches here is *in situ* creation of oxygenates via homogeneous or heterogeneous partial preoxidation of methane:

$$CH_4 + O_2 \rightarrow CH_3OH, CH_2O, \text{etc.}$$

The oxygenates obtained are much more reactive for the reduction of NO_x in comparison with hydrocarbons and are cheap because of their *in situ* production.

Under development also are catalysts for *direct* cleavage of NO_x:

$$NO_x \rightarrow \tfrac{1}{2}N_2 + \tfrac{x}{2}O_2$$

At temperatures below 1100 K such reactions are thermodynamically allowed for all the nitrogen oxides but unfortunately proceed very slowly, even in the presence of the best-known catalysts. Thus, a lot of work is still needed to put this simple chemist's idea into practice.

Recently under discussion as a serious pollutant is N_2O, although this gas is widely used in medicine and for many years was considered not to be toxic. Moreover, in most countries in the past there were no national regulations against its emission. The reason for the current attention to N_2O is its possible global environmental impact through depletion of the ozone layer.

Removal of N_2O from flue gases can be performed in traditional ways through either catalytic reduction or catalytic cleavage into N_2 and O_2. Now considered as only a harmful by-product of some energetic or chemical processes, N_2O will be paid much more attention in the future because it can be used instead of O_2 or H_2O_2 as an excellent and very selective oxidant for providing important organic synthesis, e.g. the production of phenol from benzene, etc. [1].

2.3 *Eliminating soot, carbon monoxide and other products of incomplete combustion from the main components of fuels*

The main way to lower pollution with CO and other products of incomplete combustion of carbon-containing fuels is either to improve the flame combustion or, according to the chemist's approach, to use new catalytic methods or catalysts of combustion (see Chapter 3).

As shown in Chapter 3, modern catalytic technologies allow the incineration, completely or in a controllable manner, not only of high-grade fuels but also even of low-quality solid feedstocks such as coal, wet biomass, etc.

2.4 *Prevention of creation of harmful by-products*

Incomplete oxidation of carbon-containing fuels may be accompanied by the creation of harmful aromatic compounds such as carcinogenic benzopyrenes, chlorinated compounds of dioxins (in the presence of some chlorine-containing impurities), etc.

The main tool to prevent the formation of such harmful compounds is to increase the completeness of combustion (oxidation to CO_2 and H_2O). Thus, for their removal, it is necessary either to increase the temperature of combustion to ensure destruction of these compounds or, vice versa, to diminish the temperature of combustion to provide complete oxidation of the fuels over catalysts.

Of the newest non-traditional technologies, which are now under laboratory-scale test, one can mention photocatalytic treatment of flue gases with small concentrations of some contaminants, especially of an organic nature [2,3]. Based on the utilization of very expensive UV light, these technologies can be economically suitable for treating very dilute gas mixtures because the light quanta are able to initiate highly efficient activation of (photo)catalysts even at very low temperatures, thus competing with other technologies in energy efficiency.

2.5 *Challenges between precleaning of fuels, more complete and clean but more expensive technologies of combustion and treating the flue gases*

Evidently, sulfur, arsenics, chlorine and heavy metals can be removed before the use of the chemical energy carriers, which is why in many countries much attention is paid to the production of high-quality clean fuels, with serious limitations being introduced on the purity of the fuels for large-scale thermal energies. Under especially strict limitations now are the contents of sulfur and heavy metals in coal, gasoline and diesel fuel [4]. Such limitations are being adopted more and more in many countries, but make fuel more expensive owing to the need for high levels of pretreatment. A radical but even more expensive way to avoid such contaminants in the first place is the production of *synthetic fuels* from methane or clean syn-gas (see Chapters 6 and 7).

In contrast to the above-mentioned contaminants, harmful compounds such as nitrogen oxides, carbon monoxide and carcinogenic hydrocarbons appear to be a consequence of the combustion process itself because they are created during combustion of the fuels.

For this reason, it would be natural to ask: what is more advanced from the economic, environmental and social points of view — preliminary cleaning of fuels to achieve nearly total removal of harmful compounds, or cleaning of the smoke gases after combustion of the fuel? Today, there are no unambiguous answers to this question. Also, according to the opinion of some experts, sometimes it is cheaper to use a more expensive incineration technique than to add complicated systems to treat flue gases.

3 Reducing pollution of the atmosphere with carbon dioxide

Carbon dioxide is considered now as one of the main 'greenhouse' gases. Accumulation of such gases leads to shielding of the Earth's surface from the escaping long-wave infrared radiation, which is analogous to what happens in real greenhouses covered with glass. According to some estimates (but these are still not very well adopted by the decisive majority of the critically thinking part of scientific communities), this accumulation and shielding will result in a sufficient non-reversible increase of the average global temperature of the Earth, resulting in catastrophic changes of the planetary climate. Data supporting such a scenario of climate change on Earth are those on the tight relation between the content of CO_2 in the Earth's atmosphere and the average global temperature of the Earth: the higher the CO_2 concentration, the higher the average temperature. Currently, the world scientific community has accepted the data on the steady increase of the average temperature of the Earth, which seems to be *c*. 0.5 °C higher now than several decades ago.

The annual emission of CO_2 into the atmosphere is huge, at over 3×10^9 tons (see Chapter 2), and provided mostly by both large-scale centralized and distributed small-scale local thermal sources of energy [5,6]. Clearly, it will not be possible to diminish the emission of CO_2 into the atmosphere very quickly. Indeed, even in the case of modernization of the sources of energy in the industrially developed countries, the developing countries, which consume mostly moderate-quality coal and biomass as fuels, will continue to emit CO_2 in large amounts. However, there are several obvious ways to diminish pollution of the atmosphere with CO_2.

Under consideration are numerous, perhaps curious, ideas on how to stockpile the CO_2 produced by thermal power plants, e.g. by pumping CO_2 underground or to the bottom of deep seas, freezing and stockpiling CO_2 as a solid, etc. However, these methods do not seem to be very realistic owing to their impracticability or large energy demands.

Under consideration is also the absorption of CO_2 from chimneys before it is emitted to the atmosphere, followed by further chemical transformations to produce valuable chemicals, etc. (see Section 4, Chapter 7). However, this way of reducing atmospheric pollution with CO_2 also seems to be very expensive from the point of view of energy consumption: the amount of energy that has to be introduced is equivalent to that gained during the resultant fuel combustion.

An important and much more realistic way to avoid pollution of the atmosphere with CO_2 seems to be a shift of the large-scale sources of energy to nuclear power plants or power plants based on renewable sources of energy (solar, wind, tidal or geothermal energy, renewable biomass, etc.) [4–6]. Some of these approaches related to chemistry are discussed in Chapters 9–13.

A second realistic way is to increase the output of work or electricity from the same amount of traditional energy carriers consumed by power plants and thus to diminish the specific fuel consumption. The most radical chemical tools here are extremely efficient fuel cells (see Chapter 8), as well as new energy conversion technologies based on ideas of 'chemical heat recuperation' (see Chapter 3).

A third very promising way is to use fuel either without carbon at all or with a diminished quantity of carbon. The best fuel for the distant future should be pure hydrogen: its incineration leads to the formation only of water (see Chapter 5). However, for the near future, the most realistic step will be increased use of natural gas (methane) as a widely available hydrogen-rich fuel [4,5]. Indeed, the emission of CO_2 on using this fuel is much less than when using coal or fuel oil (see Table 4.1).

Table 4.1. Comparison of amount of CO_2 emitted from different types of natural fuels on complete combustion according to the amount of heat produced.

	Emission of CO_2	
Fuel	mol MJ^{-1}	kg $Gcal^{-1}$
Coal	2.75	506
Gasoline	3.14	577
Wood	1.47	270
Natural gas	1.38	254
Hydrogen	No	No

From [6].

A possible way to upgrade an available fuel for diminishing its CO_2-forming ability could be its pre-enrichment with hydrogen via chemical treatments, e.g. by partial or complete pyrolysis with removal of solid carbon for disposal (see Chapter 5).

Note that even in the case of total removal of CO_2 emissions from the thermal energy industry, the next largest emitter of CO_2 (transport) will still remain (see Chapter 2). To prevent emission from transportation vehicles would require the radical method of substitution of motor fuels by hydrogen.

4 Chemistry in reducing heat pollution of the Earth and its atmosphere

Anthropogenic heat pollution is still not very dramatic on a planetary scale. However, it does cause local climate changes in very urban areas such as the huge urban complexes in many countries. The future development of civilization will obviously increase the anthropogenic heat pollution, which could result in much more serious consequences.

When producing work or electricity via the intermediate production of heat, some heat loss is inevitable. Moreover, complete dissipation of work and/or electricity produced by thermal power plants should result in heat emissions that are equivalent to the total calorific value of the fuels consumed. Thus, the only way to reduce heat pollution is either to increase the efficiency of conversion of the calorific ability of fuel into work (see Chapter 3) or to utilize some of the heat loss to satisfy human needs without the use of additional fuels (such as in heat pumps, heat storing materials, etc.). The possible role of chemistry in the latter approaches is discussed in Chapter 12.

An important way of reducing or, in principle, totally removing heat loss on a global scale is possible by shifting the thermal sources of energy from fossil fuels to renewable kinds of fuels such as biomass or biomass derivatives. Indeed, combustion of biomass will not change the thermal balance of our planet because the calorific ability of biomass originates from solar energy incident on the Earth and thus is included in the heat balance of the planet from the beginning. The role of chemistry in biomass utilization is discussed in Chapters 10 and 11.

5 Conclusions

As one can see from the above discussion, the role of chemistry in improving the environmental safety of traditional large-scale thermal sources of energy is crucial. The most evident and already widely exploited implementation of chemistry here is in the improvement of combustion processes to reduce atmospheric pollution with toxic compounds or in cleaning the flue gases to remove these compounds.

For the more distant future, one can expect another important role of chemistry capable of reducing pollution of the atmosphere with not only toxic gases but also CO_2, the main constituent of 'greenhouse' gases. This reduction in pollution should occur first of all through *enrichment* of the main mass of the 'integrating' chemical energy carriers with *hydrogen*. This enrichment may proceed by shifting the nature of the main fuels not only to pure hydrogen or hydrogen-containing fuel mixtures but also, at the first stages of amendment of the technologies, mainly to hydrogen-enriched hydrocarbons such as methane. An expected cost of such large-scale shifts is extremely high and comparable with that claimed in Section 4 of Chapter 2.

The next task for chemistry in creating a sustainable future will be, of course, creation of the ability to involve the widespread use of *renewable* fuels such as biomass, which completely avoids pollution of the Earth's atmosphere with both CO_2 and anthropogenic heat.

6 References

1 McCoy M. *Chem. Mark. Rep.* 1996; **250**: 1; *Chem. Week.* 1997; **1/8**: 11.

2 Ollis DS, Al-Ecabi H (eds). *Photocatalytic Purification and Treatment of Water and Air.* Amsterdam: Elsevier, 1993.

3 Vorontsov AV, Savinov EN, Barannik GB, Troitskÿ VN, Parmon VN. *Catal. Today* 1997; **39**: 207.

4 Courty PR, Chauvel A. *Catal. Today* 1996; **29**: 3.

5 Anonymous. Discussion of the 1997 Kyoto Protocol. *Oil Gas Journal* 1997; **95** (50): 17.

6 Gamburg DYu, Dubovkin NF (eds). *Handbook on Hydrogen: Properties, Production, Storage, Transportation, Application.* Moscow: Khimiya, 1989; 37.

5 Challenge of Hydrogen as an Energy Carrier of the Future

H. TRIBUTSCH[1] and V.N. PARMON[2]

[1] Freie Universität Berlin and Hahn-Meitner Institut, Dept. Solare Energetik; 14109 Berlin, Germany

[2] Boreskov Institute of Catalysis, Siberian Branch of the Russian Academy of Sciences, Prospekt Akademika Lavrentieva 5, Novosibirsk 630090, Russia

1 Introduction

From the viewpoint of practicability of chemical energy carriers, technology has seen a clear development from heavy and difficult-to-handle energy carriers, such as wood and coal, to light and easy-to-handle fuels, such as oil and natural gas. The use of hydrogen, which is the lightest gas, would be a natural consequence for energy technology of the future. Whereas wood and coal require much manual work, oil and gas can be handled easily by automatic installations. The main use of oil in the field of energetics is still its direct burning for energy generation, but hydrogen can be used easily without direct burning in fuel cells and in catalytic energy converters. Since Captain Nemos' hint in Jules Verne's story (1869) at the energy being available from water, many personalities involved in research and technology have contributed to the concept and to the science and technology of a hydrogen energy economy. In such an economy, hydrogen is assumed to be the excellent and environmentally clean, integrating energy carrier combined with a very high potential for other industrial applications.

The elegance with which hydrogen can be generated through electrolysis, its efficient and cheap transport as a gas and its energy-liberating reaction with readily available oxygen to form water have inspired much technological imagination. The hydrogen energy system is clean and non-polluting. A minimum of mass is transported during the energy transfer over long distances and the product of energy conversion is water, which is not only harmless but may also be useful as a by-product in arid regions. For a community consuming 10 kW of energy per capita, there would be an approximate daily generation of 14 gallons of fresh water per person. This would be sufficient for life in desert regions. This shows that a hydrogen economy system powered by clean solar energy would not only lead to sustainability, but would also help to develop desert regions that could not, up to now, be included in the world economy systems. Additional advantages of a solar-powered hydrogen system would be a clean-up of polluting metallurgical and refining processes.

Many metallurgical and chemical processes could be carried out much cheaper with hydrogen than with other chemical reducing agents such as carbon monoxide or pure carbon. Hydrogen would also guarantee that the processes would become inherently non-polluting. The same is true for many other types of chemical technologies. Hydrogen and oxygen, which are generated as products of water electrolysis, could also be used for upgrading effluents or industrial or household waste. Thus, hydrogenation of chemical wastes would be a very reasonable long-term strategy for recycling such wastes. Instead of oxidizing the waste and increasing its volume many times, hydrogenation would transform waste back to basic chemical energy carriers comparable with oil or natural gas.

Additionally, hydrogenation is the main tool for refining the heavy oil and carbon-containing species of oil sands and shales, as well as coal, into common hydrocarbon fuels. Conversion of renewable plant biomass into such fuels can also be done through the hydrogenation of biomass.

The advantage of hydrogen, with its high energy density and its light weight for air transport, has frequently been discussed and has also led to some pilot projects. The same is true for road and sea transport, where hydrogen in combination with fuel cells that power electro-motors could revolutionize the present propulsion system and would also result in great progress in pollution control.

Many aspects of a hydrogen economy of the future have been discussed in different works, especially those by J.O'M. Bockris [1] and by T.N. Vesiroglu (see, for example, [2,3]). A hydrogen economy would have a surprising analogy to the photosynthetic energy system evolved in nature. In both cases water would be split with non-exhaustible regenerative (renewable) solar energy, and water would then form again. The difference would mainly be that the hydrogen species in biological systems are readily attached to carbon-containing molecules for further transport and utilization of energy.

Mankind is aiming at a hydrogen economy based on the use of hydrogen as a gas but, in analogy with biological systems, one could gradually learn to use chemical energy carriers to which hydrogen is attached. Such energy carriers may be more easy to handle than hydrogen gas and could also provide additional advantages. For example, lipids (fats) used as energy storage compounds in biology have a slightly higher energy density than conventional fossil fuels, and nature has succeeded in generating and reactivating this type of energy storage apparently without major complications. Mankind may, on the basis of a hydrogen economy, gradually develop additional hydrogen-containing fuels that would allow a diversification of energy use. Below, a few basic principles and technical challenges of hydrogen generation, storage and conversion into other energy forms will be discussed.

2 Production of hydrogen

2.1 *Traditional production of hydrogen from organic substances*

The present scale of world hydrogen production is large, achieving more than 10 million tons per year. The major part of this hydrogen is produced via catalytic steam reforming of natural gas (methane) or other light hydrocarbons, followed by a water–gas shift reaction [2–6]:

$$CH_4 + H_2O \rightarrow 3H_2 + CO$$
$$CO + H_2O \rightarrow H_2 + CO_2$$

The scale of application of these well-elaborated processes is comparable with that of the other main large-scale chemical technologies and satisfies the world production of ammonia, methanol, etc. However, from the point of view of chemical engineers, the above processes of hydrocarbon reforming with steam have many disadvantages owing to their endothermicity. In other words, one must supply a large amount of thermal energy inside the reforming reactors in order to carry out the target chemical reaction. Such energy supply is presently obtained mostly through the thermo-conducting steel walls of the reactor, which greatly restricts its specific power loading and thus results in the chemical plants, etc. having to be enlarged.

For this reason, many attempts continue to be made to improve the technology of hydrogen production from light hydrocarbons [4–6]. One of the promising approaches here is to add oxygen to the reaction mixture and thus evolve heat (of oxidation), which is needed for the reforming, just inside the reactor without necessitating any heat transfer through the walls. The real success in this area, achieved only recently [7], is providing a 'dry' process of nearly 100% 'partial' oxidation of natural gas with oxygen or air into hydrogen-containing syn-gas, thus avoiding a deep oxidation with the formation of CO_2:

$$2CH_4 + O_2 \rightarrow 4H_2 + 2CO$$

This 'partial' oxidation occurs either at very high temperatures in devices resembling the nozzles of space missiles or at somewhat lower temperatures on sophisticated 'honeycomb' catalysts of special chemical composition. No doubt, such 'dry' oxygenative reforming is technologically much more convenient than traditional steam reforming technologies and is expected to be capable of substituting conventional reforming with steam, especially in the case where technologies of pure oxygen generation appear also quite cheap and easy to handle.

However, the partial oxidation also has an evident disadvantage in comparison with conventional reforming because the final amount of hydrogen produced is lower due to the fact that there is no addition of hydrogen from water as in conventional reforming. For this reason, some improvements of traditional steam reforming are also under consideration. Among these, a direct supply of energy in the form of highly wall-penetrating radiation (either microwaves or ionizing radiation, etc.; see Chapter 12), with its consequent absorption by the reforming catalysts, seems to be of most interest for advanced technologies of the future.

Production of hydrogen from various, especially solid, organic substrates (coal, biomass, etc.) can be performed simply via a variety of gasification technologies that process the substrate with oxygen and/or steam. The chemical problems here are nearly the same or even more complicated than in the above examples. However, much work is under way on the gasification of solids, especially because of the possibility of using renewable substrates such as biomass as an inexhaustible and environmentally benign raw material for hydrogen production.

2.2 *Electrochemical production of hydrogen from water*

Fossil fuels are presently the most economical source for hydrogen but they are, of course, exploited at the expense of environmental pollution. In a sustainable energy economy, hydrogen has to be produced from renewable energy sources, e.g. solar energy, which can, for this purpose, be made available either as photovoltaic electricity or as high-temperature heat. The water-splitting reaction:

$$H_2O = H_2 + \tfrac{1}{2}O_2$$

is characterized by an enthalpy change $\Delta H^\circ = 58.14$ kJ mol^{-1} and a corresponding free energy change $\Delta G^\circ = 52.99$ kJ mol^{-1} (these values are given for gaseous water at 423 K). At 1 atmos pressure, ΔG° becomes negative at *c.* 4700 K. Here, water would be spontaneously split into hydrogen and oxygen. At 2300 K, *c.* 1% hydrogen will be in equilibrium with water at 1 atmos. The very high temperature required for thermolysis of water makes direct thermal decomposition of water by concentrated solar heat rather unfeasible (see also Section 3.4.1, Chapter 13).

The same is true for nuclear heat, which is available only at 1000 K. However, electricity can conveniently be used for decomposing water into hydrogen and oxygen. Because the Carnot factor is involved, electricity generation in a nuclear power station already involves a significant energy loss. For photovoltaic cells, theoretical maximum efficiencies for electricity generation of 31% (one solar absorber) to 68% (cascade of solar absorbers) can be expected, but it will take significant development time until reasonable efficiencies can be reached technically (maximum efficiency for one photovoltaic absorber, silicon, is presently *c.* 24%).

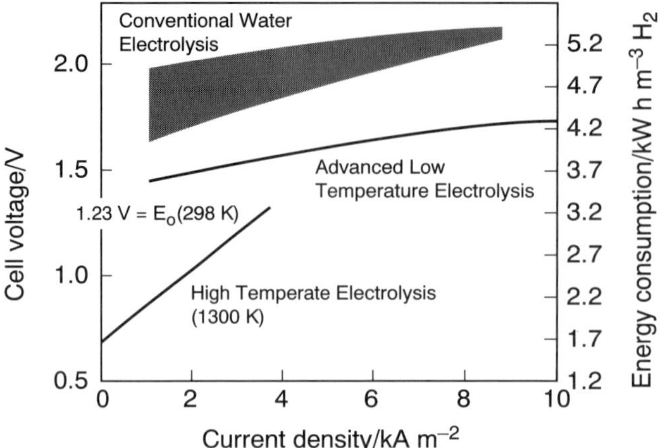

Figure 5.1. Cell voltage required and current density obtained for different water electrolysis technologies.

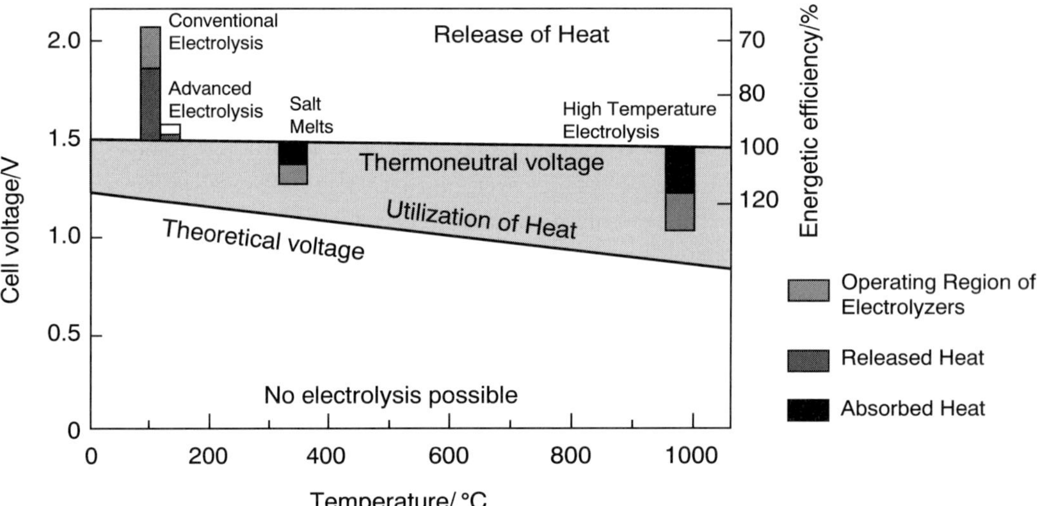

Figure 5.2. Cell voltage against temperature of operation. Visualization of thermoneutral voltage and operation region where heat can be converted into electricity.

Even though the principles for electrosynthesis of hydrogen are established, the technology for large-scale operations still needs much improvement. Under a pressure of 1 atmos, the thermodynamically needed voltage for the electrochemical water splitting is 1.23 V. This value is valid at 298 K, and with every degree of temperature rise the needed voltage is reduced by 0.25 mV. The older literature reported electrode potentials of *c.* 2 V for a current density of *c.* 500 A m^{-2}. Systematic technological research has decreased this value to *c.* 1.4 V (Fig. 5.1).

The free energy of reaction ΔG°, which is the minimal energy required for electrolysis, decreases with temperature. It is therefore of advantage to perform water electrolysis at a higher temperature. At 1.47 V, a 'thermoneutral' voltage can be defined where no heat is turned over: it is neither absorbed from the surroundings nor released into the environment (Fig. 5.2). Here, the overall enthalpy of the process due to the contribution of the entropy term $T\Delta S$ is $\Delta H = \Delta G + T\Delta S = 0$. High-temperature water electrolysis has to be performed below this potential, so that the heat can be absorbed from the environment.

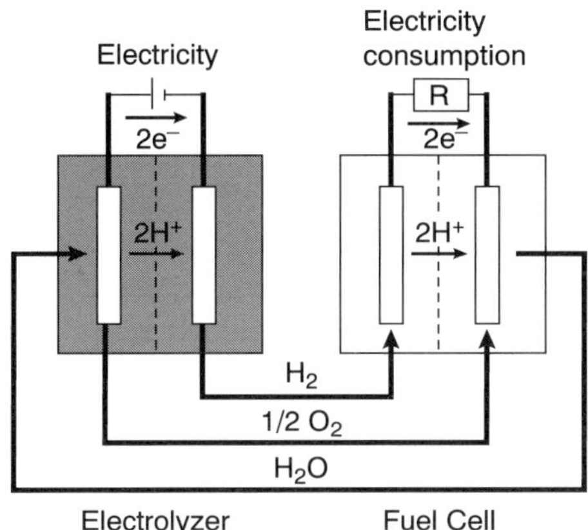

Figure 5.3. A hybrid system for coupled electrolysis and fuel cell operation.

Typically, electrolysis is performed under three different conditions: water-containing electrolytes can be used up to 420 K; salt-melts operate up to 620 K; and solid-state electrolytes up to 1300 K. At low temperature, below 420 K, the electrolyte can either be selected to be acid or alkaline. Because of corrosion problems with acid electrolytes, alkaline electrolytes are preferred. Most typically, a temperature of 362 K (85 °C) is selected in the presence of potassium hydroxide. Salt-melts operating at 620 K have the advantage of allowing lower potentials and higher electrode kinetics. At temperatures around 1300 K, the best thermodynamic and kinetic advantages can be obtained because heat can be extracted from the high-temperature environment. Electrical efficiencies of up to 120% can be achieved. At elevated temperatures, however, technological problems are also much more significant.

Because a fuel cell is, in principle, the reverse of the electrolysis cell (see Chapter 8), attempts have been made to build a bifunctional system that can both electrolyze water and generate electricity from hydrogen and oxygen (Fig. 5.3). The development of such a system turned out to be practical and, while being still in progress, has already yielded a quite satisfactory performance [2].

The most attractive source of hydrogen would be seawater, but it contains chloride anions and the problem of a parallel polluting chlorine evolution has been investigated [1]. The conclusion was reached that chlorine evolution can be neglected as long as the applied electrode potential at the electrolyzer can be kept below 1.8 V. This is possible but at high current density requires a careful design of electrodes and reaction chambers in order to facilitate efficient mass transport.

2.3 *Innovative processes for hydrogen production from natural gas and other hydrocarbons*

As mentioned above, production of hydrogen from natural gas and other available hydrocarbons through catalytic steam or other kind of reforming remains the cheapest source of hydrogen. However, this methodology not only has technical problems (see above) but also has chemical shortcomings in its advanced application for the hydrogen energy economy, typically related to the needs of extremely pure hydrogen (e.g. for low-temperature fuel cells, etc.). Such chemical shortcomings of conventional hydrocarbon reforming result in the production of hydrogen mixed with carbon oxides, which leads to sometimes very difficult and energy-consuming extraction and cleaning of hydrogen from the syn-gas obtained.

Thus, improvements or new conceptions in technologies of hydrogen production from hydrocarbons and methane (especially to avoid the above-mentioned disadvantages) appear to be very important. Recently, a few new non-traditional approaches were developed. These are based on hydrogen production by either high- or medium-temperature pyrolysis of natural gas and/or light hydrocarbons into hydrogen and valuable carbonaceous materials, as well as by partial oxidation of natural gas into syn-gas in high-temperature solid oxide fuel-cell-like systems. The evident advantages of the nearly developed processes against conventional ones lie in the possibility of production of hydrogen completely free of carbon monoxide. This is of considerable importance for designing the low-temperature fuel cells fueled with hydrogen, as well as the possibility of co-generating hydrogen with other valuable materials such as chemicals or even electricity.

2.3.1 CARBON OXIDES — FREE CATALYTIC TRANSFORMATIONS OF METHANE AND LIGHT HYDROCARBONS INTO HYDROGEN AND VALUABLE CHEMICALS OR MATERIALS

The main idea of this approach is to transfrom hydrocarbons into hydrogen and valuable carbon-containing substances or materials via direct and highly controllable pyrolytic processes. Two main processes of this type are under study. These are a high-temperature catalytic pyrolysis of methane into H_2 and C_2-hydrocarbons:

$$CH_4 \rightarrow H_2 + C_2H_4, C_2H_2 \quad (1)$$

and a high- or medium-temperature pyrolysis of CH_4 and higher hydrocarbons into hydrogen and a valuable carbonaceous material via reactions such as:

$$CH_4 \rightarrow 2H_2 + C \quad (2)$$

Reaction (1) is a highly endothermic process that has been well known for many decades in conjunction with either acetylene production in an electric arc or hydrogen production in non-equilibrium microwave plasma discharges. According to thermodynamics, predominant formation of H_2 and C_2-hydrocarbons should be achieved at temperatures above 1300–1400 K. The main reason for the lack of wide application of this plasma chemical technology is an intensive co-production of soot as a by-product via rapid pyrolysis of methane according to the competing reaction (2), which is completely shifted to carbon at temperatures above *c.* 1000 K. Thus, to achieve a high yield of C_2-hydrocarbons, one should overcome the very important problems of super-rapid heating of the initial reagents and super-rapid quenching of the products of the high-temperature (above 1400 K) pyrolysis of methane. Numerous attempts to reach the required conditions of such quenching in plasma chemical and electric arc devices have had only moderate success so far owing to the large geometric size of conventional types of plasma discharges and to their being non-penetrative for the feeding gas.

A controversial approach to solve the problem of super-rapid heating and quenching of the product mixture is to use specially designed catalytic reactors with a very thin bed of catalyst (or, more strictly, a catalyst–heat exchanger) heated by some sophisticated method such as microwave or IR and/or visible light irradiation. This unusual approach has only just started [8], and its development could be of interest for future hydrogen-producing technologies.

Another approach is to follow a reaction such as (2) and produce carbon-oxide-free hydrogen together with a carbonaceous material that is easy to separate from hydrogen [8,9]. At high temperatures (*c.* 1500–2000 K), such pyrolysis occurs very rapidly and does not need any catalyst; however, the carbonaceous material formed at these conditions appears to be only carbon black (soot) with poorly controllable properties [9]. Thus, to improve the parameters of methane pyrolysis and produce a much more valuable carbon material rather than conventional soot, a medium-

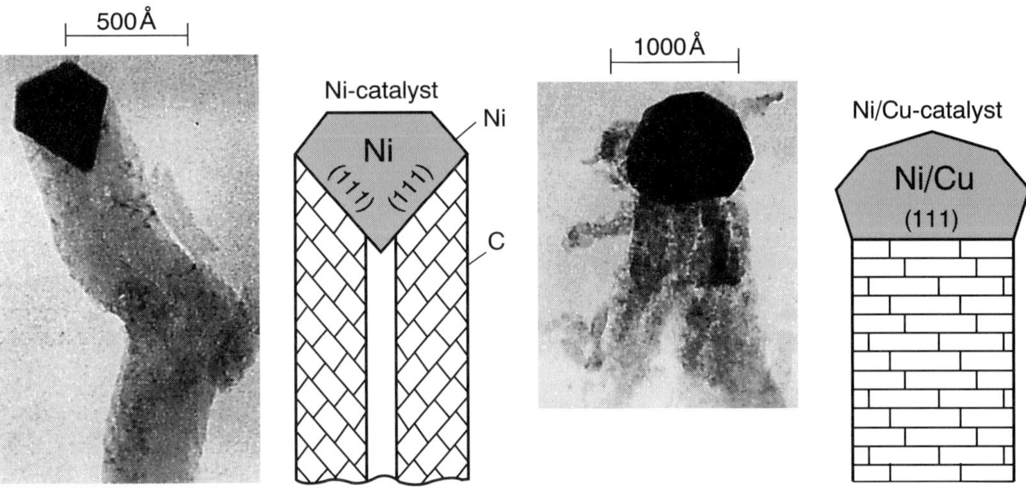

Figure 5.4. Formation of filamentous or nanotubular carbon obtained together with hydrogen at medium-temperature catalytic pyrolysis of methane and light-hydrocarbons according to the reaction $C_nH_m \xrightarrow[720-920\ K]{catalyst} C\downarrow + 2H_2\uparrow$. The weight of the initial catalyst can be increased by a factor of 300 due to the formation of carbonaceous filamentous material. (From [7,9].)

temperature catalytic technology was recently designed and tested on a pilot scale [7]. It is based on the pyrolysis reaction of natural gas or other light hydrocarbons to form carbon and hydrogen:

$$C_mH_n \rightarrow C + H_2 \tag{3}$$

at 720–920 K at atmospheric pressure over nickel or nickel/copper-containing catalysts.

At first glance, reaction (3) is not suitable for hydrogen production because thermodynamics restricts the formation of high yields of hydrogen at temperatures lower than 1000 K and the catalysts are rapidly deactivated, namely by the carbon produced. Indeed, until recently, processes (2) and (3) were usually considered only with regard to the problem of how to prevent carbon deposition on the reforming catalysts [5]. However, the newly proposed catalysts appear to be very stable and not deactivating, owing to the formation of *filamentous* carbon materials, with carbon deposited *behind*, not over, the active metallic nanoparticles of the catalyst. The structure of the filament resembles graphite (Fig. 5.4) [10].

Of importance is that the newly designed catalysts for growing the filamentous or even nanotubular carbon allow a yield of carbon to be obtained that is 200–300 times greater than the starting mass of the pyrolysis catalyst, as well as to enrich the treated methane with 30–40 vol.% of hydrogen. Note that it is much easier to use such a methane–hydrogen-containing mixture for fueling low-temperature fuel cells than to use syn-gas with a large content of carbon monoxide that inhibits the platinum electrodes, etc. Also, methane is much more easily separated from hydrogen than from the carbon oxides.

Pyrolytic filamentous carbon materials have some advanced properties and can be used as promising sorbents, catalyst supports, etc. Also, they can be used either for the production of carbonaceous composite materials or as an active chemical substance for some metallurgical or chemical applications. The properties of these materials can be easily controlled by the composition of the pyrolysis catalyst as well as by the process conditions [10].

In the case of co-production of hydrogen together with the pyrolytic filamentous carbon, the main task is to obtain maximum yield of the carbon material using a minimum amount of catalyst.

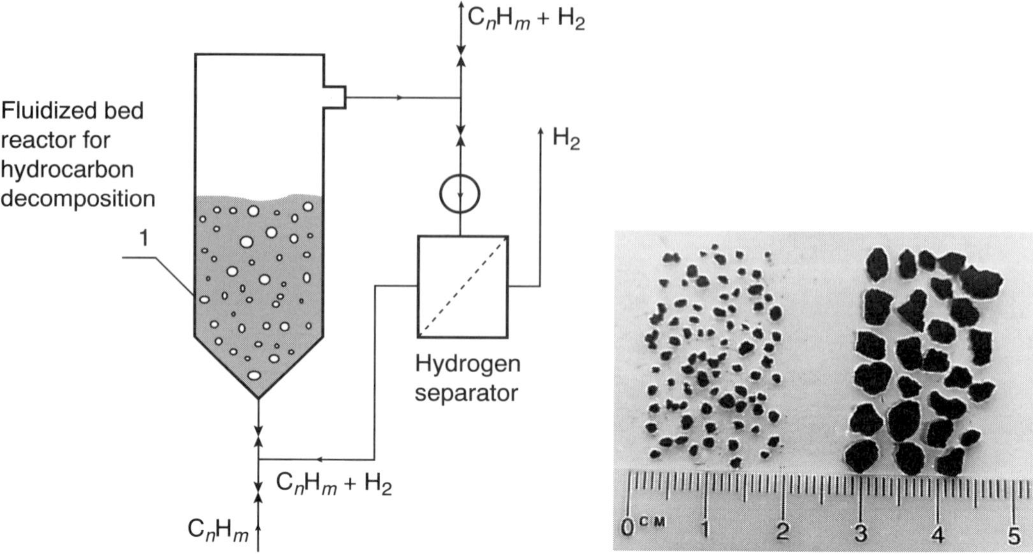

Figure 5.5. Scheme of a new process of medium-temperature (720–920 K) catalytic pyrolysis of methane and light hydrocarbons showing granules of the filamentous carbon grown in the reactor with the fluidized catalyst bed. (From [7].)

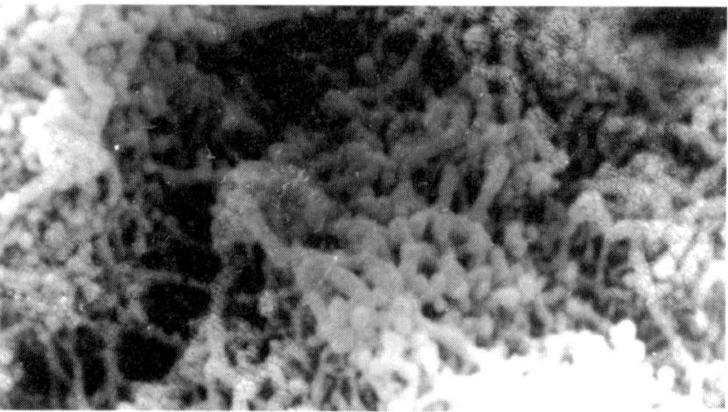

Figure 5.6. Granular catalytic carbon materials of filamentous structure obtained by the medium-temperature catalytic pyrolysis of methane in the pilot set-up shown in Fig. 5.5.

In addition, for elaborating an industrial process, one should solve also the problems of a scaling effect and of preventing agglomeration of the carbon material formed.

A principal scheme of the suggested process is shown in Fig. 5.5.

In order to prevent agglomeration of carbonaceous particles and to produce the carbon in a granulated form, the use of a reactor with a fluidized catalyst bed is suggested. Figure 5.6 gives a view of the granular carbon material obtained in the pilot reactor. The carbonaceous granules appear to be mesoporous particles of 1–5 mm in size, composed of bounded graphite-like filaments.

2.3.2 CO-GENERATION OF ELECTRICITY AND CHEAP SYN-GAS VIA OXIDATIVE REFORMING OF METHANE IN SOLID OXIDE FUEL CELL SYSTEMS

The aim of such co-generation is, first, the possibility of combining the use of high-temperature fuel cells both as an electricity generator and as a catalytic reactor for providing valuable chemicals, especially hydrogen, and, second, to avoid preliminary very expensive production of pure oxygen for oxidative reforming of light hydrocarbons.

Recently, there have been reports of several examples of electrochemical oxidation of methane to syn-gas (a mixture of H_2 and CO) over platinum and other metal electrode–catalysts in a high-temperature cell with yttria-stabilized zirconia (YSZ) as a solid oxygen-conducting electrolyte [8,11]. Schematically, such a cell is similar to the solid oxide fuel cell (SOFC; see Chapter 8):

CH_4, Pt | YSZ | Pt, air

At the site of electrochemical oxidation of methane in the SOFC system, the methane in the flow does not contain oxygen. It is fed directly into the reaction zone through a membrane of the solid oxygen-conducting electrolyte by passing an anodic current through the cell. In this case, oxygen ions (O^{2-}) are formed at the air-blown electrode (cathode). Then, these ions transfer to the electrode–catalyst (anode) through the YSZ electrolyte. On the anode they can either discharge to produce O_2 or oxidize methane directly.

These reported experiments were carried out at atmospheric pressure within the temperature range 930–1070 K and with anodic current flowing through the cell of the order of 300–600 mA. Practically, a 100% conversion of the electrochemical oxygen pumped to the reaction zone was observed, whereas the conversion of CH_4 achieved *c.* 90% and the yield of H_2 and CO reached 70% or more. Similar results were also reported for rhodium and nickel electrode–catalysts.

Note also that electricity is co-generated via syn-gas generation in the above fuel-cell-like systems. The driving force for this co-generation of electricity is the very high chemical affinity of the partial oxidation of hydrocarbons with oxygen.

The electrochemical oxidation of methane to syn-gas also has some expected advantages in comparison to the conventional partial catalytic oxidation. Thus, in addition to the possibility of producing not only syn-gas but electricity as well, the electrochemical oxidation of methane to syn-gas in SOFC systems has the following characteristics:

1 the probability of the reaction mixture to explode decreases, because the input methane and air are separated by the solid electrolyte membrane, and it is easy to control the oxygen amount in the reaction zone by simply regulating the electric current passing through the SOFC system;

2 when air is used as an oxidant, no additional nitrogen accumulates in the reaction zone, because oxygen supplied for the methane partial oxidation is separated from the other components of air just at the stage of its electrochemical supply to the reaction zone. Note that separation of oxygen from air is typically the most energy-expensive stage of any process of oxidative conversion of methane and other light hydrocarbons into hydrogen, carbon oxides and other oxygenates.

There is also a possibility in fuel cell systems to use water vapor and carbon dioxide instead of air for the electrochemical oxidation of methane to syn-gas, e.g. by arranging the following SOFC systems:

CH_4, electrode | YSZ | electrode, H_2O
CH_4, electrode | YSZ | electrode, CO_2

Clearly, the summed-up electrode semireactions expected in such systems represent the overall chemical reactions of steam and carbon dioxide reforming of methane. However, unlike conventional catalysis, their electrochemical performance allows arrangement of the required CO/H_2 ratio in the syn-gas via simple mixing of the outlet gas flows from the anodic and cathodic compartments of the SOFC system.

Taking all the above into consideration, one should expect that the electrochemical oxidation of CH_4 to syn-gas in the SOFC-like systems is also quite promising for future hydrogen technologies and thus should be a subject of further study by chemists.

3 Material aspects of hydrogen technology

Hydrogen can penetrate most metals and, because it is chemically reactive, may produce embrittlement. It was discovered that local mechanical stresses increase hydrogen solubility by several orders of magnitude. Also, welding joints are critical places of hydrogen activity. Embrittlement occurs typically in the form of cracks, even at moderate pressures of hydrogen. The effect of embrittlement is, however, significantly reduced by the addition of impurities such as oxygen to the gas. Protection of surfaces by layers of specially developed steels is a feasible alternative. The mechanism of generation of embrittlement and the technologies towards its suppression remain an interesting challenge for the future. Because hydrogen has been transported and handled for many decades successfully in many countries, the embrittlement problem is not a complication that may slow down the introduction of hydrogen as an energy carrier [2–4].

4 Safety aspects of hydrogen technology

Many documents are available that deal with up-to-date standards for hydrogen technology [2–4]. Compared to methane and gasoline, hydrogen can easily be lost and be ignited or exploded. The best prevention is to allow the lightweight hydrogen to escape upwards and to use sensors for hydrogen detection. It is not any more difficult to handle hydrogen than methane, provided that routine safety standards are applied.

5 Transport of hydrogen

Transmission of hydrogen via pipelines becomes cheaper than transmission of electricity at distances of 300–1000 km [2–4]. The present natural gas pipelines can be adapted for the transmission of hydrogen, but an advanced hydrogen economy will require improved transmission systems. Transmission of liquid hydrogen in pipes seems to be too expensive compared with transmission of gas under pressure.

A chemical way to avoid this difficulty is to bind hydrogen reversibly with other chemical substrates to produce hydrogen-enriched solids or liquids. The well-known examples of this type are hydrides of different metals, which can absorb up to several wt.% of hydrogen and give it off easily. Such hydrides were successfully tested for storage and for numerous kinds of utilization of hydrogen, e.g. for powering cars.

Another chemical approach is to bind hydrogen reversibly with, for example, organic molecules. The best-known example of this type is the well-elaborated and industrially utilized hydrogenation of aromatic compounds, e.g. benzene:

$$C_6H_6 + 3H_2 \rightleftarrows C_6H_{12}$$

This scheme allows the transportation of hydrogen as part of a liquid (cyclohexane) and the liberation of hydrogen through a dehydrogenation reaction (see also Chapter 12). The evident advantage of such a hydrogen transportation process has created a basis for considering its practical application for the transportation of hydrogen by tankers across the Atlantic from Canada to Europe.

A non-conventional, but nevertheless possible, way of 'hydrogen transportation' from the energy-abundant areas is in the production of some reactive metals, such as aluminum or magnesium, and the generation of hydrogen directly at its place of use via reaction of the metal (promoted with special kinds of catalysts) with water. Indeed, this method of 'hydrogen transportation' will hardly be reversible from the point of view of the 'hydrogen carrier'; however, it is seriously discussed for the co-generation of hydrogen and electricity in electrochemical energy sources [12].

6 Storage of hydrogen

Hydrogen is difficult to liquefy owing to the low efficiency of refrigeration at low temperatures, but the technology of using liquid hydrogen has become well developed, especially because of space flight applications. Hydrogen can, in principle, be conveniently stored in depleted gas fields and solid cavities, but also under compression in caverns.

Small-scale storage systems can be provided by hydride technology, for example, based on iron–titanium alloys that absorb a significant amount of hydrogen and release it upon heating. The above-discussed hydrogenation–dehydrogenation technologies producing liquid 'hydrides' such as cyclohexane, as well as non-reversible 'hydrogen storage' in the form of reactive metals, are also under consideration.

Good results in the storage of hydrogen at moderate pressure and temperature can be expected when using special types of hydrogen-oriented adsorbents based on oxide or carbonaceous molecular sieves [13]. Unfortunately, the data on such adsorbents are still very scarce.

7 Outlook

The main aspects of hydrogen technology are all developed so that hydrogen technology could be adopted widely provided that cheap hydrogen was available. As in many other technologies that are not widely used, there are still many details that can be improved by systematic research and development. After several decades' experience with hydrogen technology there is no technological reason against the large-scale introduction of hydrogen as an energy carrier. In fact, in the introduction to this chapter, reasons have been summarized that make hydrogen the most elegant and the ultimate energy carrier for the future world energy system. Even when hydrogen is used at a very large scale, there do not appear to be problems because the very light hydrogen can escape into space and would not accumulate in our environment. Once cheap hydrogen was readily available as an energy carrier, chemists would increasingly try to use this hydrogen for producing alternative energy carriers for specialized applications. Here, the elaborate chemical energy carriers from biology could be used as examples, such as lipids, which can be readily stored and transported and would provide a significantly different technological infrastructure compared with hydrogen gas. Prior to this, however, lipids as an emulsion of small particles could be used as a chemical energy carrier for fuel cells. Much knowledge would have to be accumulated on the catalytic interconversion of chemicals into electrochemical energy.

The above examples of recent studies on some innovative processes for hydrogen production from natural gas (methane) and other hydrocarbons are evidence of the good prospects for the development of new technologies based on traditional raw materials that will be able to compete with existing well-developed and large industrial-scale technologies such as the steam reforming of methane. The main advantages of the technologies under development are the possibility of producing hydrogen free of heavily extractable carbon oxides or the possibility of co-generating hydrogen along with some valuable materials or even electricity. Such processes are nearing readiness for commercial application.

The simplicity and environmental advantages of such medium-temperature catalytic pyrolysis of natural gas and light hydrocarbons, as well as the possibility of producing hydrogen simultaneously with a valuable carbonaceous material, indicate that this process may be the basis of a promising technology for the production of hydrogen from hydrocarbons in the future. The present advantage of this process is its possible direct application for fueling low-temperature hydrogen fuel cells. In the distant future, this process might be regarded as an alternative for even industrial steam reforming, because in principle it does not pollute the atmosphere with carbon dioxide.

8 References

1 Bockris JO'M. *Energy Options*. Sydney: Australia & New Zealand Book Company, 1980.
2 Vesiroglu TN, Winter CJ, Baselt JP, Kreysa G (eds). *Hydrogen Energy Progress. XI. Proceedings of 11th World Hydrogen Energy Conference, Stuttgart, Germany, 23–28 June, 1996*. Frankfurt: Schön & Wetzel, 1996.
3 Vesiroglu TN. *J. Hydrogen Energy* 1995; **20**: 1; 1996; **21**: 147.
4 Gamburg DYu, Dubovkin NF. *Handbook on Hydrogen: Properties, Production, Storage, Transportation, Application*. Moscow: Khimiya, 1989 (in Russian).
5 Rostrup-Nielsen JR. *Catal. Sci. Technol.* 1984; **5**: 1; *Catal. Today* 1993; **18**: 305.
6 Tsang SC, Claridge JB, Green ML. *Catal. Today* 1995; **23**: 3.
7 Parmon VN, Kuvshinov GG, Sadykov, Sobyanin VF, Parmaliana A (eds). *Stud. Surf. Sci. & Catal.* Oxford: Elsevier; 1998; **119**: 677–683.
8 Parmon VN, Kuvshinov GG, Sobyanin VA. In Veziroglu TN, Winter CJ, Baselt JP, Kreysa G (eds). *Hydrogen Energy Progress XI*. Stuttgart: Schön & Wetzel, 1996; 2439.
9 Fulcheri L, Schwob Y. *Int. J. Hydrogen Energy* 1995; **20**: 197.
10 Likholobov VA, Fenelonov VB. *Sci. Russ.* 1997; **4**: 6.
11 Galvita VV, Belyaev VD, Demin LK, Sobyanin VA. *Appl. Catal. A* 1997; **165**: 301.
12 Lehman TA, Renich P, Schmidt NE. *J. Chem. Educ.* 1993; **70**: 495.
13 Weitkamp J, Fritz M, Ernst S. *Int. J. Hydrogen Energy* 1995; **20**: 967.

6 Integrating Chemical Energy Carriers (other than Hydrogen) for the Future

V.N. PARMON[1] and H. TRIBUTSCH[2]

[1] *Boreskov Institute of Catalysis, Siberian Branch of the Russian Academy of Sciences, Prospekt Akademika Lavrentieva 5, Novosibirsk 630090, Russia*

[2] *Freie Universität Berlin and Hahn-Meitner Institut, Dept. Solare Energetik, 14109 Berlin, Germany*

1 Introduction

Civilization cannot exist without some widely adopted 'integrating' energy carriers that create links between different technologies. The most universal links of this type are created by electricity and chemical fuels. Links of analogous types based on compounds such as adenosine triphosphate (ATP) and reduced nicotinamide adenine dinucleotide ($NADH^+$) also have been widely exploited in bioenergetic mechanisms, which serve as a basis for life. Among the chemical fuels, liquid hydrocarbons such as gasoline, diesel fuel, kerosene or jet fuel, etc. are the most universal at the current state of civilization. Also, biological systems use carbon-containing compounds reduced with hydrogen originating from photosynthetic water cleavage.

There are a lot of advantages of liquid hydrocarbon fuels: they are only moderately toxic, they are convenient to use, they have sufficiently large energy content, their properties are not very sensitive to a change in the fuel composition, etc. These advantages served as the basis for choosing the liquid hydrocarbons as the virtually universal 'integrating' energy carrier. The main distinguishing feature of an 'integrating' fuel is the possibility of its use not only in stationary conditions, such as for driving thermal power plants, but also as a motor fuel.

In the distant future, when there will be depletion in the available sources of hydrocarbon fossils, it will be necessary to choose new integrating energy carriers. These could be either the still available fossil organic materials or some artificial or synthetic fuels.

It was shown in Chapters 2 and 5 that the most promising integrating chemical energy carrier for the distant future is indeed hydrogen. The respective activity in the development of the technologies needed for arranging large-scale employment of hydrogen is in progress and was reviewed in Chapter 5.

However, for the near future, one could expect an increase in the use of more available, cheap or energy-rich chemical energy carriers, which could also serve as prototypes of fuels for integrating different technologies.

The most evident examples of available energy carriers are methane (natural gas), coal (carbon) suspensions in water or in a combustible liquid and syn-gas (a mixture of hydrogen and carbon monoxide). With regard to methane and syn-gas, these have properties analogous in some respects to those of hydrogen and can be considered as an intermediate step in the transition from natural to synthetic fuels.

At last, there are also some synthetic fuels with convenient or even advanced properties. Synthetic hydrocarbons obtained by Fischer–Tropsch or other types of synthesis are the best-known examples of this kind. Methanol and dimethyl ether are other different examples. There are also some more exotic synthetic fuels (ammonia, hydrazines, etc.) that are, nevertheless, widely and successfully used in some military fields of application and, for this reason, obviously should also be considered as possible candidates for high-density chemical energy carriers of the future.

Below, we will not discuss the problems and recent trends in the technologies of production of the mentioned energy carriers. The aim of this chapter is to concentrate attention on those substances that are the most plausible candidates for the role of integrating energy carriers. Under consideration here are mostly the fuels that are used in a conventional manner, i.e. when being oxidized with oxygen from air.

However, in the future, a wide application can be expected for 'monofuels', which do not need coupling with external reagents for liberating the stored energy. The most attractive could be 'reversible' monofuels, which allow repetitive use, being 'charged' or 'discharged' without a separation of their components.

With regard to solid energy carriers, this chapter concerns only finely suspended coal or carbonized biomass to create liquid-like fuels with properties similar to crude or fuel oil, etc., despite significant and well-known efforts to elaborate different kinds of synthetic solid fuels, including highly reactive metals.

2 Methane as a new integrating and widely available chemical energy carrier

Methane (CH_4) is the main component of natural gas, along with the so-called methane hydrates, which are very peculiar crumbly solid minerals of gray color precipitated in permafrost conditions and at the bottom of seas at high latitudes (see also Section 2, Chapter 7). The natural resources of both fossils are enormous: estimated reserves of fossil natural gas exceed the value of 100 billion tons; and reserves of the methane hydrates are estimated by a value at least one order of magnitude higher. There are also some artificial renewable sources of methane, e.g. anaerobic fermentation or digestion of biomass or municipal wastes with the production of so-called biogas, which is a mixture of mostly CH_4 and CO_2 (see also Chapter 10). Thus, there is no doubt that, even in the distant future, after depletion of resources of natural fossil oil, methane will still stay available in large amounts and will attract attention for its use not only for direct combustion in thermal power plants or as a convenient raw material for its conversion into valuable chemicals or synthetic fuels, but also as an integrating energy carrier for multipurpose needs. The main, and not very traditional, area of the last application seems to be the use of methane as a motor fuel.

Evident advantages of gaseous methane are its high volume energy density (calorific value), exceeding that even for hydrogen (see Table 6.1), as well as its very high quality in terms of ecological demand. Indeed, the chemical composition of methane is very favorable: it has double the H/C ratio of conventional liquid hydrocarbons and, of course, an H/C ratio much higher than that for coal. Thus, when using CH_4, pollution of the atmosphere with 'greenhouse' CO_2 is significantly less in comparison with the use of liquid hydrocarbons or coal (see Table 4.1, Chapter 4).

However, the danger arises from the fact that CH_4 itself is a 'greenhouse gas' that is ten times more efficient than CO_2. Pipeline or gas field leaks of 10–20% may thus neutralize the greenhouse advantage of CH_4. Leakage of CH_4 from high-pressure transcontinental pipelines (now in Russia constituting up to 20% of the pumped natural gas), as well as from the natural gas wells and permafrost regions, will continue overheating our planet. If production and transport technology are not improved, and if the temperature increase on our planet is not halted (avoiding release of methane from permafrost), CH_4 will add to the problems. Methane should therefore only be considered to become a transient large-scale integrating energy carrier before the arrival of hydrogen energetics.

Table 6.1. Comparison of some properties of gaseous substrates that are considered now as possible large-scale gaseous 'integrating' energy carriers for the future.

Substrate	Calorific value $Q_{p,298}$ (kJ mol^{-1})	Calorific value (MJ kg^{-1})	Calorific value (MJ Nm^{-3})	Fluidization temperature at 1 bar (°C)	Density of fluidized substrate (kg m^{-3})
H_2	285.8	142.9	12.8	−252.77	70.78
CH_4	882.0	55.1	39.4	−161.49	415
Syn-gas (H_2/CO = 2:1)	854.7	26.7	12.7	–	–
NH_3	474.9	27.9	21.2	−33.42	790
CO (for reference)	283.0	10.1	12.6	−191.5	–

From [1,2].

A remarkable environmental advantage of methane is that its cleaning to eliminate sulfur compounds is typically much easier than that in the case of liquid hydrocarbons or coal. In fact, application of methane as a high-quality fuel is very wide and tending to become wider, especially in natural-gas-rich countries such as Russia. Moreover, according to an official scenario of development of large-scale energetics in Russia, for the next 20–30 years in this country one should expect a so-called 'natural gas pause' [3], which means a predominant development of all fields of large-scale energetics based on the use of methane as the main energy carrier for thermal power plants. Consumption of natural gas for these needs is very favorable from many points of view. Indeed, emission of CO_2 is reduced by a factor of two or more in comparison with the use of coal or fuel oil (see Table 4.1, Chapter 4). Emissions of SO_2 and fuel-originated NO_x also can be significantly reduced. There is also a further possibility to improve the technologies of methane combustion, mainly through its efficient deep catalytic oxidation. Note, however, that much more active and advanced catalysts are needed for this in comparison with the catalytic oxidation of higher hydrocarbons, owing to the much lower chemical activity of methane.

Of importance also is the possibility of the direct use of natural gas and methane in high-temperature fuel cells, demonstrating a much higher efficiency of chemical-energy-to-electricity conversion than the thermal power plants (see Chapter 8).

However, as shown in Chapter 2, of most interest for chemists is the possibility of substituting the 'unavoidable components' of the chemistry vector of energetics by methane, where motor fuels constitute the main part of consumption of chemical energy carriers.

Note that wide application of methane meets the same difficulties as application of hydrogen: one should condense gaseous methane to improve its properties for transportation and usage in autonomous vehicles. The main approaches here are nearly the same as the approaches in hydrogen energetics: compression of methane in high-pressure tanks and its cryogenic liquefaction.

Note that large-scale adaptation of the above technologies should be considered as a very important and, possibly, a necessary step before achieving wide-scale hydrogen energetics because it will resolve nearly the same, but nevertheless easier, problems as those for hydrogen. Indeed, both compression and cryogenic liquefaction of methane are much easier to perform than those for hydrogen (see Table 6.1).

The most advanced utilization of methane (or non-rectified natural gas) as a motor fuel seems to be in Russia at the moment, where a large fleet of trucks, buses and limousines fueled with very cheap compressed natural gas is in operation. Also, at the beginning of the 1990s, there were successful tests of large passenger jet aircraft ('Tupolev-156') and helicopters fueled with cryogenic methane. There are some serious expectations that compressed and even cryogenic natural gas will be utilized on a large scale in the Nordic zone of Russia or transported by ships from these areas to Europe, possibly in the coming decade.

Other European countries are also searching for the possibility to use methane as a good-quality motor fuel.

From the viewpoint of chemists, an interesting contribution could be made to the direction of new technologies of condensing methane with the use of adsorbents. Indeed, according to numerous available data, the use of adsorbents such as special kinds of active carbon allows the compression pressure of methane in 200 bar tanks to be reduced by a factor of three or more for achieving the same density of this substrate even at room temperature. Indeed, this gives a good prospect for decreasing the weight of high-pressure tanks for transportation of the same amount of methane.

One may hope that in the future other new ways of mild chemical bonding of methane could be elaborated to develop principally new approaches for efficient storing of methane.

Because methane is easily usable as a fuel in fuel cells, which can generate at least twice as much energy compared to combustion engines, mobile transport via methane-powered fuel cell trucks or buses is an attractive alternative. The low carbon dioxide emission in combination with a high energy density and the comparably easy handling of methane make this fuel vector especially suitable for controlling traffic pollution until the time when totally clean (hydrogen) energy systems are economically feasible.

3 New synthetic energy carriers for the future

3.1 *Syn-gas*

Syn-gas was widely used at the end of the last century and later under the name of 'city gas' or such like, taking the same place in the energy industry as natural gas now takes for stationary energy devices (mostly for the production of heat in ovens, etc.). Syn-gas and 'city gas' are easily produced from coal, wood, bio-gas or other low-quality organic raw materials by various gasifying or steam reforming techniques, etc. The content of 'city gas' depends very much on the source of the gas and the technology of its production and can vary on a large scale from the 'real' syn-gas (a mixture of H_2 and CO in different ratios) to syn-gas with a large inclusion of CH_4, CO_2, N_2, etc. The quality of syn-gas in terms of energetics using simple conventional combustion technologies is not so good as that of hydrogen and methane (see Table 6.1) because of the large content of CO with moderate calorific ability.

An evident disadvantage of syn-gas as a universal gaseous fuel for 'open' combustion is indeed the toxicity of CO, which cannot be detected by the human sensory organs. This seems to narrow the expected application of syn-gas (or 'city gas') and bio-gas to small centralized energy systems supplying the gas to small towns, villages or separate industrial or agricultural facilities. Also, the ability of CO to poison the noble metal cathodes of low-temperature fuel cells narrows its possible application in an advanced electricity production industry.

However, one should expect a much wider use of syn-gas as an 'integrating' fuel for large-scale commercialization of some thermochemical technologies with closed-loop cycles of operation of 'reversible' fuels (so-called 'monofuels') not needing separation of their components after creation of the 'energy-enriched' form of the fuel (see Chapter 12). Indeed, the ability of the components of syn-gas to participate in very exothermic and highly reversible reactions of methanation type:

$$CO + 3H_2 \rightleftarrows CH_4 + H_2O + Q_1$$
$$2CO + 2H_2 \rightleftarrows CH_4 + CO_2 + Q_2$$

(thermodynamic parameters of these reactions are given in Chapter 12, Table 12.1) avoids oxidation reactions of syn-gas for the evolution of the energy accumulated in this fuel and thus avoids pollution of the atmosphere with CO_2 due to the possibility of recycling the fuel in thermochemical cycles (see also Chapter 12). Note that this ability of syn-gas to serve as a recyclable fuel was suggested during the 1970s as the so-called ADAM-EVA scheme for the long-distance transportation of energy from nuclear plants after its thermochemical conversion (see Chapter 12). For this reason, a lot of interest is paid nowadays to improving the production technologies of syn-gas by reforming methane with steam or carbon dioxide, which is needed to complete the loop of thermochemical cycles (see [4] and Chapter 12).

Note also that in the absence of air, as in the case for closed-loop thermochemical cycles, syn-gas is not explosive at all and can be stored for any amount of time before use. Syn-gas will obviously be used in the future as a universal and easily available feedstock for the synthesis of other synthetic fuels.

3.2 *Methanol and other synthetic alcohols*

Since the early 1970s, numerous papers have been published discussing synthetic methanol as the first candidate for the role of a universal liquid fuel for the future. Indeed, methanol is synthesized usually from syn-gas of various origins and obtained from any carbon-containing raw materials:

$$CO + 2H_2 \rightarrow CH_3OH$$

Evidently, methanol has satisfactory properties as a liquid fuel (Table 6.2).

World production of chemical methanol in 1995 reached 29 million tons [6] and shows a tendency for steady growth.

The synthesis of methanol is much easier and cheaper in comparison with that of liquid hydrocarbons (see Section 3.4). In some countries that are developing new liquid fuels, e.g. in the USA, attempts were made for the stepwise implementation of methanol as a large-scale motor fuel, starting from the preparation of methanol–hydrocarbon mixtures. Unfortunately, only modest results have been obtained so far in this field. The main difficulties for the wide utilization of methanol as a conventional fuel are: its high and well-known toxicity and its corrosivity (which is of primary importance for its use in conventional metal-made engines of cars). Moreover, recent data show some suspicious properties of methanol and other oxygenates (including the well-known fuel additives ETBE (ethyl-*t*-butyl ether) and MTBE (methyl-*t*-butyl ether)) when analyzing the toxicity of engine exhausts: in their presence in the exhausts there is the possibility of a remarkable rise of hazardous components such as formaldehyde, etc. Also, a serious disadvantage of methanol is its poor solubility in hydrocarbons in the absence of special stabilizers.

Nowadays, there is vivid discussion of methanol usage as a large-scale liquid motor fuel. The most challenging option is to use methanol in combination with a fuel cell for transport systems. In this case, only CO_2 should be a by-product and the energy efficiency should be more than twice that of a combustion engine. Prototypes of cars using a high-temperature converter to generate

Table 6.2. Comparison of some properties of synthetic liquid fuels that are considered as possible candidates for the role of liquid 'integrating' energy carriers in the future.

Substrate	Calorific value $Q_{p,298}$ (kJ mol^{-1})	Calorific value $Q_{p,298}$ (MJ kg^{-1})	Density (10^3 kg m^{-3})	Melting temperature at 1 bar (°C)	Boiling temperature at 1 bar (°C)
CH_3OH	715	22.3	0.8	−97.9*	64.5
C_2H_5OH	1370.7	29.8	0.8	−114.2*	78.4
Iso-C_4H_9OH (for reference)	2633	35.5	0.8	−108*	108.1
$(CH_3)_2O$†	1454.3	31.6	0.67	−135.5	−23.7
N_2H_4	622.2	19.4	1.0	2	113.5
$(CH_3)_2N{=}NH_2$	1983.4	35.4	0.8	−57	63
Iso-C_8H_{18} (for reference)	5456.1	47.8	0.7	−107.4	99.2
Coal C (for reference)	≤ 390	≤ 30	*c.* 1.0	–	–

From [1,2,5].

* Absolutely dry.

† Dimethyl ether is typically not considered as a gaseous fuel being suggested for use in a liquefied form like propane–butane mixtures.

syn-gas for a fuel cell are already on the road. Low-temperature direct methanol fuel cells are expected to become feasible in the near future (see Chapter 8).

The satisfactory properties of methanol as a liquid fuel that can be conveniently stored, etc., allow us to expect its possible large-scale use for stationary thermal power plants, either in new 'energochemical' (i.e. combined) technologies of utilization of coal or other available low-quality combustibles (see Chapter 3) or in 'cheminuclear' concepts of future nuclear energetics. In the first concepts an intermediate step of co-production of methanol and heat (electricity) from the initial raw fossils is assumed (see Section 5, Chapter 3). In the last concepts the primary heat of nuclear fission or fusion reactions is converted through thermochemical cycles into the energy of syn-gas (see Chapter 12) and then into that of methanol, which in comparison with syn-gas is indeed much more suitable for long-term storing and transporting.

Note that large-scale and/or long-distance transportation of methanol could also create significant environmental problems. Indeed, owing to the high toxicity of methanol and its good solubility in water, damage of a pipeline or ship transporting methanol is much more hazardous in comparison with that for natural oil or more sophisticated hydrocarbons, which do not mix with water and can easily be removed from the surface of water, etc.

From this point of view ethanol has more advanced properties. Nowadays it is available both synthetically and produced by fermentation of sugar or other kinds of renewable biomass (see Chapter 10). Production of ethanol is gradually increasing in the world, reaching *c.* 27 million tons of pure ethanol in 1995 [7]. The main raw material for its production is high-quality renewable green plant biomass with a high sugar content (see Chapter 10).

The largest producer of ethanol at the moment is Brazil, with production of 17.3 million tons in 1995 mainly for domestic use as an important constituent of motor fuel [7]. Ethanol is much less toxic than methanol and mixes better with hydrocarbons, which is the reason for its large-scale use as an important component of 'gasohol'. 'Gasohol' is a gasoline–ethanol mixture widely used as a high-quality motor fuel in countries such as Brazil that have large quantities of sugar-cane from which to produce ethanol. In the USA, the utilization of 'gasohol' was recognized by special governmental acts promoting the use of renewable raw materials for motor fuel production.

From the viewpoint of chemists engaged in the synthesis of alcohols from syn-gas, it would be much more efficient to synthesize not pure alcohols but so-called 'fuel alcohols', e.g. methanol with certain, not so severely controlled, combustible admixtures of higher alcohols, hydrocarbons, etc., or even 'higher alcohol mixtures' containing C_1–C_6 alcohols with the properties of good liquid fuels. Note that the synthesis of higher alcohols can be considered as a 'spoiled' version of both pure methanol synthesis and the Fischer–Tropsch synthesis of hydrocarbons, and thus can be managed in catalytic reactors more easily than methanol or Fischer–Tropsch synthesis. Note also that the production of fuel alcohols could be performed in much milder and cheaper conditions than those for 'chemical grade' alcohols; for example, some energy parameters (e.g. energy content of the fuel) in the resulting mixture should be optimized, in contrast to the demands of the conventional chemical industry to optimize the content of methanol, etc.

One can expect synthetic alcohols to be satisfactory components of the synthetic fuel base in the not too distant future.

3.3 *Dimethyl ether*

From the beginning of the 1990s, dimethyl ether, $(CH_3)_2O$, has been widely discussed as a possible candidate for the role of a very promising synthetic motor fuel. The main reasons for this are: better calorific characteristics in comparison with its closest artificial analog, methanol (see Table 6.2); the possibility of its production by simple technologies that could be complementary to those of methanol production; and recent positive experimental tests in the use of dimethyl ether as a satisfactory and environmentally benign fuel for efficient diesel engines [8]. These features appear to be very attractive and will dominate in the tests of this fuel to distinguish its real potential.

Typically, dimethyl ether is produced either by simple catalytic dehydrative coupling of methanol:

$$2CH_3OH \rightarrow (CH_3)_2O + H_2O$$

over some solid acid-type catalysts in a single pass, or by direct synthesis from syn-gas:

$$4H_2 + 2CO \rightarrow (CH_3)_2O + H_2O$$

The latter is considered to be cheaper and, moreover, needs much milder conditions than conventional synthesis of methanol (no high pressure of 80–120 bar is needed). Industrial technology of dimethyl ether production is not very difficult and can be scaled up easily if necessary. Of importance is the fact that the 'fuel dimethyl ether' can contain some impurities of alcohols that make its production cheaper in comparison with a 'chemical grade' ether. Also, one can expect the elaboration of a very cheap synthesis of dimethyl ether–gasoline mixtures from syn-gas over some acid catalysts; an expected advantage of such a synthesis is the easy separation of these mixtures and thus the possibility of producing both gasoline and diesel fuel in one reactor and in a controllable ratio.

An evident disadvantage of dimethyl ether as a motor fuel is indeed its gaseous state at ambient conditions. This necessitates condensing the dimethyl ether in special tanks under a pressure of several bars for use in autonomous vehicles. However, there are expectations that this procedure will be no more difficult than operation with liquid propane–butane mixtures, which are now very convenient in many countries and are widely used as a cheap and rather efficient motor fuel. Moreover, it will be possible to use the same infrastructure, which will diminish very significantly the cost of wide commercial acceptance of this synthetic fuel.

The interest in dimethyl ether is now expanding rapidly and, certainly, very soon will reveal unknown features of this substance such as its real both applicability and toxitity at large scale use as the motor fuel, etc.

3.4 *Synthetic liquid hydrocarbons*

Synthetic liquid hydrocarbons are obtained usually from syn-gas via the well-known Fischer–Tropsch synthesis in rather mild conditions (pressure below 50 bar, temperature 470–620 K) over iron- or cobalt-supported catalysts or over some types of high-silica zeolites:

$$CO + H_2 \rightarrow C_mH_n + (CO_2, H_2O)$$

The quality of the hydrocarbons of moderate molecular mass (so-called middle distillates) produced by the Fischer–Tropsch synthesis corresponds to 'premium' diesel fuels or kerosenes with the highest possible parameters; mostly, these are linear paraffins. Simultaneously, at the same synthesis, when changing the nature of the catalyst or parameters of the process, one can obtain also heavy slightly branched paraffins — so-called synthetic 'waxes' — which can be used either as themselves as high-quality hydrocarbons or be split into middle distillates via hydrocracking processes, etc.

Hydrocarbons obtained from syn-gas over zeolites (sometimes with an intermediate step of methanol formation) have more complicated composition and usually contain a large portion of aromatic compounds. Thus, without special treatment, these hydrocarbon-containing mixtures correspond mostly to gasoline but with some properties beyond modern environmental standards.

Production of synthetic hydrocarbons from syn-gas can be considered now as being well elaborated. For example, in South Africa, there are large-scale facilities in operation producing more than 1 million tons of synthetic hydrocarbon fuels per year [9]; and a large facility producing *c.* 0.5 million tons per year was constructed in Malaysia and has been exploited commercially since the mid-1990s [10].

Of course, the costs of synthetic fuels produced by such technologies are much higher than those of liquid hydrocarbons produced by conventional well-elaborated processing of natural fossil oil. However, in some particular conditions, especially in the case of the fossil-oil-poor but coal- or natural-gas-abundant countries such as South Africa or Malaysia, the cost of synthetic hydrocarbons can compete with those imported from remote areas. The technology in South Africa is operated economically due to the production of a well-selected spectrum of valuable products.

Moreover, nowadays under very vivid discussion is the possibility of a large-scale chemical liquefaction of Russian natural gas obtained at the giant, but very remote, natural gas fields at the far northern part of Siberia (Yamal Peninsula, etc.). The main reason for this is the very remarkable cost reduction of the transportation of hydrocarbons over long distances (e.g. over more than 4000–5000 km toward the western borders of Russia) when transporting liquid instead of a gas, the latter demanding large energy expenditure for pumping and compressing (estimated as 30–40% of the initial energy content of the natural gas at those fields).

The technologies of industrial synthesis of synthetic hydrocarbons are still far from optimal, even when good catalysts are used for the process. The main obstacle to be overcome by chemical engineers in this area is suitable handling of the process at rather mild conditions but evolving a huge amount of heat; it is well known that there are huge difficulties in the removal of large heat fluxes under restricted conditions of small temperature gradients. Thus, one can expect that the real need for synthetic hydrocarbons will promote the invention of new chemical technologies for carrying out some particular catalytic process. An example of such novel technology is the Fischer–Tropsch synthesis in a slurry of dispersed catalysts [9].

To have high-quality liquid hydrocarbons on a large scale and as universal energy carriers, no principal problems are expected for the future except the cost of their production and the availability of the raw materials for syn-gas production.

Another way of synthesizing liquid hydrocarbons is by chemical liquefaction of coal. This method is under extensive research and development in many countries, but practical results seem to be modest in comparison with hydrocarbon synthesis via syn-gas.

A real breakthrough in the field of synthetic fuel production is expected by co-processing oil and/or coal with methane, which under certain conditions can lead to alkylation of long paraffins with methane molecules directly, owing to so-called 'thermodynamically coupled processes' [11,12]. According to preliminary data, such coupling can substitute up to 20–30% of oil carbon with that from methane, which could result in enormous savings of crude oil as feedstock and thus changes in the raw material balance of the existing oil refinery industry.

Indeed, of prime interest for chemists is the possibility of synthesizing artificial hydrocarbons directly from available inorganic carbon-containing substrates such as CO_2 or limestone ($CaCO_3$). In principle, there are no fundamental difficulties in arranging such synthesis, but the most serious disadvantage would be the very large consumption of energy in comparison with the production of H_2 (see Chapter 2 and Section 4 in Chapter 7).

3.5 *Carbon or coal suspensions*

An evident deficiency in high-quality liquid hydrocarbon fuels in many countries resulted in attempts to 'invent' their liquid substituents. In recent decades, great interest has been generated by fine suspensions of coal or other solid carbonaceous materials in water or, preferably, in combustible liquids such as low alcohols or their water solutions. The main interest in such suspensions was due to the possibility of substituting these suspensions for more 'noble' fuels such as fuel oil, etc. to be used first of all at conventional stationary thermal power plants.

Indeed, the production of such suspensions is very cheap due to the use of a simple procedure of dry or wet crushing and dispersion of cheap coal; and the rheological and calorific parameters of the suspensions obtained are satisfactory for their transportation by pipelines and for many applications, such as incineration in flame, etc.

Japan is now importing large quantities of such suspensions from China, and the possibility of large-scale applications of such suspensions is being looked at in other countries, e.g. Russia. In the late 1980s, an attempt was made to put into operation a long-distance (200 km) pipeline for transporting coal suspension between two large cities in Siberia. Unfortunately, this very important experiment was interrupted by the known events in the former Soviet Union.

An evident disadvantage of the coal suspension is, indeed, its lower calorific value in comparison with conventional fuel oil, etc. Also, crushing of natural coal in the absence of chemical or other pretreatment of the coal dispersion will pollute the suspensions with sulfur and mineral residues usually abundant in the natural coal, which can create some environmental problems. Because sulfur is typically present as iron sulfide (FeS_2), sulfide-oxidizing bacteria (*Thiobacillus ferroxidans*) may be used to convert the sulfide to sulfate, which remains unreactive during combustion.

Indeed, one can avoid some of the mentioned uncertainties of the suspensions when using soot or dispersions of artificial carbonaceous materials (see Section 2.3.1 in Chapter 5 for example) for preparation of the suspensions.

Another serious problem is stabilization of the coal suspensions from sedimentation. This problem can be solved easily by adding some available and cheap surface-active components such as alcohols, preferably 'higher' alcohols (C_3–C_{5+}). In the latter case, some other properties of the suspensions are simultaneously improved: the rheology of the suspension is improved, the calorific value increases and the melting point decreases. Note that modern technologies allow the production of cheap 'fuel alcohol' mixtures instead of coal excavation, through complete or partial gasification of coal:

$$C + O_2, H_2O \xrightarrow[\text{Temperature}]{} CO + H_2$$

with subsequent synthesis of alcohols from the syn-gas obtained (see Section 3.2).

One can expect a continuous increase of interest in such fine carbon suspensions in the future, especially when approaching depletion of natural fossil oil resources: the reserves of coal are much larger and coal is more uniformly distributed throughout the world.

A good source for high-quality water–carbon suspensions could be finely dispersed carbonized biomass as well.

4 Carbon-free and high-density chemical energy carriers

Along with conventional synthetic chemical energy carriers of a simple hydrocarbon or alcohol nature, one should take into account also less traditional kinds of fuels, especially those with the absence of carbon atoms (see Chapter 2). One can also expect interest in more energy-dense chemical fuels than common hydrocarbons. The needs for such fuels are accelerated mostly by a continuous progress in the design of new kinds of aircraft, etc.

It is of interest that current civilization already has experience in a wide range of such fuels, experience that originates mostly from military fields of technology.

4.1 *Hydrogen peroxide*

Hydrogen peroxide (H_2O_2) is a very well-known and environmentally clean chemical energy carrier that can be produced either by partial water oxidation or, vice versa, by partial chemical or electrochemical reduction of oxygen. In electrochemistry, the respective overall electrode reactions are written as:

$$2H_2O - 2e^- - 2H^+ \rightarrow H_2O_2$$

or

$$O_2 + 2e^- + 2H^+ \rightarrow H_2O_2$$

Thus, in principle, no other raw materials apart from air (oxygen) and water are needed to synthesize this environmentally reasonably benign and energy-rich substrate (however, there are some well-known oxidation problems when H_2O_2 comes into contact with the skin or biological substrates or is distributed via aerosols).

Table 6.3. Some properties of liquid 'monofuels' at their direct decomposition.

Substrate and reaction*	ΔH°_{298} (kJ mol^{-1})	ΔG°_{298} (kJ mol^{-1})	ΔH°_{298} (MJ kg^{-1})	Fuel density (10^3 kg m^{-3})
$H_2O_2 \rightarrow H_2O + \frac{1}{2}O_2$	−98.0	−116.8	2.88	1.45
$N_2H_4 \rightarrow N_2 + 2H_2$	−50.5	−149.2	1.58	1.00

From [1].
* The data are given for liquid H_2O, H_2O_2 and N_2H_4.

The most advanced use of H_2O_2 as an energy carrier is its direct decomposition:

$$2H_2O_2 \rightarrow 2H_2O + O_2$$

in the presence of special catalysts due to its intrinsic thermodynamic instability, with the release of a significant amount of heat (see Table 6.3). Thus, when being used in such a way, H_2O_2 can be considered as a 'monofuel' (i.e. fuel not needing additional chemical components for energy release).

The mentioned property of H_2O_2 was used in the first generation of the German World War II missiles 'V-2' when decomposition of H_2O_2 was used for the rotation of a turbine pumping the main fuel components (ethanol and liquid oxygen). Well known also are the H_2O_2-driven Walter turbines widely used in submarines. Since the 1960s there have been demonstrations of the use of H_2O_2 for driving small jet engines for transporting individuals, for example.

Large-scale use of H_2O_2 as a common fuel was never considered seriously, probably due to its poor energy capacity in comparison with 'binary' fuels (the pair 'oxidizer + fuel (reducing agent)'). However, the future of H_2O_2 as a fuel seems to be still only under initial discussion.

4.2 *Ammonia and hydrazines*

Ammonia and hydrazine are the most evaluated and large-scale-produced carbon-free compounds that have the properties of a fuel.

Hydrazine (N_2H_4) is similar to H_2O_2 in its carbon-free origin and intrinsic thermodynamic instability (see Table 6.3). In principle, the initial reagents for N_2H_4 could be water and N_2 from air:

$$N_2 + 2H_2O \rightarrow N_2H_4 + O_2$$

However, until now the synthetic path has been much more complicated.

Nevertheless, simple hydrazine has found wide application in space technologies as a 'monofuel' for small jets, being decomposed according to the reaction:

$$N_2H_4 \rightarrow N_2 + 2H_2$$

over thermostable catalysts such as metallic iridium. Nowadays, N_2H_4 is frequently used for space vehicles on long-term interplanetary flights as a convenient propellant for the 'correction' jet engines.

With regard to the Earth's surface, the large-scale application of N_2H_4 is restricted due to its significant toxicity, which also restricts its use as a combustible component of 'binary' fuels of conventional type (see Table 6.2).

Ammonia (NH_3) is produced now in quantities of several tens of millions of tons per year by simple catalytic synthesis:

$$N_2 + 3H_2 \rightarrow 2NH_3$$

and, in principle, can also be used as a component of conventional 'binary' fuels with moderate properties (see Table 6.1). For its use as a motor fuel it should be liquefied by cooling and compressing. Incineration of NH_3 with oxygen is typically accompanied by a large emission of hazardous nitrogen oxides, which nevertheless can be sufficiently decreased by using appropriate catalysts.

Nowadays, ammonia is not considered as a fuel, mostly due to its intrinsic toxicity and NO_x formation on oxidation. However, one can expect a rise of the interest in the application of ammonia in the future, especially in conditions of a shortage of carbon-containing raw materials for fuel synthesis.

Note also that the large-scale production of ammonia is now based mainly on cheap hydrogen produced from natural gas. Thus, in the case of a shortage of hydrocarbons, hydrogen should be produced from water; this indeed will follow from comparative economic studies of the predominance of hydrogen energetics over those based on ammonia or other hydrogen-derived fuels.

Alkylated hydrazines, mostly mono- and dimethylhydrazine, are well known as widely used liquid propellants for military missiles and space vehicles. These hydrazines have good properties of 'binary' fuels (see Table 6.2) and, for this reason, one can expect an interest in them in the future for civil application. However, the most serious obstacle for speeding up the application of synthetic fuel of this type is again its enormous toxicity.

4.3 *High-density hydrocarbons*

High-density synthetic hydrocarbons are rather new and are a very interesting kind of hydrocarbon fuel developed in recent decades mostly to fuel the military 'wing flying missiles'. The main feature of these fuels is their peculiar chemical composition, because they are composed mostly of strongly 'stretched' hydrocarbons such as norbornadiene derivatives, having thus more advanced calorific value per unit mass than 'common' hydrocarbons (see Table 6.2). Also, their density can approach 1.2 kg dm^{-3}, which is of prime importance for reducing the weight or number of tanks storing such fuels.

No doubt this type of fuel, especially when the technologies of synthesis are simplified, will attract wider interest in the future.

4.4 *Biomimetic energy carriers*

Biological evolution has led to excellent chemical energy carriers, which, with progressing chemical technology, may also become integrating for future artificial energy systems. The rubinethroat humming bird (*Archilochus colubris*) requires only 2 g of fat for an 18 h non-stop flight across the Gulf of Mexico with 3.2×10^6 wing strokes. Fat is indeed the most energy-dense chemical compound in biology and, with a specific energy of 39 kJ g^{-1}, is slightly superior or identical to that of conventional fossil fuel and twice as high as that of glucose or starch. This is the reason why most animals use fat for long- and medium-term energy storage, proving that regenerative energy cycles can generate and handle fuels with optimum energy density. Biological systems also use carbohydrates, which are easier to mobilize and therefore are used when smaller energy quantities are needed repeatedly (e.g. by harvesting bees, which can eat frequently).

In living organisms, fat is transported in the form of small droplets of 1 μm or less, which are stabilized by a protein film (α- and β-globulins). Approximately 40% of its energy is recovered in the form of the more reactive energy carrier ATP, which carries only a small fraction of the energy stored in fat (typically 1 mol of fat is similar in energy to 400 mol of ATP), as needed in the microscopic biological energy-containing structures. For ATP, utilization energy efficiencies of typically 79% have been estimated. Adenosine triphosphate is a universal energy carrier like electricity, which is easily converted into other energy forms (light, mechanical, chemical). It cannot easily be compared with conventional fuels because energy release is not essentially due to the broken phosphate bond but due to entropic energy released during hydrolysis.

Biological energy storage systems will remain attractive model systems for future energy strategies for years to come. However, the challenge for chemists is significant. A lot has to be learned about catalytic generation and turnover of intermediates. In addition, low-temperature (fuel-cell-type) mechanisms for energy conversion cycles have to be implemented.

The fact that nature has succeeded in evolving an efficient regenerative energy system tells us that the task is feasible within the laws of chemistry.

5 Conclusions

The above non-complete list of promising (especially when adding hydrogen as the first candidate; see Chapter 5), integrating chemical energy carriers for the future shows the wide choice for our civilization in this direction.

One can expect that progress in the adaptation of new energy carriers will be stepwise and follow firstly the particular demands of some selected countries as well as the demands of ecological purity, safety, etc.

As to the particular properties of the listed fuels, it is evident that, unfortunately, the well-known liquid hydrocarbons were, are and will be the best energy carriers among 'binary' fuels with air (oxygen) as the oxidizer. However, in the future, one can also expect much more serious interest in 'monofuels', especially those that are ecologically pure and recyclable.

Indeed, a very important step towards the large-scale implementation of hydrogen could be the preliminary large-scale adaptation of methane as motor or fuel cell energy carrier, because the adaptation necessitates solving nearly the same technical problems of condensing and transportation as for hydrogen.

Nearly all the problems that have arisen in the large-scale adaptation of new integral energy carriers are related in many aspects either to the chemistry and physical chemistry of these energy carriers themselves or to the chemistry and some chemical engineering problems of their synthesis, including those of the more available and easier processed raw materials.

6 References

1 Rabinovich VA, Khavin ZY. *Kratkij Khimicheskij Spravochnik (Short Chemical Handbook)*. Leningrad: Khimiya, 1991 (in Russian).

2 Gamburg DYu, Dubovkin NF (eds). *Handbook on Hydrogen: Properties, Production, Storage, Transformation, Application*. Moscow: Khimiya, 1989 (in Russian).

3 Shafranik JF (ed.). *New Energy Policy of Russia*. Moscow: Energoatomizdat, 1996 (in Russian).

4 Tsang SC, Claridge JB, Green ML. *Catal. Today* 1995: **23**: 3.
5 *Khimicheskaya Entsiclopediya (Chemical Encyclopedia)*, Vol. 1. Moscow: Sovetskaya Entsiclopedia, 1988 (in Russian).
6 Crocco JR. *Chem. Ind.* 1996; **March**: 215.
7 Ivanov A. *Chem. Complex Russia* 1995; **10**(12): 6 (in Russian).
8 Rouhi AM. *Chem. Eng. News* 1995; **May**: 37.
9 Jager B, Espinoza R. *Catal. Today* 1995; **23**: 17.
10 De Jong KP, Van Wechem HMH. *Int. J. Hydrogen Energy* 1995; **20**: 493.
11 Parmon VN. *Catal. Today* 1998 (in press).
12 Lee S. *Methane and its Derivatives*. New York: Marcel Dekker, 1997.

7 Production of Traditional Carbon-containing Energy Carriers from Alternative Non-renewable Raw Materials (other than Oil, Natural Gas or Coal)

V.N. PARMON

Boreskov Institute of Catalysis, Siberian Branch of the Russian Academy of Sciences, Prospekt Akademika Lavrentieva 5, Novosibirsk 630090, Russia

1 Introduction

A comparison of the properties of different candidates for the role of integrating chemical energy carriers for the future shows unambiguously, as mentioned in Chapter 6, that the best fuel of the future should be either certain synthetic carbon-containing compounds (preferably liquid hydrocarbons such as our common fuels) or, in some respects, pure hydrogen. With regard to the most promising raw materials for the large-scale production of hydrogen, these are inexhaustible because common water serves as the main raw material. The situation is not so apparent when discussing the large-scale production of carbon-containing materials in a situation of complete depletion of the main traditional fossil hydrocarbon resources, such as oil, natural gas and coal. Indeed, renewable green plant biomass or municipal solid wastes that contain a high proportion of organic substances can be used as raw materials.

However, there are some additional or alternative and practically non-exhaustible sources of raw materials for the production of traditional or synthetic carbon-containing energy carriers. In this chapter, a brief overview of the expected ways of production of such carbon-containing fuels from some proposed non-renewable raw materials is given.

2 Production of fuels from oil sands and shales, methane hydrates and other non-traditional hydrocarbon fossils

Although these carbon-containing fossil raw materials were not mentioned in Chapter 6 as being those for the future, their natural deposits are enormously large and exceed those of oil, natural gas and coal by several orders of magnitude. For this reason, they may serve as important sources of hydrocarbons at least for the coming century, provided that the Earth's atmosphere can sustain the CO_2 input. The processing of these raw materials has some peculiarities and has not been developed for commercial use as yet.

2.1 *Oil sands and shales*

Oil sands and shales are present on the Earth in enormous quantities. Their total resources are estimated as 5×10^{22} J [1]. They are located all over the world, the main deposits probably being located in the USA.

The main peculiarity of virgin oil sands and shales as raw materials for valuable hydrocarbon production is the very high content of inorganic inclusions. Also, hydrocarbon constituents of the oil sands and shales are represented mostly by fractions of heavy hydrocarbons with low H/C ratio. Thus, the main path for conversion of this raw material to high-quality hydrocarbon fuels should be efficient separation of inorganics (sand or other mineral inclusions) from hydrocarbons of the sands or shales, followed by reforming of the extracted hydrocarbons with hydrogen or other hydrogen carriers to improve the H/C ratio.

Under discussion are numerous ways of separating inorganics and subsequently upgrading the cleaned hydrocarbons of the sands and shales with hydrogen. However, despite the enormous resources of hydrocarbons in the sands and shales, their widespread use for creating the basis of large-scale energetics of the future still remains very questionable.

Indeed, with regard to the separation of inorganic inclusions, a huge amount of energy, in addition to that for excavation of the sands or shales, is needed when their pre-processing occurs in a common manner (e.g. by treating with superheated steam, etc.). Thus, one should be very careful in verifying that these energy expenses will be less than the final content of energy in the chemical carriers produced.

As for true chemists, of much higher interest are the expected ways of increasing the H/C ratio of the organic constituents of the oil sands and shales. Note that when using pure hydrogen as a hydrogen donor for this need we lose a huge amount of energy in advance for the production of hydrogen (see Section 3.1, Chapter 2). This provides a disadvantage for this method of production of fuels in comparison with direct use of H_2 as the chemical energy carrier. For this reason, more advanced technologies of upgrading the heavy hydrocarbons seem to be based on the use of steam or, better still for the transient period of civilization, natural gas as the donors of hydrogen for the heavy oil-reforming processes.

Note that the mentioned technologies do not yet exist and this is a very wide field of basic and applied research for chemists.

2.2 *Methane hydrates*

Methane hydrates are not well known even for many chemists. Nevertheless, they are the most rich and practically inexhaustible source of methane deposited in huge amounts (more than 10^{18}–10^{21} tons [2–4]) in the Arctic zone of the Earth. The amount of methane in the solid hydrate form is considered to be at least one order of magnitude higher than that of 'free' methane in natural or oil-assisting gases. Recently, large amounts of gas hydrates have been found also at the bottom of the sea near India and Japan, i.e. in areas considered earlier as very poor in fossil organic energy carriers.

Methane hydrates are unstable gray solids of clathrate type with composition $CH_4{\cdot}nH_2O$, where n is no less than 5.67 [4–6]. The intrinsic chemical structure of methane hydrates resembles that of zeolites, composed of water molecules with light hydrocarbon molecules captured inside the cavities of this zeolite-like framework (see Fig. 7.1).

Methane hydrates have very peculiar thermodynamic properties: their phase diagram shows that the region of hydrate stability is very restricted and needs elevated pressure and coexistence with liquid or frozen water [4–6]. Because of this instability, hydrates are located mostly in permafrost layers and on the bottom of the sea.

It is of interest that methane hydrates are well known by those employed in the excavation of natural gas or gas condensates, because at the expansion of wet hydrocarbon gases these hydrates form corks in the valves, pipelines or other units of the natural gas- or light hydrocarbon-producing equipment.

The chemical methods for the decomposition of methane hydrates are very simple and based usually on the addition of methanol to dissolve the hydrate cages, and so decomposing the

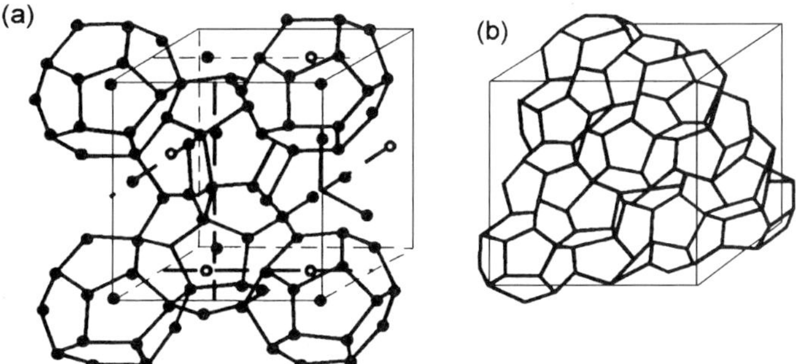

Figure 7.1. Simplified view of the structure of two of the most common types of methane hydrates. (a) The framework of cages is composed of water molecules, (b) with hydrocarbon molecules captured inside the cages. (According to [5].)

hydrates to release their methane (and some other light hydrocarbons) constituents is expected to be quite simple.

Much more difficult and still under discussion is a suitable way of excavating the hydrates. Indeed, even being completely saturated with methane, the solid hydrates contain only very small amounts of hydrocarbons: note that the calorific value of a unit mass of saturated methane hydrate does not exceed 15% of that for pure methane. For this reason, the allowed methods of excavation of methane hydrates have to be very cheap and consume much less energy than that for the excavation of coal, etc. Such technologies, possibly including underground or underwater chemical destruction of the hydrates or heating *in situ* to overcome the thermodynamic conditions of hydrate stability, have not yet been invented. Moreover, the particular thermodynamic properties of the methane hydrates might prevent their easy decomposition underground or underwater. If these problems were to be successfully resolved, further processing of the extracted hydrocarbons would certainly be the same as that of conventionally produced natural gas.

2.3 *Methane-containing vent gases of coal mines and gases of coal beds*

A curious but, nevertheless, enormous resource for fossil methane in the future could appear in the ventilation gases of coal mines. Indeed, diffusion of methane into the mines strecks from the surrounding coal beds requires continuous ventilation of the mines in order to prevent explosions. Much richer in methane are the virgin gases of coal beds themselves.

According to some estimates, the total calorific value of methane in the coal deposits approaches 10% of that of the coal. Emission of such methane through the ventilation is also huge: only for the countries of the former Soviet Union do the respective estimates give a value of over 10 billion m^3 annually.

The main problem of utilization of this easily available source of methane is its small content in the vent gases. Indeed, the concentration of methane there is typically lower than 0.2–0.3 vol.% and only sometimes reaches 2–3%. This creates large difficulties in direct utilization of such methane, because concentrating it according to existing technologies is energy consuming. Attempts have been made to develop more efficient technologies of enrichment of such methane through, for example, isothermal pressure swing adsorption (the so-called PSA technologies) on suitable carbon molecular sieves, etc. As for direct combustion of lean methane-containing air mixtures without their enrichment, it is not so easy to accomplish due to the low temperature of 'adiabatic heating' of the mixtures (e.g. at 0.3% methane in air, the mixture is able to be heated

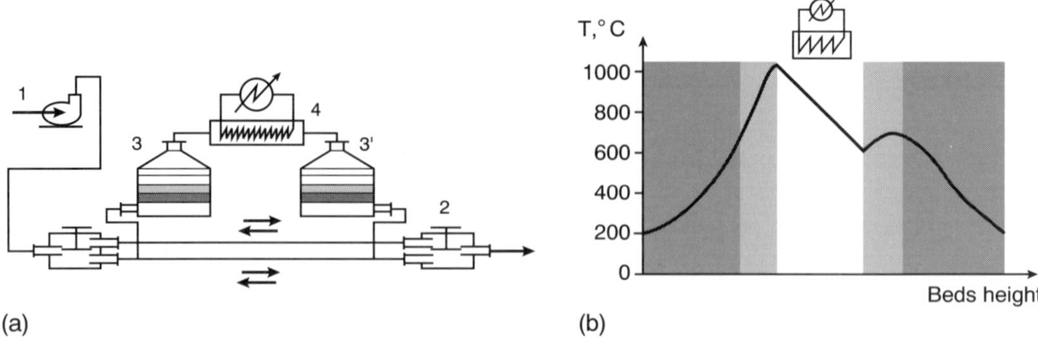

Figure 7.2. Production of heat and electricity via utilization of low-calorific methane-containing vent gases of coal mines. (a) Scheme of a catalytic heat generator based on the 'reverse' technology: 1, fan for supply of the vent gases; 2, switching valves; 3,3′, catalytic reactors with the catalyst beds; 4, heat utilizer (e.g. combined with a steam turbine for generation of electricity). (b) Profile of the 'captured' temperature in the catalyst beds of the heat generator. (From [7].)

adiabatically by no higher than 150 °C even in the case of complete oxidation of methane). For this reason, even sophisticated catalysts are typically able to support direct oxidation of methane from only the most concentrated vent gases.

From this viewpoint, it is of interest that now under large-scale pilot tests is the newest catalytic technology based on the so-called 'reverse' operation of catalytic oxidation reactors, which allows moderate- or high-temperature heat to be obtained even from very lean fuel mixtures (see Fig. 7.2).

The idea of this technology is to maintain the hot zone of the reactor by changing, periodically, the direction of the reagent flow [8]. Indeed, in some regimes, such a procedure allows the oxidation catalyst bed to be used simultaneously as the internal recuperative heat exchanger, thus considerably increasing the temperature inside the hot zone over the mentioned temperature of adiabatic heating.

With regard to the commercial utilization of the above technology for the methane-containing vent gases of particular mines, it should be possible to produce high-temperature technological heat capable of driving steam turbines for the production of electricity.

3 Waste and used polymers: recycling versus production of fuels

Nowadays, annual production of polymeric materials in the world is expressed by an enormously large figure approaching 100×10^6 tons that tends to increase rapidly. At the same time, most of the polymers produced quickly appear in the form of used plastic containers, bags, vessels, etc., which forms a large part of the solid municipal wastes contaminating the environment. Note that from the chemist's point of view waste polymers can be considered as very valuable solid raw materials resembling the best-quality fossil hydrocarbon resources; much more valuable than coal of the best quality.

For this reason, the wastes of used plastics and polymers should obviously be considered as a large-scale and very promising carbon-containing technogeneous raw material for the future.

Indeed, a lot of work is being done in developing the most suitable and cheap technologies of utilization of such valuable wastes. Among the most attractive approaches are: recycling of spent plastics, which means either their compounding into new plastics or depolymerization into monomers with subsequent polymerization; direct incineration of a good- or moderate-quality solid fuel (for the new technologies of such incineration, see Section 2.3, Chapter 3); and utilization of used plastics as a high-quality feedstock for the production of liquid fuel hydrocarbons.

The potential of each mentioned approach depends on many particular technological and economic issues and is still to be carefully evaluated.

To use polymer waste for the production of liquid fuels, the main approaches here are:

1 Pyrolysis or gasification of the wastes with oxygen and/or steam, with the formation of volatile olefins, alkanes, etc. that can be upgraded on zeolite or other types of acid catalysts into hydrocarbons. The existing technologies result in the production of a large fraction of aromatics.

2 'Methanopyrolysis' (co-processing with natural gas), which can lead, in principle, to liquefaction of the plastic waste.

Also, plastic waste, especially waste polyolefins (polyethylenes, polypropylenes), is considered an excellent carrier of hydrogen that can be used for co-processing with coal, biomass, heavy residuals of oil, etc. in order to liquefy them. The available data show that, for example, static pyrolytic transformation of dry biomass into liquid organics at 700–900 K occurs much more easily and deeply in the presence of waste polyolefins, producing a liquid fuel of much better quality than in the absence of polymers.

4 Production of fuels from atmospheric gases and inorganic sources of carbon

Some of the chemist's approaches for the production of hydrogen and hydrocarbons from inorganic substrates have been mentioned and discussed in Section 3 of Chapter 2. Here, we discuss in more detail only the production of carbon-containing valuable chemicals or fuels from CO_2 and solid carbonates. Nature is very experienced at efficient fixation of very dilute CO_2 during green plant or bacterial photosynthesis, and has been for billions of years. Intensive works are under way to elaborate numerous artificial chemical processes of analogous fixation; the reliable fundamental background for such transformations of inorganic carbon-containing materials is still to be created. Also, one should take into account that, when estimating the real perspectives for large-scale transformation of CO_2 or carbonates into valuable chemicals or energy carriers, one should obviously consider the energy cost of these transformations. If this cost is very high, some other approaches for the production of chemical energy carriers would appear to have more potential. Note also that in all the examples considered below, CO_2 is assumed to be concentrated from the beginning, the technologies of such concentration and their energy cost being outside the scope of this book.

4.1 *Conventional high-temperature catalytic hydrogenation of CO_2*

Carbon dioxide is known to be a chemically rather active substrate at moderate and high temperatures. For example, CO_2 molecules rather than those of CO are considered by some experts to be the main participants in the industrial catalytic process of production of methanol from syn-gas (a mixture of H_2 and CO) over copper-containing catalysts. However, the direct catalytic hydrogenation chemistry of CO_2 is not sufficiently advanced, except for the so-called 'shift reaction':

$$CO_2 + H_2 = CO + H_2O$$

and the methanation reaction:

$$CO_2 + 4H_2 = CH_4 + 2H_2O$$

In addition, only a few large-scale processes based on CO_2 as a reagent are known. Among them are the synthesis of urea via a combination of CO_2 and ammonia, etc.

A possible reason for the non-developed situation with industrially adopted CO_2 transformations via conventional medium- or high-temperature catalytic processes seems to be the failed competition with the much easier and variable chemistry of CO. Note that CO can be produced easily from CO_2 so one can expect that the 'C_1 chemistry' will remain based mostly on CO in the near future.

Of much more fundamental interest for chemists dealing with future energetics are, indeed, the processes of CO_2 reduction into valuable organic compounds at mild conditions (low temperature and pressure), mimicking the fixation of CO_2 in plants and bacteria.

4.2 *Direct low-temperature reduction of CO_2 with hydrogen*

The knowledge on such processes is still very scarce, except for the 'wet' analogs to the 'water gas shift reaction'. Only recently, a more complicated direct hydrogenation reaction was reported to occur in water solutions of bicarbonate ions on Pd catalysts at elevated hydrogen pressure, providing sufficient amounts of formates [9]. In spite of the obvious potential of this reaction of direct transformation of CO_2 into oxygenated organics, the main problem here remains the well-known difficulties in subsequent transformations of non-stable formates into valuable chemicals or chemical energy carriers.

One should expect that further progress in such studies will depend on the progress of formate chemistry.

4.3 *Electrochemical and electrocatalytic reduction of CO_2*

Direct electrochemical reduction of CO_2 seems to be very promising. Such reactions are well known to occur in liquid electrolytic cells on some metals or electrodes modified with transition metal complexes, the most attractive results being obtained over copper-containing electrode–catalysts or over some other electrolytically 'passive' metals. Note that copper appears to be typically the most favorable chemical element for CO_2 activation in numerous situations [10]. Various compounds, such as CO, alcohols, methane, ethylene, etc., have been reported as the products of this direct electrocatalytic reduction.

However, this kind of electrosynthesis is typically considered as not a very stable process, because the electrode is gradually deactivated. Periodic anodic activation improves the activity but only for a limited time. Unfortunately, most of the results cannot be considered to be of any practical interest due to the very small electrochemical reduction or the negligible current density (typically no more than 10 mA cm^{-2}) of the processes. The main obstacle for increasing the reduction or the current density originates from heavy competition between the desirable processes, with much simpler electrochemical reduction of water into molecular hydrogen. Thus, at the liquid-phase electrolysis of water solutions of CO_2 we again usually obtain hydrogen instead of other energy carriers.

A much more fruitful approach appears to be the *gas-phase electrocatalytic* reduction of CO_2 on gas-diffusion electrodes that resemble the electrodes of many types of fuel cells (see Chapter 8) and improve the mass transfer of CO_2 in the electrode zone with a diminished concentration of water. The scheme of a possible set-up for such a reduction is drawn in Fig. 7.3. Among the most active catalysts for the electrocatalytic reduction of CO_2 in this mode appear to be immobilized

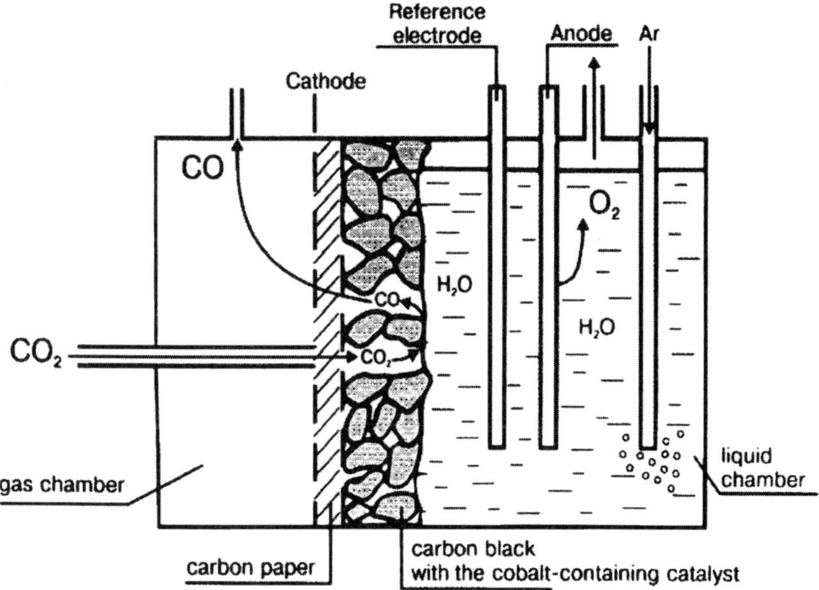

Figure 7.3. Schematic diagram of an electrochemical cell for CO_2 reduction on a carbon gas-diffusion electrode impregnated with cobalt–phthalocyanine complexes. The cathodic reaction is $CO_2 + 2H^+ + 2e^- \rightarrow CO + H_2O$. The reported CO current yield at $T = 293$ K and pH 6 is 97% at a current density of *c.* 100 mA cm^{-2} and the CO_2 conversion is 30–90%. (According to [11].)

cobalt–phthalocyanine complexes, which allow a high electrochemical reduction of CO_2 with a current density of practical interest even under mild conditions.

Unfortunately, in the mentioned set-up, the main product of CO_2 reduction appears to be only CO rather than more complicated organic compounds. One can expect, however, that further studies in this direction will allow much more advanced systems to be developed for the gas-phase reduction of CO_2.

One should mention that the main problem of activation of CO_2 under mild conditions seems to be 'screwing' of the rigid rod-like CO_2 molecule before its reduction, because virgin rod-like CO_2 hardly undergoes a chemical reduction.

Indeed, in nature during photosynthesis the CO_2 molecule is chemically bound by non-saturated organic molecules, thus dramatically transforming hybridization of its chemical bonds; the process of reduction of the carbon atoms of this chemically bound CO_2 molecule occurs only as the next step of its transformation. This strategy is especially pronounced in desert plants, which allow CO_2 to enter the leaf openings only at night (to avoid water loss) and then store it for photosynthesis during the day (crassulacean acid metabolism). Thus, one can expect that artificial systems for efficient reduction of CO_2 under mild conditions also have to mimick the well-elaborated natural photo- and chemosynthesis. Experimental works in this direction are still to be done.

4.4 *Photocatalytic conversion of CO_2*

Natural photosynthesis serves as a prototype for the photocatalytic conversion of CO_2 and may be regarded as a branch of the electrochemical (in the liquid phase) or catalytic (in the gas phase) transformations of CO_2 discussed above. In the first case, the desirable reaction is initiated by light-generated reducing equivalents (either movable electrons in light-sensitive semiconductors or strong molecular reducing agents in molecular photocatalytic systems; see Section 3 of Chapter 13 and [12,13]). In the second case, the nature of the CO_2-activating center can be, for example, a metal ion in a low oxidation state. Indeed, the products of photocatalytic transformations of CO_2 are typically the same as in electrocatalytic reduction.

Being very attractive from the scientific point of view, the well-known photocatalytic systems for CO_2 fixation are still very non-efficient in terms of their cost and light energy consumption [12,13]. For this reason, much should be done by researchers before seriously considering artificial photocatalytic systems for the fixation of CO_2 as a tool for large-scale energetics of the future.

However, no doubt the elaboration of such photocatalytic systems (which will hardly be able to compete with natural photosynthesis in terms of integral parameters) is a real challenge for chemists, who have to reproduce the most sophisticated tools for the production of chemical energy carriers created by nature.

4.5 *Conversion of solid carbonates*

Direct chemical conversion of solid carbonates, which are easily available as fossil limestones, marbles, etc., into chemical fuels is not very well elaborated. There are a few possible ways of chemical utilization of carbon from solid carbonates: for example, preliminary calcination of limestone with the evolution of gaseous CO_2 and its subsequent chemical treatment, or dissolving the limestone in acids with the formation of gaseous CO_2 or more reactive carbonate or bicarbonate ions.

No doubt, of the approaches for the utilization of fossil carbonates that are attractive practically, the mentioned pretreatment with subsequent activation of CO_2 in the processes discussed in Sections 4.1–4.4 will be preferable. It is evident also that the consumption of energy for chemical conversion of solid carbonates is much greater than that for the conversion of concentrated CO_2 (see Section 3.2, Chapter 2). Thus, in spite of inexhaustible resources of fossil carbonates, their wide use for the large-scale production of artificial carbon-containing energy carriers will be very questionable in the future.

5 Conclusions

The above discussion shows that there are now several possible ways to produce traditional carbon-containing energy carriers from largely available and practically non-exhaustible raw materials. However, in terms of energy consumption, they are rather costly. With regard to the available organic raw substrates, future technologies of co-processing methane with other carbon-containing feedstocks, such as coal, heavy oil, oil shales or biomass, appear promising. During such co-processing, methane can serve as an efficient hydrogen donor, upgrading the mentioned carbon- or oxygen-rich substrates. Also, in some cases, one can expect the possibility of 'alkylation' of different kinds of hydrocarbons with methane [14]. Evidently, the latter reactions can increase the yield of valuable liquid hydrocarbons when embedding extra carbon atoms of methane into the existing hydrocarbon molecules, thus diminishing the consumption of deficient oil.

With regard to the use of available inorganic carbon-containing compounds, extended studies in this area are now very much appreciated. However, one should expect any practical implementation of the elaborated technologies of CO_2 fixation or carbonate transformations for energy needs to be hindered by the extremely high energy cost of such transformations. For environmental reasons, only solar energy may be safely considered for large-scale CO_2 fixation in analogy to natural photosynthesis.

6 References

1 Skinner BJ. *Earth Resources*. New Jersey: Prentice-Hall, 1986.
2 Makogon YuF. *Doklady Akad. Nauk SSSR* 1971; **196**: 197 (in Russian).
3 Collet TS. In *The Future of Energy Gases*, USGS Prof. Paper 1580. Washington: USGS, 1993; 299.
4 Sloan ED, Jr. *Clathrate Hydrates of Natural Gases*. New York: Marcel Dekker, 1997.
5 Byk SSh, Makogon YuF, Fomina BI. *Gas Hydrates*. Moscow: Khimiya, 1980.
6 Davidson DW, Garg SK, Gough SR *et al*. *J. Inclus. Phenom.* 1984; **2**: 231.
7 Noskov AS, Klenov OP, Godin LL, Lakhmostov VS. In *Proceedings of the First European Congress on Chemical Engineering — ECCE1, Florence, Italy, May 4–7, 1997*, Milan: AIDIC S.r.l., Vol. 4. 1997; 3059.
8 Matros YuSh (ed.). *Unsteady State Processes in Catalysis*. Utrecht: VSP, 1990.
9 Wiener H, Blum J, Feilchenfeld H, Sasson Y, Zalmanov N. *J. Catal.* 1988; **110**: 184.
10 Sullivan BP, Krist K, Guard HE. *Electrochemical and Electrocatalytic Reactions of Carbon Dioxide*. New York: Elsevier, 1993.
11 Savinova ER, Yashnik SA, Savinov EN, Parmon VN. *React. Kinet. Catal. Lett.* 1992; **46**: 249.
12 Halmann M. In Grätzel M (ed.) *Energy Resources through Photochemistry and Photocatalysis*. New York: Academic Press, 1983; 549.
13 Ford PC, Fridman AF. In Serpone N, Pelizzetti E (eds) *Photocatalysis: Fundamentals and Application*. New York: Wiley, 1989; 541.
14 Lee S. *Methane and its Derivatives*. New York: Marcel Dekker, 1997.

8 Electrical Energy Carriers, Chemical Technologies and Electricity

H. TRIBUTSCH[1], W. PLIETH[2] and D. RAHNER[2]

[1] *Freie Universität Berlin and Hahn-Meitner Institut, Dept. Solare Energetik, 14109 Berlin, Germany*

[2] *Institute of Physical Chemistry and Electrochemistry, Bergstraße 66b, 01062 Dresden, Germany*

1 Introduction

The discovery that chemical and electrochemical energy can be interconverted is older than the discovery of the principle of the combustion engine. When chemicals such as hydrogen and oxygen, which can combine under energy gain, are interacting with electrodes, dipping into an electrolyte and connected with an electron-conducting wire, electrochemical energy can be generated (Fig. 8.1). This reverse process of electrolysis is theoretically described by the relation $\Delta G = -zFE_0$, where ΔG is the energy turned over, z is the number of current equivalents transferred, F is the Faraday constant and E_0 is the electromotive force. Whereas the development of combustion engines was very dynamic and successful, the research and technology of fuel cells for the generation of electrochemical energy from chemicals was very sluggish and time-consuming. Convincing estimates have been made for the entire fuel cell research and development showing that no more money has been spent than is typically spent on the development of a new car model by larger companies.

The main attraction of fuel cells compared to combustion engines governed by 'Carnot' efficiencies is that much higher energy conversion efficiencies can be obtained. Another very important advantage is the fact that energy conversion can be accomplished without noise and mostly without chemical products that are toxic and unpleasant for the environment.

There is one very important argument that should convince us of the practicability of fuel cells and the need to develop this technology with more effort than in the past: energy conversion of living systems essentially follows electrochemical principles. Explained in a simplified way: light energy or chemical energy is converted into the electrochemical potential of membranes. This electrochemical potential then is used to drive ions, especially protons, and these protons perform work. Metabolic energy is thus first converted into electrochemical potentials across membranes that are generating ion currents. These are driven through membrane-bound macromolecules, which act as 'working resistances', that are converting the protonic energy into the desired new energy form (Fig. 8.2). Biological energy conversion systems in this way are working like electrical energy conversion systems consisting of an electrical generator and an electrical circuit with an external working resistance that generates light, heat or mechanical work. The main difference between biological and technical electrical systems is that electronic conductors are replaced by ionic conduction in biological systems. This is a consequence of the fact that biological systems convert energy in wet aqueous environments, whereas technical electrical systems are typically solid-state systems based on electronic conduction.

In biological systems, the efficiency with which the energy involved in electron transfer reactions is converted into electrochemical potential gradients and the efficiency with which these potential gradients transfer ions for the purpose of synthesis of chemicals or the performance of

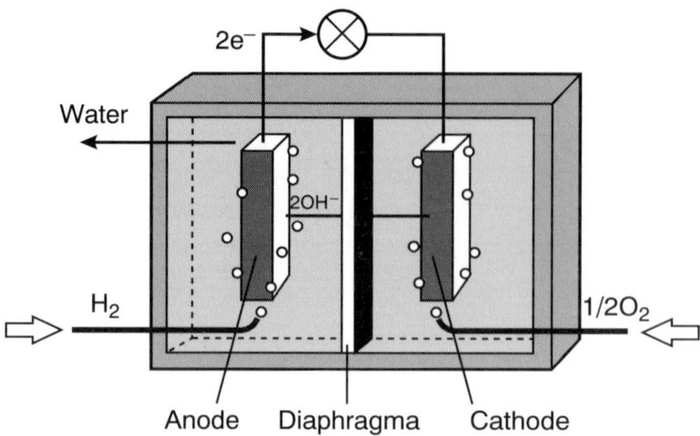

Figure 8.1. Principle of a polymer electrolyte membrane fuel cell.

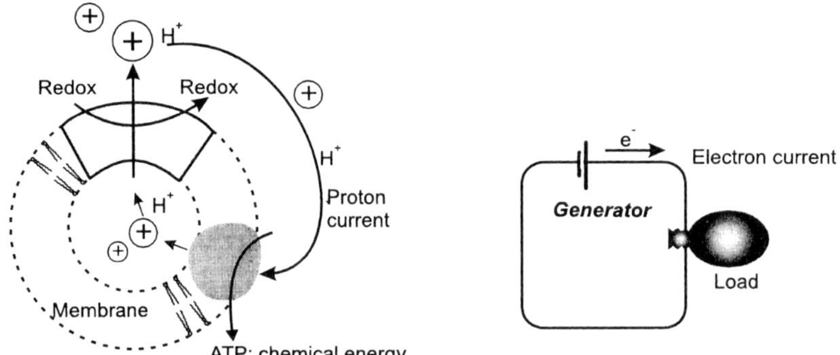

Figure 8.2. Comparison of electrochemical conversion in biology and electrical energy conversion in technology.

mechanical or electrical work is really astonishing. Biological electrochemical conversion, which typically operates between the environmental temperature and 40 °C, reaches an energy conversion efficiency of 40–80%. This requires not only a high degree of topological organization, i.e. closed membrane vesicles that allow the separation of anodic and cathodic electrochemical processes, but also very efficient electrochemical catalysts that, in many cases, have to provide efficient multi-electron transfer at very moderate operation temperatures. Biological systems have succeeded in optimizing both electrolysis and the reverse reaction of conversion of chemical energy into electrochemical energy.

Biological systems of course take advantage of self-organization, which allows the build-up and maintenance of complicated membrane structures. Technological efforts to put the principle of the fuel cell into technical practice have had to deal with many scientific and technological problems, which in the following will be analyzed to pinpoint unsolved or challenging problems for the future. Technological fuel cells have basically evolved into systems in which thin-layer electrodes and thin-layer electrolytes in combination with semipermeable thin membrane layers are used to both increase the effective electrode surface and to decrease the production costs of systems that should be assembled easily in a modular way for a convenient scale-up. The working principle of a hydrogen/oxygen fuel cell is shown in Fig. 8.1: a fuel cell with anode, cathode, ion-conducting separating membrane and electrocatalyst that is catalyzing the electrode reactions. The ion-exchange membrane in this case is a proton-exchange membrane.

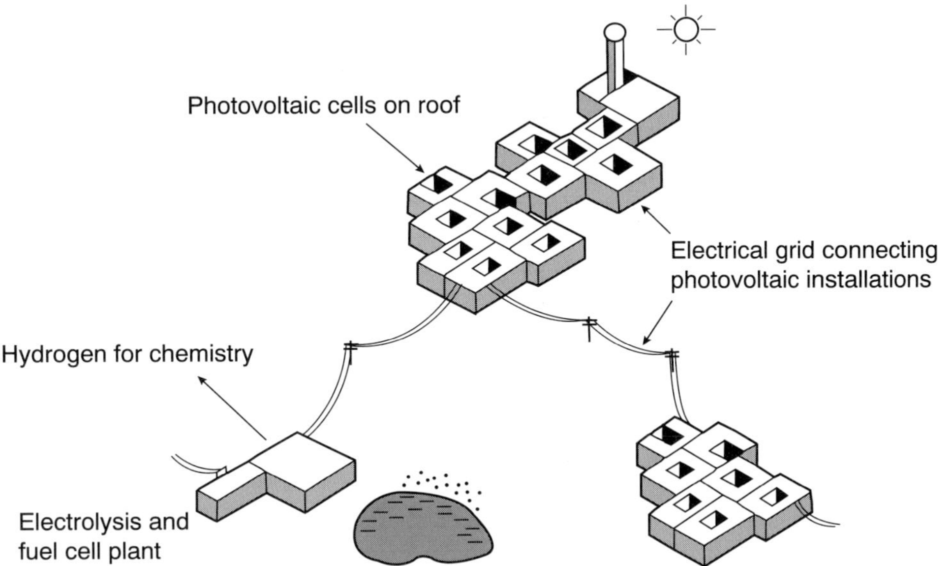

Figure 8.3. Visualization of a solar hydrogen energy economy based on photovoltaic electricity generation and transmission into the public grid and hydrogen generation via electrolysis, hydrogen storage and transformation of hydrogen again into electricity via fuel cells.

It is interesting to compare a polymer fuel cell with the biological fuel cell cycle (Fig. 8.2). Because, in both cases, protons are maintaining electrolytic conduction, the technical cell requires metallic electronic conductors to connect anodes with cathodes but the biological cell uses electron transfer protein chains linking anodic and cathodic sides of the biological membrane.

The development of technical fuel cells is a big challenge towards the reduction of environmental pollution and savings in energy [1,2,3]. Many scenarios for our energy future already consider a wide application of fuel cells, including more efficient utilization of fuels generated for energy storage, i.e. hydrogen in a future solar hydrogen economy. As shown in the simplified scheme of Fig. 8.3, photovoltaic electricity generated on the roofs of houses is transferred into the public electrical grid from where excess electrical energy is used for electrolysis and the liberation of hydrogen. This hydrogen is stored until energy is needed and reconverted into electricity via fuel cells. The energy efficiency of such electricity generated in fuel cell facilities may be twice as high or higher compared to the efficiency of present thermal power plants. A further interesting use for fuel cells will be in traffic; trucks and buses may be constructed to provide sufficient space for fuel cells. In this way, they would at least become twice as energy efficient and, in addition, non-polluting and essentially quiet. A further use that has already started to become reality is the use of fuel cells for the combined generation of electricity and high-temperature heat. Introduction of a fuel cell technology on a larger scale will basically depend on the ability to decrease the costs of fuel cells, to make them more compact and to ensure a longer operational life.

2 Chemistry and catalysis for the production of electricity — high-output fuel cells

Fuel cells are typically classified by the types of electrolyte they use. When all practical requirements are met, the choices are relatively limited. In principle, any electrolyte with a sufficient ionic conductivity may be used in a fuel cell. However, to avoid concentration gradients of the electrolyte, active conduction should occur via the ions that are turned over during the electrode reactions. This means that the ions generated and consumed at both anode and cathode should provide the anionic and cationic components. In aqueous fuel cells that operate on hydrogen and oxygen, high concentrations of H^+ and OH^- may be used, which means that either strong acids or strong bases can serve as electrolytes. These species are the only ones that are turned over at the anode and the cathode. In the acid fuel cell, H^+ is carried via the H_3O^+ ion or higher water complexes of the proton. The alkaline fuel cell works basically with OH^- ions, which are the reaction product of the oxygen reduction at the cathode. Phosphoric acid, on the other hand, which is also used in certain types of fuel cells, is a self-ionizing amphoteric system: a molten acid (H^+) salt. In molten salts, either H^+ or O^{2-} may be used as a conductor. Carbonate melts or aqueous carbonates use carbonate carrier ions. Solid oxides, on the other hand, directly conduct O^{2-} ions. Many fuel cells are operated at high or very high temperatures because catalysis is not as important at high temperature and technically can be handled more easily, but only certain temperature regions are sufficiently well handled.

2.1 *The efficiency of interconversion of electrical and chemical energy*

A fuel cell directly converts the Gibbs free energy of a chemical reaction into electrical energy using isothermally controlled electrochemical processes. Theoretically, the energy conversion may occur with an efficiency $\eta = \Delta G/\Delta H$ (ΔH is the change in the fuel enthalpy) of 83–124% for typical fuel cell reactions. More than 100% efficiency can be obtained theoretically because energy can be drawn from the environment, but only if the system is perfectly reversible, which is generally not the case. For this reason, a part of the chemical energy is liberated in the form of high-temperature heat. Typically, electrical efficiencies of 50–60% are reached with fuel cells, the rest being liberated as high-temperature energy. The two chemical compounds involved must react unmixed at the electrodes in contact with the electrolyte by providing an electron stream through the electrodes and the connecting wires under a voltage difference between the two catalytic electrodes. An ion flow in the electrolyte completes the reaction. The electrical resistance of the entire system has to be kept as low as possible to avoid ohmic losses.

A fuel cell may be considered to be a primary battery with ideal electrodes and stable electrolytes. Ordinary primary batteries, on the other hand, contain consumable electrode materials that are turned over during the conversion of chemical energy into electrochemical energy.

The selection of practical fuel cell systems is basically determined by the catalytic properties that can be obtained with known catalytic materials. At a low temperature, hydrogen is one of the few systems that can be converted into electricity at a high rate (Fig. 8.1). For this reason, hydrogen fuel cells are the most used fuel cells at low temperature. Carbon-containing fuels, on the other hand, react at much lower rates and are, therefore, typically used in high-temperature fuel cell systems (Table 8.1). Common fossil-based fuels must be converted to hydrogen and carbon monoxide by steam reforming before being used in fuel cells. Between these two extremes, methanol fuel cells should be mentioned because they can be used at lower temperatures. Methanol oxidation fuel cells are, however, still relatively ineffective. As an electron acceptor, most present-day fuel cells use oxygen from air which is also the case in biological systems. Possible concentration gradients that lead to efficiency losses are avoided by using concentrated electrolytes and by raising the temperature, which not only makes the catalytic reactions proceed more easily but also facilitates ion conduction.

The ion reactions used in practical hydrogen–oxygen fuel cells are described below:

Table 8.1. Theoretical efficiencies and operating potentials of possible fuel cells at 298 K.

Fuel	Reaction	E°_{rev}	Efficiency (%)
Hydrogen	$H_2 + 0.5O_2 \rightarrow H_2O$ (l)	1.229	82.9
Methane	$CH_4 + 2O_2 \rightarrow CO_2 + 2H_2O$ (l)	1.060	91.8
Carbon dioxide	$CO + 0.5O_2 \rightarrow CO_2$	1.066	90.8
Carbon	$C + 0.5O_2 \rightarrow CO$	0.712	124.1
	$C + O_2 \rightarrow CO_2$	1.020	100.2
Methanol	$CH_3OH + 1.5O_2 \rightarrow CO_2 + 2H_2O$ (l)	1.214	96.6
Formaldehyde	CH_2O (g) $+ O_2 \rightarrow CO_2 + H_2O$ (l)	1.350	93.0

E°_{rev}, 'reversible' electromotive force for the given reaction with the 'standard' concentration of reagents.

$$2H_2 \rightarrow 4H^+ + 4e^- \quad (1)$$
$$O_2 + 4H^+ + 4e^- \rightarrow 2H_2O \quad (2)$$

and in alkaline electrolyte:

$$2H_2 + 4OH^- \rightarrow 4H_2O + 4e^- \quad (3)$$
$$O_2 + 2H_2O + 4e^- \rightarrow 4OH^- \quad (4)$$

The only efficient choice of fused salt electrolytes for fuel cells using hydrogen-containing fuels and oxygen are molten carbonates. The reaction on the cathode is:

$$O_2 + 2CO_2 + 4e^- \rightarrow 2CO_3^{2-}$$

The reaction with hydrogen at the anode is:

$$2H_2 + 2CO_3^{2-} \rightarrow 2H_2O + 2CO_2 + 4e^-$$

It is apparent from these reactions that the CO_2 at the anode exit gas stream must be collected and recycled to the cathode.

It is realized that only a few systems can meet the many requirements needed to make a fuel cell efficient and stable in the long term.

2.2 *Various types of fuel cells*

As we have seen, fuel cells may conveniently be classified by the types of electrolytes that may apply: strong acids, strong alkaline solutions (purified of CO_2), molten alkali carbonates and solid oxide electrolytes (Fig. 8.4). They operate in specific temperature windows adjusted according to the availability of suitable materials and the need for accelerating electrode reactions that are not sufficiently catalyzed to proceed well at low temperatures. Depending on the type of fuel cell, the temperature range may extend from 340 K to 1400 K.

2.2.1 ALKALINE FUEL CELL

This fuel cell operates typically between 340 and 470 K, is characterized by high efficiency and high power density but has a drawback that, due to the alkaline electrolyte, CO_2 has to be excluded

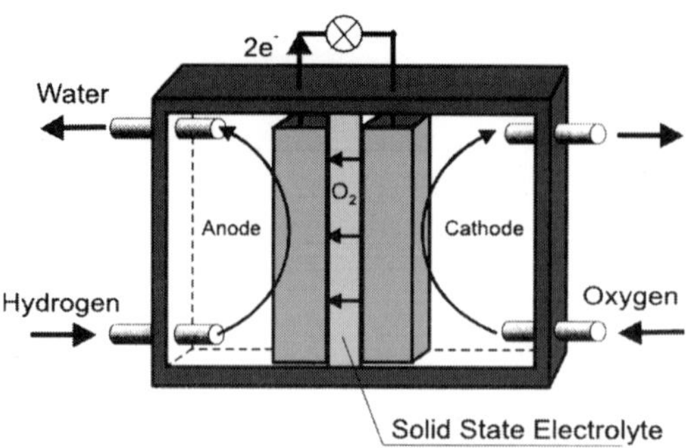

Figure 8.4. High-temperature solid oxide fuel cell (SOFC).

from the system. The alkaline fuel cell reaches the highest efficiencies of all fuel cells and, besides the electrodes of noble metals, needs only relatively inexpensive materials. The chemical fuel, however, is expensive. Pure oxygen has to be used to avoid CO_2 and it is not feasible to produce H_2 from methane or natural gas because this would result in contamination by CO_2.

2.2.2 PROTON-EXCHANGE MEMBRANE FUEL CELL

This type of fuel cell uses ion-exchange membranes, such as perfluorosulfonic acid membranes (NAFION®), which make operation very practical and efficient. The temperature range for operation is between 350 and 380 K. High power density and a long operation life can be obtained with this system. The most advanced ion-exchange membranes satisfy the criteria for successful performance of both stationary and automotive applications. However, the high cost of these membranes is a significant drawback and cheap alternatives are not yet in development. Additional problems are the noble catalysts and water management for this fuel cell.

2.2.3 PHOSPHORIC ACID FUEL CELL

This fuel cell operates between 420 and 480 K, is technologically very advanced and commercially available, but has a relatively low efficiency, a limited lifetime and needs quite noble catalysts. Nevertheless, prototypes of a few hundred kilowatts are already operating. The overall efficiency for electrical energy production is only approximately 40%, but high-temperature heat can be gained and an overall efficiency of 90% is feasible if this high-temperature heat is used to heat buildings.

2.2.4 MOLTEN CARBONATE FUEL CELL

This is a fuel cell type that operates at a temperature between 820 and 920 K and offers a high efficiency and high-grade waste heat. The operation life is, however, relatively short because there are still problems with the electrolyte and material stability.

2.2.5 SOLID OXIDE CERAMIC FUEL CELL

This fuel cell operates at 1200–1400 K and typically uses zirconium oxide as the electrolyte (Fig. 8.4). At a high temperature it becomes conductive for oxygen ions. The electrodes for this fuel cell consist of nickel and conductive oxides. The high operating temperature is a big challenge for the materials applied. A modular production of this kind of fuel cell is feasible and laboratory testing on the kilowatt scale is under way. Commercial implementation of these solid oxide ceramic fuel cells is especially favored by the possibility of a direct reforming of natural gas as an integral part of the fuel cell operation. In addition, a long operation life and potentially inexpensive production technology favor the application of this type of fuel cell.

2.3 *Prospects for the advancement of fuel cell technology*

Fuel cell technology is still immature so advances of the fundamental scientific understanding can still lead to significant progress for engineering. The biggest challenges in fuel technology appear to be:

1 The design of fuel cells that can operate at a very low temperature. This would require considerably improved catalysts and improved transport for ions and chemical reactants and products.

2 The development of fuel cells for certain temperature ranges that are not yet accessible due to the lack of suitable materials.

3 A decrease in the cost of proton-exchange ion membranes. At the moment, there is no realistic replacement for NAFION® (films of a sulfonated fluorocarbon), which is too expensive for broad commercialization.

4 A reduction of the material costs for electrocatalysts. Platinum but also other expensive noble elements are used intensively in fuel cell technology. It has to be remembered that biological catalysts accomplish all difficult energy conversion reactions with organic catalysts containing relatively abundant transition metals only (e.g. Fe, Cu, Mn, Mo).

5 Selectivity in the reactivity of catalytic electrodes is highly desirable. In this case, membranes would not be needed to separate the anode and cathode in methanol fuel cells. Such selective catalytic materials have indeed been identified with ruthenium-cluster compounds and should be tested on an industrial scale [4].

6 Topological studies are required on the optimum distribution of catalysts in electrodes and the development of porous catalysts with time.

Controlling the aging process in the fuel cell will, in part, depend on understanding the migration and degradation of small particles. Research has to be conducted to develop polymer and solid oxide membranes with substantially increased conductivities and decreased potential losses. Further research is necessary to shrink the stacks that build the fuel cell, in order to increase the power density. The tolerance for fuel stream impurities has to be increased and the reliability of the overall system improved. Industrialized countries are presently spending less than 1% of total funds for energy research and development for the purpose of fuel cell design. This is not sufficient, considering the great potential of fuel cells for increasing the efficiency and environmental compatibility of energy systems. Co-generation of electricity and heat for buildings with small- and medium-sized units is one example, and methanol fuel cells for trucks and buses is another example of the useful application of fuel cell technology. It is definitely a field where many different branches of science and technology can contribute. Public funding agencies must take care that a sufficiently large momentum is provided to push fuel cells along a steep learning curve towards a broad technological application.

As indicated at the beginning of this chapter, life has efficiently exploited the possibility of interconverting chemical and electrical energy. It has achieved this at ambient or very low body temperature, which proves that fuel cell technology is an intelligent and practical approach;

however, it has to be realized that significant scientific obstacles still have to be surmounted: a much better understanding of electrocatalysis is needed; and improved mechanisms for ion transport and chemical transport need to be developed. If these obstacles are overcome, then low-temperature fuel cells with high efficiency and high power density should be achieved in the long term.

3 Electrochemical production of hydrogen

According to present technology, the most efficient technique for the production of hydrogen from water is electrolysis. As discussed in Chapter 5, the decomposition of water into oxygen and hydrogen is achieved by passing an electrical current through an electrochemical cell containing two electrodes dipping into salt-containing water. Today, only approximately 1% of the hydrogen is generated via electrolysis, usually where cheap hydroelectric power is available. Hydrogen generation occurs typically in power ranges up to 150 MW, corresponding to 33 000 m^3 of hydrogen per hour. Technical electrolysis units typically consist of a series of single cells with peripheral support units to maintain the electrolysis circuit and for gas deposition and cooling. The cross-section of the single cell may amount to up to 4 m^2 and a single cell unit may produce up to 225 m^3 of hydrogen per hour. The alkaline electrolysis cell is the most developed, but the highest efficiencies are provided by high-temperature electrolysis cells, which have already been tested as prototypes.

Significant fundamental knowledge is available on the electrolysis of water [5]. The industrially applied electrolysis systems typically do not yet apply the significant progress achieved in improving the catalysis of oxygen evolution. More modern prototypes, on the other hand, still show problems with both the stability of catalytic electrodes and with the more advanced system technology. It was especially found that electrolysis does not easily tolerate strongly varying current supply, which occurs when photovoltaic systems are generating electricity. This problem could, however, be solved reasonably well by providing an auxiliary potential during periods where energy supply was below a critical limit [5].

Presently, the best catalyst for oxygen evolution from water, RuO_2, equally evolves chlorine from water (this is actually its principal technical application). Seawater therefore cannot be used directly for electrolysis due to chlorine pollution of the environment or only under strict control of operation conditions and catalysts (see Chapter 5). Research would be needed to find a selective catalyst or to separate ions with cheap technology. Additional research is needed to improve the long-term stability of catalysts for electrolysis. System technology has to be improved and simplified for mass production.

4 Battery systems for energy storage

Batteries are the oldest type of electrical energy carriers. The lead–lead dioxide accumulator (lead acid battery) was introduced in 1854, and the Leclanché (zinc–manganese dioxide battery) goes back to the year 1868. Both systems are based on the same electrochemical principle, i.e. the storage of energy in a metal and in a metal oxide of higher valence state:

$$Me + Me^{4+}O_2 + 4H^+ \rightleftarrows 2Me^{2+} + 2H_2O$$

It depends on the reversibility of this reaction whether the battery is a primary or a secondary accumulator. The secondary battery as a rechargeable device is a fast and convenient energy storage system. In renewable energy systems (solar energy, wind energy) it can act as a direct storage device, or as a buffer if storage in other energy forms is considered. Typically, the capacity of these systems is limited because the energy density is low compared to other chemical forms of energy storage, such as hydrocarbons or hydrogen. The energy density achieved is between 30 and 120 Wh kg^{-1}. Intensive developments of advanced systems might lead to an increase to 200 or 250 Wh kg^{-1}, but no landslide change can be expected.

Nevertheless, the simplicity of handling and the reliability of most devices will give secondary batteries a permanent position in energy storage systems. Below, the state of the art and the expected developments in secondary systems are briefly considered.

4.1 *The present status of rechargeable battery systems*

The available battery systems with rechargeable properties are listed in Table 8.2. Some of them go back to the years of the introduction of the lead acid battery. Only a few systems are of more modern origin, the most spectacular example being the lithium battery, which as the first commercial secondary cell is only a few years old. Table 8.2 highlights two problems for the developers of secondary batteries: the limited number of theoretical systems, simply based on the few combinations of chemical elements of the Periodic Table that can be coupled to a working system; and the large discrepancy between theoretical energy density and practically achieved results.

Table 8.2. Some parameters of available rechargeable batteries.

System	Negative electrode	Positive electrode	Electrolyte	Operational temperature (°C)	Achieved energy density (Wh kg^{-1})	Theoretical energy density (Wh kg^{-1})	Cycles
Lead acid	Pb	PbO_2	4.5 M H_2SO_4	–40 to 60	30–50	160	500
Nickel/cadmium	Cd	NiO(OH)	4 M KOH	–30 to 45	50	210	2000
Nickel/metal hydride	$H_2(LaNi_5)$*	NiO(OH)	4 M KOH	–20 to 50	60	Max 390	1000
Zinc/manganese dioxide	Zn	MnO_2	9 M KOH	–20 to 55	120	790	300
Zinc/bromine	Zn	Liquid Br_2 complex	$ZnBr_2$	–10 to 50	70	440	200
Zinc/air	Zn	Air	KOH	30	200	1350	–
Sodium/sulfur	Liquid Na	Liquid S	β-Alumina	290 to 390	150	790	250
Sodium/nickel chloride	Liquid Na	$NiCl_2$	β-Alumina/ $NaAlCl_4$	220 to 450	100	788	500
Lithium ion	Li(C)	$LiMn_2O_4$ $LiCoO_2$ $LiNiO_2$	Li^+ (polymer) Li^+ (organic, liquid)	–40 to 75	500–1500	1000–2000	Up to 1500
Lithium/iron sulfide	Li(Al)	$Li(FeS_2)$	LiCl/KCl eutectic melt	400	150	490	800

* Base type.

This discrepancy is not a lack of experimental skill but is based on the principal restrictions of reversibility: the higher the density of energy, the smaller the reversibility between the charged and discharged level. From this point of view, the development of these systems marks a point of optimization that can be improved but will not lead to a substantial increase of energy density. An increase of at least one order of magnitude is necessary to develop a real alternative to the present high-energy storage systems.

4.2 *Ongoing developments for improved performance*

4.2.1 LEAD ACID BATTERY

The oldest type of accumulator is based on the reversible electrochemical reaction:

$$Pb + PbO_2 + 2H_2SO_4 \underset{\text{charging}}{\overset{\text{discharging}}{\rightleftarrows}} 2PbSO_4 + 2H_2O$$

The electrolyte is 20% sulfuric acid. With its typical performance data (Table 8.2) it combines high current discharge and, with adapted models, fast recharging with the excellent cyclability of more than 1000 cycles. It is a very cheap system and its recycling rate approaches nearly 100%. The only disadvantage is its low energy density, especially in terms of the energy/mass ratio (30–40 Wh kg^{-1}). It is remarkable that more than 100 years of development have led only to a ratio of 25% of the theoretical energy density (160 Wh kg^{-1}). The reason for this misfit is shown in Fig. 8.5. In addition to the active mass (17%) with its limited utilization, mechanical components such as grits, housing, etc. are necessary for a functional device.

Over the years, continuous system improvements have been introduced. The latest development is geared towards a bipolar system. If successful, this development might increase the energy density to more than 50 Wh kg^{-1}.

4.2.2 NICKEL-BASED BATTERIES

Nickel/cadmium is also an old system, dating from 1899. It is based on the following principal reaction:

$$Cd + 2NiO(OH) + 2H_2O \rightleftarrows 2Ni(OH)_2 + Cd(OH)_2$$

The electrolyte is a 40 wt.% potassium hydroxide solution. The development of this system has brought a number of improvements during the last two decades, increasing the energy density to 50 Wh kg^{-1} (theoretical value 210 Wh kg^{-1}). Typical for the performance of the system is its high

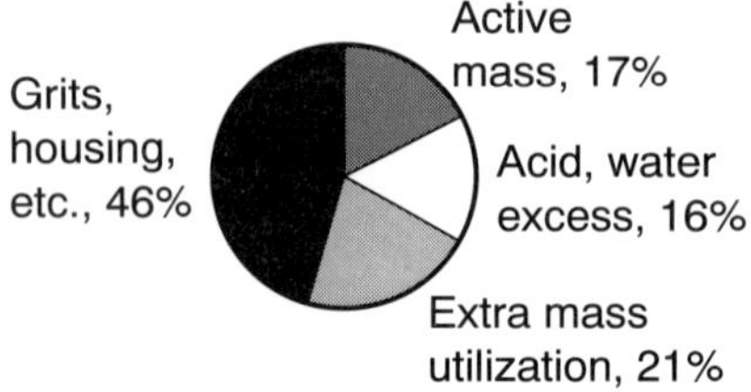

Figure 8.5. Contribution of various system subunits to the overall mass of lead acid batteries.

stability (cycle numbers up to 2000). Big problems are the toxicity of cadmium and the price, which is nearly 5–10 times higher than for the lead system.

The disadvantage of this system is that it contains an extremely toxic metal, cadmium. To overcome this disadvantage, nickel/metal hydride batteries have been developed. Progress in the synthesis of metal hydrides as efficient hydrogen storage devices, e.g. the $LaNi_5$-based structure, has made possible a battery system where the cadmium is substituted by the non-toxic metal hydride. The reversible electrochemical reaction is described by the following schematic overall reaction:

$$LaNi_5H_6 + 6NiO(OH) \rightleftarrows LaNi_5 + 6Ni(OH)_2$$

The performance data achieved (energy density *c.* 70 Wh kg^{-1}, theoretical value 390 Wh kg^{-1}) make this system suitable for many high-performance applications in the near future.

4.2.3 ZINC-CONTAINING SYSTEMS

The group of primary and secondary batteries containing zinc represents also a classical battery system. Primary cells dominate the market (the alkaline zinc/manganese dioxide system where the electrolyte is 9 M potassium hydroxide). The classical primary zinc/manganese dioxide system was the Leclanché battery.

Recently, attempts have been made to make the system rechargeable. This has led to a cell that can be reloaded up to 20–30 times. These are data that are acceptable for a consumer battery but are unacceptable for a larger energy storage system.

Another noteworthy development is the zinc/air battery. This system is described by the equation:

$$2Zn + O_2 + 2H_2O \rightleftarrows 2Zn(OH)_2$$

Oxygen is converted at a catalyst. While many attempts to make the system rechargeable have not yet been successful, a recent development made the zinc electrode exchangeable and reprocessed the zinc hydroxide on a larger scale in a factory, thus separating charging from discharging. The theoretical energy density is 1350 Wh kg^{-1}, but the present performance data are far from this theoretical value (*c.* 200 Wh kg^{-1}). A great problem is the energy loss caused by the high overvoltage of the zinc ion reduction.

Another development using metallic zinc is the zinc/bromine system. This is a secondary battery with a liquid bromine complex stored in a separate storage device. The energy density in Table 8.2 (theoretical energy density 440 Wh kg^{-1}) is not as high as the zinc/air battery.

4.2.4 SODIUM/SULFUR AND SODIUM/NICKEL CHLORIDE SYSTEMS

In the last three decades, high-temperature systems (300–350 °C) have undergone intensive development. The first system was the sodium/sulfur battery and the remarkable performance data (150–200 Wh kg^{-1}) have achieved considerable attraction. The anticipated use for electrical vehicles has suffered severe setbacks in terms of safety problems, which appeared if the systems were tested in cars under real conditions, such as spontaneous combustion in an accident.

In stationary systems, as with solar or wind energy, the advantages will dominate. The electrochemical reaction is given by the equation:

$$NaS_x + Na \rightleftarrows NaS_{x-1}$$

The liquid sodium and the liquid sulfur are the critical components that prevent use in electrical transport systems. To avoid the possible risks of this combination, sulfur was substituted by $NiCl_2$ and the solid β-alumina electrolyte was modified by adding sodium tetra-aluminum chloride. The achieved energy density of 100 Wh kg^{-1} is still low if compared with the theoretical values of 790 (Na/S) and 788 (Na/$NiCl_2$). It is obvious that this system can be improved and it should be possible to achieve 200–250 Wh kg^{-1} in the near future.

4.2.5 SECONDARY LITHIUM BATTERIES

Secondary lithium batteries can be considered as the great breakthrough in secondary battery technology. Considering the high energy density (Table 8.2), the non-rechargeable battery already provides an energy source superior to conventional systems.

Regarding secondary systems, there is a general trend towards the 'lithium ion battery' with lithium intercalation compounds for both electrodes. The negative electrode is typically a lithium graphite (LiC_6) or a lithium carbon system and the positive electrode consists of a lithium metal oxide.

The electrolyte can be a thin layer of liquid organic electrolyte (e.g. Li^+ in a mixture of alkyl carbonates or related solvents) or a solid polymer electrolyte (e.g. Li^+/modified polyethylene oxide). Systems are still under rapid development and energy densities vary between 100 and 200 Wh kg^{-1} (theoretical energy density between 500 and 1500 Wh kg^{-1}, depending on the system).

4.2.6 THE LITHIUM/IRON SULFIDE SYSTEM

This is a high-temperature system working at 400 °C in a eutectic melt of LiCl/KCl, which is still in an experimental stage. It was mentioned as a possible system for electric transport but this will need more time for development.

4.2.7 THE DEMANDS FOR ENERGY CONVERSION SYSTEMS

The present research and development in the area of secondary batteries concentrates on four strategies: supply of energy to mobile systems, such as computers, communication devices, etc.; supply of energy for mobile tools demanding high power density; supply of energy for transport systems; and energy storage for load leveling and isolated energy supply stations.

The supply of energy for transport systems was stimulated by political decisions, especially in California, USA, demanding a certain percentage of cars running with zero emission. Will this lead to a battery-powered vehicle? Other systems, such as fuel cells, may finally provide the better solution.

Future expectations from secondary batteries may deviate from the present ideas. In converting solar energy or wind energy into electricity, larger immobile systems will be necessary as a buffer system, which compensates for the irregularities of the regenerative energy sources. One possible solution is the use of fuel cells. Fuel cells have the advantage of a separation between the conversion system (the fuel cell) and the storage system (the hydrogen, for instance). Developments in secondary batteries, e.g. the zinc/air battery or the zinc/bromine battery, may proceed in a similar direction.

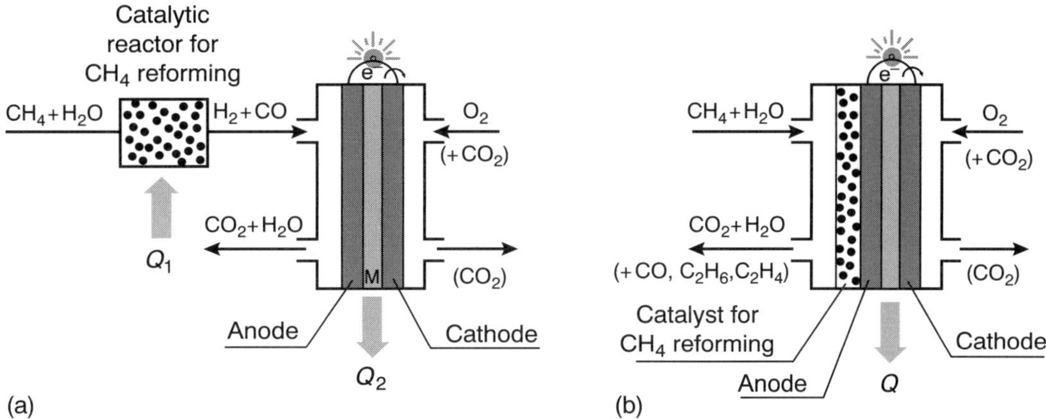

Figure 8.6. Operational scheme of a high-temperature fuel cell with the traditional scheme of external methane reforming (a) and with internal methane reforming in the integral fuel cell (b). Thick arrows indicate heat fluxes Q in such devices; M is the high-temperature oxygen-conducting membrane. Oxygen is transported through the membrane in the form of oxygen anion (for ceramic oxide membranes) or carbonate anion (for molten carbonate membrane). Possible products of gas-phase electrocatalytic partial oxidation of methane are shown in parentheses at the gas outlet of the integral fuel cell.

5 Catalysis for the production of electricity and co-generation of electricity and chemicals

In fuel cells, the electricity is produced as a result of catalytic oxidation of a fuel at one of the electrodes (anode) and reduction of an oxidant (usually oxygen) at the other electrode (cathode). In order to maintain the equilibrium of charges during redox transformations and the electric current flow between the working electrodes, the ion-conducting membranes are usually placed in the cell. For a long time chemical processes in fuel cells attracted mainly chemists, who were indeed well aware of the fact that the electrodes of the fuel cell must be good catalysts of the corresponding heterogeneous redox processes. However, experts of 'pure' catalysis have shown recent interest in fuel cells too. This is primarily due to marked intensification of research work aimed at the creation of medium- and high-temperature fuel cells where alcohols, methane and other hydrocarbons can be used as fuels [7]. Application of high temperature in new fuel cells requires the operation of a working electrode–catalyst system in the gas phase. The latter is well known to chemists dealing with 'pure' catalysis.

It should be noted that the initial scheme of hydrocarbon use in fuel cells assumed their endothermic reforming with steam or carbon dioxide to syn-gas in a separate conventional catalytic reactor (so-called external reforming); then the syn-gas obtained was fed to a unit of a hydrogen–oxygen fuel cell, which is very familiar to electrochemists. Although heat supply is needed for operation of the subunit for external hydrocarbon reforming, operation of the fuel cell subunit produces heat in much larger amounts (see Fig. 8.6a). This leads to the suggestion that an integrated system could be developed instead of one consisting of spatially separated subunits. This can be done, for example, by arranging the hydrocarbon-reforming catalyst and the fuel cell unit into a single multipurpose structure, which is called an integral fuel cell or fuel cell with internal reforming (Fig. 8.6b and [6,7]). The successful operation of such integral fuel cells would be impossible without the development of special catalysts, because the specific exacting requirements of the properties of ion-conducting membranes and other components of fuel cells impose

Table 8.3. Examples of gas-phase electrocatalytic reactions accomplished over ion-conducting electrolytes (membranes) in electrochemical devices.

No.	Reactions*	Electrolyte membrane†	Electrode–catalyst
Oxygen-conducting membranes			
1	$H_2 + [O] \rightarrow H_2O$	YSZ, MC	Various metals
2	$CO + (O_2) + [O] \rightarrow CO_2$	YSZ, MC	Various metals
3	$C_2H_4 + (O_2) \pm [O] \rightarrow$ ethylene oxide, CO_2	YSZ	Ag, Pt
4	$C_3H_6 + (O_2) + [O] \rightarrow$ propylene oxide, CO_2	YSZ	Pt
5	$C_4H_8 + [O] \rightarrow CO_2$	YSZ	Pt
6	$CH_3OH + (O_2) + [O] \rightarrow CO_2, H_2CO$	YSZ	Pt
7	$CH_4 + [O] \rightarrow C_2H_2, C_2H_4, C_2H_6, CO, CO_2, H_2O$	YSZ, MC	Ag, $LiNiO_2$, Cu, Ni, Ag–Pd, etc.
8	$C_6H_5C_2H_5 + [O] \rightarrow$ styrene	YSZ	Pt
9	$CH_4 + NH_3 + [O] \rightarrow HCN$	YSZ	Pt
10	$NH_3 + [O] \rightarrow NO_x$	YSZ	Pt
11	$SO_2 + [O] \rightarrow SO_3$	YSZ	Pt
12	$CH_3OH \pm [O] \rightarrow H_2CO, CO, H_2, CH_4$	YSZ	Ag, Pt
13	$H_2 + CO + CO_2 - [O] \rightarrow CH_4$	YSZ	Ni, Co, Pt, Fe
14	$H_2O - [O] \rightarrow H_2$	YSZ	Various metals
15	$CO_2 - [O] \rightarrow CO, C$	YSZ	Various metals
Hydrogen-containing membranes			
16	$CH_3OH - [H] \rightarrow$ methylformate, dimethoxymethane, CO_2	Solid heteropolyacid, phosphoric acid, NAFION	Pt
17	$CO_2 + [H] \rightarrow CH_4$	NAFION	Cu
18	$CO, CO_2 + [H] \rightarrow CH_3OH, C_2H_5OH$	Solid heteropolyacid, Sb_2O_5	Cu, Ni
19	$C_2H_4 + (H_2) \pm [H] \rightarrow C_2H_6$	$CsHSO_4$	Ni
20	$C_2H_6 + O_2 + [H] \rightarrow C_2H_5OH, C_2H_4O$	NAFION	$Fe(OH)_x$?

From [6,8].
* Symbols [O] and [H] with the corresponding signs '+' or '–' indicate the direction of transferring oxygen and hydrogen atoms through the ion-conducting membrane. Symbols (O_2) and (H_2) indicate a possible presence of these substances in the gas phase as an admixture to the principal reagents.
† YSZ, oxygen-conducting membrane based on solid yttria-stabilized zirconia; MC, molten carbonates; NAFION, proton-conducting film of sulfonated Teflon.

certain extra requirements on the catalysts used in these systems. These catalysts, for example, must have a larger resistance to poisoning by carbonates, etc. Problems of this kind have been solved already and one can expect further studies to result in substantial improvement of the existing fuel cells and the development of new ones.

As the development of integral high-temperature fuel cells progressed, it was found that such fuel cells can also synthesize organic compounds via gas-phase electrocatalysis with the aid of ion-conducting membranes. Typical examples of gas-phase electrocatalytic reactions that have been accomplished with the aid of ion-conducting membranes are listed in Table 8.3. The published results are still quite sparse, but they suggest that in future the electricity production (generation) in fuel cells can be combined with gas-phase electrocatalytic synthesis. For example, with methane as a fuel, one can achieve operating conditions in the high-temperature fuel cell such that methane will oxidize not to CO_2 but to CO or ethane and ethylene or even syn-gas (see Example 7 in Table 8.3 and Section 2.3.1 in Chapter 5). The oxidative dehydrogenation of methane to C_2-hydrocarbons or syn-gas is accomplished in the fuel cell along with electricity production.

These reactions can be controlled by electric potential variations on electrodes. The process selectivity with respect to C_2-hydrocarbons or syn-gas turns out to be comparable to that of the currently developed conventional process of catalytic oxidative dehydrogenation of methane.

Solid zirconium oxide with the addition of yttrium at working temperatures (*c.* 870–1070 K) or molten metal carbonates at working temperatures (*c.* 870 K) are commonly used as oxygen-conducting membranes in high-temperature fuel cells. Such membranes are commonly used in the processes of gas-phase electrocatalytic oxidation (Table 8.3). Some processes of gas-phase electrocatalytic reduction have been reported to occur also over oxygen-conducting membranes, although proton-conducting membranes operating at low temperatures seem to be more suitable for such purposes. For example, CO_2 is reduced at room temperature to methane over polymer proton-conducting electrolytes such as 'NAFION' [9]. Reduction of CO to methyl and ethyl alcohols was observed at the same temperature over proton-conducting membranes of heteropolyacids or antimony oxides (see Table 8.3).

Recently, cathodic partial oxidation of ethane on NAFION-like membranes under mild conditions has been reported, the process being able to proceed in the fuel cell mode, i.e. with co-generation of electricity (see Reaction 20 in Table 8.3 and [8]). Another efficient version of carrying out electrocatalytic processes with gaseous reactants is by using gas-diffusion electrodes, which were also developed for fuel cell application. An example of CO_2 reduction to CO on a gas-diffusion electrode is mentioned in Section 4.3 of Chapter 7.

Incidentally, such oxidation and reduction processes may be interesting for chemical synthesis irrespective of whether or not they are coupled with electricity production. Moreover, the processes of gas-phase electrocatalytic synthesis over ion-conducting membranes are also of great interest for fundamental studies of catalysis, because they allow us to study the chemical processes on the catalyst surface under the unique conditions of supplying one of the reactants (e.g. oxygen or hydrogen) not from the gas phase but directly from the volume of the catalyst, with simultaneous control of its reactivity by applying an appropriate voltage to the electrode–catalyst [9,10].

6 Electrochemical cells for energy production from the sea, underground and wastewaters

Also of interest for energy production may be electrochemical cells that utilize various electron-donating organic and inorganic compounds dissolved in wastewaters or in the sea and underground water.

There are specialized geological environments where reducing chemicals are accumulated: for example, the possible perspectives of utilization of the energy potential of H_2S dissolved in deep water layers of the Black Sea have been repeatedly discussed. In this sea, the water below 200 m depth is enriched with H_2S. Also, water near vulcanic sites frequently has high concentrations of H_2S. There are also many deposits of metal sulfides, which may supply reducing power after catalytic activation. The reducing environments may serve as electron donors for batteries in which the electron flow is maintained to air as a suitable oxidant. A similar electrochemical energy exploitation could be attempted using wastewater containing H_2S, alcohols or phenols. In this way, cleaning of these liquids would be a favorable side-effect. It is evident that any progress in this direction will depend on the development of catalysts capable of providing electrooxidation of electron donors over anodes — in the case of the particular example, sulfide ion to elemental sulfur — under appropriate very specific conditions with durable preservation of electrode activity. The possibility of developing such fuel cells has been demonstrated by the well-known advancement in the development of selective sensor electrodes, with enzymes as electrocatalysts. However, much work remains to be done to develop electrochemical devices of practical interest for this type of energy production.

It may also be worthwile investigating whether underground coal and oil shale gasification may be coupled directly with underground electricity generation. It may be possible in this way to keep the final product, CO_2, underground. Such a technology may also be attempted with natural gas deposits, which typically already provide high pressure and elevated temperature for improved reactivity. In all cases, confinement of CO_2 to deep underground layers may be a significant chance for energy technology in times of increasing 'greenhouse' problems.

7 Conclusions

Interconversion of electrical and chemical energy has been a fundamental strategy of biological systems. Evolution has proved that this technological strategy is reasonable, efficient and can be an integral part of a sustainable energy system that nature has evolved for the utilization of solar energy.

Technical systems for producing electrical energy from chemicals (fuel cells) and for generating chemicals from electrical energy (electrolysis systems) are both working on the basis of technology that is relatively little tested and established. Both technological developments therefore cannot easily compete with energy technology based on fossil fuel utilization. Both technologies have to be improved by systematically working on many detailed technological problems that make up these complicated systems. In both technologies, the lack of understanding of multi-electron transfer catalysis is still a major problem. It forces us to either work at relatively high temperatures (fuel cells) or to accept relatively large potential losses (electrolysis cells). Basic research will definitely lead to improvements over the years, but the most important factor is the continuing support of these technologies so that they can gradually become commercially feasible. This will then automatically provide a basis for continuing the improvement of these valuable technologies. It may be mentioned that the combination of electrolysis and fuel cell operation in the same system is being developed and is an attractive technological concept because it allows considerable savings in materials. On the other hand, the challenge of developing an efficient reversible oxygen electrode should not be underestimated.

The present situation in the development of secondary batteries is determined by the limited availability of systems comparable with the energy density of fossil energy carriers. Because the general limits are known, one cannot expect a real jump in energy density, reversibility, etc. Another limiting factor is the market price, which excludes some systems. A continuous improvement in the systems available at present will be sufficient for some applications but not all. In particular, the expectations on batteries for electrical cars might be too high and other solutions may determine the transportation systems of the future. Fuel cells might be the more realistic choice.

For further improvement of the existing energy storage systems, one needs research and development into new materials. Such circumstances make the investigation of new promising electrochemical systems for rechargeable batteries both actual and timely.

8 References

1 Steck AE. In Savadogo O, Roberge PR, Viziroglu TN (eds) *New Materials for Fuel Cell System I, Conf. Proceedings: Montreal, Quebec, Canada, July 9–13, 1995*. Montreal: Edition de L'Ecole Politechnique de Montreal, 1995; 74.

2 Kartha S, Grimes P. *Phys. Today* 1994; **Nov**: 54.
3 Bockris JO'M. *Energy Options*. Sydney: Australia and New Zealand Book Company, 1980; 311.
4 Alonso Vante N, Giersig M, Tributsch H. *J. Electrochem. Soc.* 1991; **138**: 639.
5 Grasse W, Oster F (eds). *Hysolar, Solar Hydrogen Energy, Results and Achievements 1985–1989*. University of Stuttgart: DLR, 1990.
6 Parmon VN, Ismagilov ZR, Kerzhentsev MI. In Thomas JM, Zamaraev KI (eds) *Perspectives in Catalysis*. Oxford: Blackwell–IUPAC, 1992; 337.
7 Hutchings GJ, Burstein GT (eds). *Catal. Today* 1997; **38**: 391.
8 Kuzmin A, Savinova ER, Parmon VN. *React. Kinet. Catal. Lett.* 1999; **66**: 351, 359.
9 Neophytides S, Vayenas CG. *J. Catal.* 1989; **118**: 147.
10 Politova TI, Sobyanin VA, Belyaev VD. *React. Kinet. Catal. Lett.* 1990; **41**: 321.

9 Biomass: a Future Renewable Carbon Feedstock for Energy

D.O. HALL and F. ROSILLO-CALLE

Life Sciences Division, King's College London, Campden Hill Road, London W8 7AH, UK

1 Overview

This chapter deals mostly with biomass energy but chemicals from biomass are also considered. Biomass energy, or bioenergy (the chemical energy stored in organic matter and derived from solar energy via photosynthesis), is experiencing a surge in interest. This is due to: greater recognition of its current role and future potential contribution as a modern fuel; its availability, versatility and sustainability; its global and local environmental benefits; the existing and potential development and entrepreneurial opportunities; technological advances and knowledge that have recently accumulated on many aspects of biomass energy, etc. However, despite these advantages and advances, bioenergy still faces many barriers stemming from economic, institutional and some technical factors: for example, it is striking that in many countries where biomass energy plays a significant role, so few resources (either human or financial) are allocated to biomass energy. It is essential that the contribution and potential of biomass, both as a modern and traditional energy source, are fully recognized and that resources are accordingly made available for its further development.

1.1 *Potential sources of biomass for energy and chemicals*

Biomass resources are potentially the world's largest and sustainable energy source — a renewable resource comprising 220 billion oven-dry tons (about 4500 EJ) of annual primary production. The annual photosynthetic storage of energy in biomass is about ten times more than the present energy use from all sources (4000 EJ). The annual bioenergy potential is about 2900 EJ, although only 270 EJ could be considered available on a sustainable basis and at competitive prices [1]. The problem is not availability but the sustainable management and delivery of energy to those who need it. Most major energy scenarios recognize bioenergy as an important component in the future world's energy [2,3], ranging from 10% to 33% of primary energy supply by around 2050, with modern biomass providing between 17% and 33% of the world's electricity. In the Intergovernmental Panel on Climate Change (IPCC) [4], in all its five scenarios, biomass takes an increasing share of total energy over the next century, rising to 25–46% in 2100 (see Table 9.1).

Biomass is already the fourth largest source of energy (about 14%) in the world [12]. Developing countries derive 38% of their energy from biomass, but with large variations from almost zero to over 90%, mostly in the form of traditional fuels such as fuelwood, straw and dung, but with increasing use in modern applications. Industrial countries are also increasingly using bioenergy, e.g. the USA derives nearly 4% of its total energy from biomass (about 1.4 million barrels of oil equivalent (b.o.e.) per day, equivalent to hydro and nuclear contributions), has over 7600 MW of biomass power generation and possibly over 13 000 MW by 2010, and produces over 4 billion liters of fuel ethanol from corn. In the EU, biomass currently supplies the equivalent of 45 million tons of oil equivalent (t.o.e.) (3% of primary energy consumption) and is planned to increase to about 135 million t.o.e. (8.5%) by 2010.

Table 9.1. The role of biomass in future global energy use (EJ)01—present biomass energy use is about 55 EJ annually.

Scenario	Year 2025	2050	2100
IEA [5]	60	–	–
IIASA/ WEC [6]	82	153	316
Shell [7]	85	200–220	–
IPCC [4]	72	280	320
Greenpeace [8]	114	181	–
Johansson *et al.* [2]	145	206	–
WEC [3]	59	94–157	132–215
Dessus *et al.* [9]	135	–	–
Lashof and Tirpak [10]	130	215	–

From [11].

Almost every petroleum-derived chemical currently being produced could be produced from biomass. Biomass is already used as a source for various large-scale industrial chemicals, such as dimethyl sulfoxide, rayons, tall oil, tannins, specialty chemicals, etc. A major difficulty is price and feedstock supply for large-scale production of chemicals, e.g. the petrochemical industry is very capital intensive and normally operates all year around but this is not normally the case with biomass-based feedstocks. Considerable research has been done in past decades to find suitable alternatives for petrochemical feedstocks but a major barrier continues to be the cost of alternative feedstocks.

1.2 *Main sources of biomass energy*

Biomass energy can be obtained from a very large number of biological-based sources containing sugars, starches or cellulose. There is an enormous untapped biomass potential, particularly in improved utilization of existing resources, forest and other land resources, higher plant productivity and efficient conversion processes using advanced technologies. Forests and agroforestry residues are currently the main sources of biomass energy. A large number of energy-dedicated crops/forestry have been investigated over the past decade. The most promising in the short to medium term are sugar-cane and short-rotation woody crops (SRWCs), e.g. willows and poplars. Significant increases in productivity and cost reduction have been achieved, e.g. in Brazil productivity from sugar-cane has increased from 2630 l ethanol ha^{-1} $year^{-1}$ in 1977 to 5100 l in 1996, and in Sweden productivity from willow increased from about 8 t ha^{-1} $year^{-1}$ (dry wt) to over 10 t ha^{-1} $year^{-1}$ over the past decade.

1.3 *Land uses*

Land availability and bioenergy production are intrinsically interlinked. Many studies have been carried out to determine how much land is available globally for non-agricultural purposes. Estimates range from 150 to 1200 Mha, reflecting the lack of adequate criteria for classifying degraded and abandoned lands. What these and other studies seem to demonstrate is that the perceived constraints on land are not well founded, notwithstanding local priorities of land use. Factors such as food production and consumption, distribution patterns, hunger, lack of purchasing power, inequality, land and grains used for livestock, under-utilization of agricultural land, lack of appropriate investments, export of crops, land tenure, wars and political policies are major factors.

1.4 *Costs and commercial opportunities of bioenergy*

Costs are very specific, depending on a large number of variables including raw material, management practices, type of technology and environmental considerations. Most biomass energy technologies have not yet reached a stage where market forces alone can make the adoption of these technologies possible and they need some kind of support in the form of taxes, subsidies, fiscal incentives, etc. Conventional energy sources (fossil fuels and nuclear) still continue to receive large subsidies estimated at over $300 billion already. The more established technologies, e.g. combined heat and power, ethanol fuel and bio-gas, have seen important cuts by about 4% annually for the last decade and a further 23% would be possible with relatively minor investment (see Table 9.2).

The most immediate commercial prospects are in co-generation (heat and electricity), probably spearheaded by the pulp, paper and timber industries using wood wastes and the sugar-cane industry using bagasse. Co-generation from sugar-cane bagasse is becoming a commercial reality in various sugar-cane-producing countries, e.g. India and Mauritius. In Brazil, the world's largest producer of sugar-cane, the main barrier for co-generation is the low cost of hydroelectricity, which has many hidden costs not reflected in the final price.

A major barrier to the commercialization of bioenergy is that current energy markets mostly ignore the social and environmental costs and risks associated with conventional fuel use, the direct and hidden subsidies, the long-term costs of depletion of finite resources and the costs associated with securing reliable supplies from foreign sources. However, growing environmental and ecological pressures, combined with technological advances and increases in efficiency and productivity, are making biomass feedstocks more economically attractive in many parts of the world, with some technologies reaching commercial maturity, such as wind power.

1.5 *Management practices*

Management practices are an important factor in biomass production and use. Although we can provide considerable data on traditional forms of management practices, particularly in the pulp/paper, cellulose and sugar-cane industries, by comparison little research and development (R&D) has been carried out to modify or develop forestry/crops specifically tailored to meet specific energy characteristics, e.g. basic density, low ash content, etc. However, significant R&D funding is being channeled to address many of these issues, particularly in the USA and the EU, with encouraging results.

1.6 *Future prospects*

It is notoriously difficult to forecast energy production and use. Nevertheless, all indications point to a more complex and varied future energy supply matrix in which renewable sources will have a major and increasing share of the market. Energy demand will continue to grow, although the pace will depend on population, economic growth and technological advances. A more decentralized and diversified energy supply system would allow greater control at national, regional and local levels. Fossil fuels will continue to be used on a large scale well into the next century, but with a diminished role on a percentage basis, whereas alternative energy sources, the so-called

Table 9.2. Ethanol production costs in Brazil and further potential for reduction (based on an autonomous distillery with a capacity of 120 000 l year^{-1}).

Item	(1990 US$ l^{-1}, average)
Direct costs	
Labor	0.006
Maintenance	0.004
Chemicals	0.002
Energy	0.002
Other	0.004
Interest (working capital/financial)	0.022
Sugar-cane	0.127
Fixed costs	
Capital (6% discount rate)	0.03
Capital (12% discount rate)	0.051
Others	0.011
Total	
6% discount rate	0.208
12% discount rate	0.229
Potential for reducing costs	Cost reduction (%)*
Sugar-cane production (agriculture)	
Selection of varieties & handling	9.8
Lime application	1.6
Liquid fertilizers	0.7
Stillage application	1.0
Weed control	2.1
Transportation	0.5
Operational planning	3.4
Ethanol production (industry)	
Milling	1.3
Fermentation	3.3
Distillation	0.3
Energy	1.5
Total	23.1†

From [13].
* For each item, 'cost reduction' corresponds to the ratio between net benefit gains minus associated costs, including cost of downstream processing and the total costs of production and storage of ethanol.
† The total does not correspond to the sum of all items because some items are inter-related.

'carbon-free' energies, will see their market share increase considerably, particularly biomass energy (see Table 9.1).

2 General introduction

The growing interest in biomass, both for energy and chemicals, stems from a combination of factors: (i) awareness of the enormous potential of biomass, particularly for energy, at accessible costs; the reflection of considerable technical gains of the past decade, mainly in the industrial world; the present and future possibilities of converting biomass to convenient energy forms such as electricity, liquid and gaseous fuels and to chemical products; (ii) increasing concern with the environmental implications of fossil fuel use combined with the potential local and global environmental and ecological benefits arising from bioenergy when produced sustainably; and (iii) new possibilities for reducing government subsidies to farmers through economically viable bioenergy production on surplus land.

None the less, despite this growing interest, bioenergy in particular still faces difficulties because of: inadequate political, financial and institutional support; insufficient funding for R&D and demonstrations, particularly in the developing countries; exclusion of external costs and the non-monetary benefits in economic evaluations of energy that place biomass energy on an unequal footing compared to conventional energy sources; low oil prices; the varied and sometimes unpredictable nature of biomass energy sources and uses; the free or low-cost availability of biomass resources for energy, especially in developing countries, which results in little incentive to improve energy efficiency or to find alternative energy sources unless they can be provided on an equal delivered cost basis.

This chapter presents a general overview of the importance of biomass primarily as a source of energy. It pays particular attention to the biomass energy potential, some characteristic attributes of energy forestry/crops, the possibilities of obtaining biomass-based chemical products, land availability issues, costs and potential job creation, the management practices of energy forestry/crops, some environmental aspects and the future prospects of biomass energy.

3 The potential of biomass

Four broad categories of biomass use can be distinguished: (i) basic needs, e.g. food, fiber, etc.; (ii) energy, e.g. domestic and industrial; (iii) chemicals and materials, e.g. construction; and (iv) environmental and cultural, e.g. the use of fire. Biomass use throughout the course of history has varied considerably, greatly influenced by three main factors: population size and resource availability [14], and understanding how to use biomass to our own advantage. We will deal mainly with energy and also briefly with chemicals.

3.1 *Energy*

3.1.1 PRESENT POTENTIAL

Bioresources are potentially the world's largest and sustainable source of fuel and chemicals — a renewable resource comprising 220 billion oven-dry tons of annual primary production. Estimates of the total annual amount of biomass are about 2×10^{11} tons of organic matter, which is equivalent to about 4×10^{21} J of energy. The average coefficient of utilization of the incident photosynthetically active radiation by the entire flora of the Earth is only about 0.27%. The energy content of biomass on the Earth's surface is equivalent to about 36×10^{21} J compared to 3.9×10^{20} J of all the energy consumed in the world in 1991 [1,12]. The problem is not availability but the sustainable management and delivery of energy to those who need it.

It is well understood that biomass is a major source of energy in many developing countries, but perhaps what is not so well known is that a revival in the modern applications of bioenergy is being led by the industrial countries, e.g. the EU and the USA. An exception is Brazil's ethanol fuel program. For example, currently in the EU, renewable energies presently contribute to about 6% of its primary energy consumption and is planned to increase to around 12% by 2010. Biomass in the EU provides about 3% (45 million t.o.e.) but with wide variations, e.g. 12% in Austria, 18% in Sweden and 23% in Finland. By the year 2010, about 135 million t.o.e. (8.5%) of the primary energy could be derived from biomass, representing an additional 90 million t.o.e. The projected additional biomass energy use is: bio-gas, 15 million t.o.e.; agroforestry residues, 30 million t.o.e.; energy crops, 45 million t.o.e. Table 9.3 assesses some of the main potential costs and benefits of renewables in the EU, showing that a total investment of about 165 billion ECU would be needed to achieve the overall additional increased use of renewables. As a result there is: an estimated annual generation of new business of 36.6 billion ECU by 2010, the creation of a large number of jobs, avoidance of 21 billion ECU of fuel costs, a reduction of energy imports by 17.4% and a reduction of 402 million tons of CO_2 annually compared to 1997 [15].

However, despite this large potential, in practice only a few types of biomass feedstock can seriously be regarded as potential sources of energy and chemicals due to various economic and environmental constraints. Residues from forestry and agriculture are invaluable as immediate and relatively cheap energy resources to provide the initial feedstock for the development of bio-energy industries. They are also frequently an environmentally acceptable way of disposing of unwanted and polluting wastes, but their use must encompass environmental sustainability. The energy content of potentially harvestable residues (the total residues produced that could potentially be collected) is about 93 EJ year^{-1} worldwide. Assuming that only 25% of this is realistically recoverable, residues could provide 7% of the world's energy [16].

Although residues can provide an important kick-start for the bioenergy industry, the development of large-scale energy production from biomass will probably rely in the future on specifically grown energy crops, such as sugar-cane, miscanthus, switchgrass and trees (particularly SRWCs). Biomass productivities must be improved because they are generally low, being much less than 5 t (dry wt) ha^{-1} year^{-1} for many woody species without good management. It is now possible, with good management, continued research and planting of selected species and clones on appropriate soils, to obtain 10–15 t ha^{-1} year^{-1} in temperate areas and 15–25 t ha^{-1} year^{-1} in tropical countries [17]. Record yields of 40 t ha^{-1} year^{-1} have been obtained with *Eucalyptus* in Brazil and Ethiopia. High yields are also feasible with herbaceous crops; for example, in Brazil, the average ethanol yield from sugarcane has risen from 2633 l ha^{-1} in 1977 to 5100 l ha^{-1} in 1996 [13].

While the modernization of bioenergy is to be desired because it can be used with greater efficiency and produced and used in an environmentally sustainable manner, it is important to recognize the fact that traditional energy is still a key source of energy in many communities. Woodfuels are of enormous economic value, e.g. over $30 billion annually in Asia alone (excluding Japan and Siberia). Evidence shows that bioenergy is used by both low- and high-income groups in many parts of the world and that modern use of biofuels in many instances is complementary to traditional fuels. The prime concern is not the availability of woodfuels as such, but their distribution to people in need. In areas of woodfuel scarcity, other biomass-type fuels are likely to increase in importance as complementary sources of energy [18].

Recent evidence in Asian countries shows that two-thirds of all woodfuels originate from non-forest land and thus are not the cause of deforestation as was stated in the 1970s. When the 'Fuelwood Gap Theory' was proposed in the 1970s, it implied that woodfuels were consumed on a non-sustainable basis. An important conclusion was that deforestation and forest degradation were largely caused by fuelwood harvesting. Much more has been learned since then and we now know that this is not the case. Non-forest land will continue to be the main source of woodfuels. Wood energy use is not, and will not be, a general or main cause of deforestation [18].

Table 9.3. Estimated total investment costs and benefits of renewable energy in the EU, 1997–2010.

Type of energy	Additional capacity 1997–2010	Unit cost 1997 (ECU)	Unit cost 2010 (ECU)	Average unit cost (ECU)	Total investment 1997–2010 (billion ECU)	Additional annual business 2010 (billion ECU)	Benefits of annual avoided fuel costs 2010 (billion ECU)	Total benefit of avoided costs 1997–2010 (billion ECU)	CO_2 reduction 2010 (Mt year^{-1})
Wind	36 GW	1000 kW^{-1}	700 kW^{-1}	800 kW^{-1}	28.8	4.0	1.4	10.0	72
Hydro	13 GW	1200 kW^{-1}	1000 kW^{-1}	1100 kW^{-1}	14.3	2.0	0.9	6.4	48
Photovoltaics	3 GWp	5000 kWp^{-1}	2500 kWp^{-1}	3000 kWp^{-1}	9.0	1.5	0.06	0.4	3
Biomass	90 million t.o.e.				84.0	24.1	–	–	255
Geothermal (+heat pumps)	2.5 GW	2500 kW^{-1}	1500 kW^{-1}	2000 kW^{-1}	5.0	0.5	–	–	5
Solar collectors	94 million m^2	400 m^{-2}	200 m^{-2}	250 m^{-2}	24.0	4.5	0.6	4.2	19
Total for EU market					165.1	36.6	3.0	21.0	402

From [15]. Column 2: additional capacities needed to be installed to achieve the estimated renewable energy contribution. Columns 3 and 4: current unit costs by technology type and projected costs in 2010, respectively. Column 5: average reference unit costs, considering projected time frame for the deployment of each technology. Column 6: total investment needed for installations. Column 7: expected annual business. Installation rate, Operations and Maintenance (O&M) and fuel costs for biomass are also included. Columns 8 and 9: estimations of avoided fuel costs, based on coal and oil at 1997 prices. Column 10: considers mainly displacement of coal power plants in electricity generation (1 TWh produced by renewable energy sources saves 1 Mt of CO_2). For biomass, emissions during feedstock production are also included.

3.1.2 MUNICIPAL SOLID WASTE (MSW)

Municipal solid waste can be an important source of energy because considerable amounts are produced all over the world, e.g. over 200 Mt $year^{-1}$ in the USA alone, causing considerable disposal problems. However, there are a number of reasons why this source of biomass will not be considered in this chapter: (i) the nature of MSW, which comprises many different organic and non-organic materials; (ii) difficulties and high costs associated with sorting out such material, which make it an unlikely candidate for renewable energy except for disposal purposes; (iii) re-used MSW is mostly for recycling, e.g. paper; and (iv) MSW disposal would be done in landfills and incineration plants. Such complex issues require greater attention than are possible in this chapter.

3.1.3 BIOMASS ENERGY SCENARIOS

In the past few years, a number of global energy scenarios have been published, most of which include substantial roles for energy efficiency and renewable energies, but some have studied biomass in more detail and incorporated large roles for bioenergy.

The IPCC [4] has considered a range of options for mitigating climate change and increased use of biomass for energy features in all of its scenarios. In five scenarios biomass takes an increasing share of total energy over the next century, rising to 25–46% in 2100. In IPCC Working Group II's biomass intensive energy scenario, with biomass providing for 46% of the total energy in 2100, the target of stabilizing CO_2 in the atmosphere at present-day levels is approached. Annual CO_2 emissions fall from 6.2 Gt.C in 1990 to 5.9 Gt.C in 2025 and to 1.8 Gt.C in 2100: this results in cumulative emissions of 448 Gt.C between 1990 and 2100, compared to 1300 Gt.C in their business-as-usual case [19]. Estimates for future biomass use from various sources are summarized in Table 9.1.

The Renewables — Intensive Global Energy Scenario (RIGES) prepared for the Rio 1992 Conference [2] proposes a significant role for biomass in the next century. It is concluded that, by 2050, 'renewable sources of energy could account for three-fifths of the world's electricity market and two-fifths of the market for fuels used directly'. Within this scenario, biomass should provide about 38% of the direct fuel use and 17% of the electricity.

A Fossil-Free Energy Scenario (FFES) was developed [20] as part of Greenpeace International's study of global energy warming. They forecast that, in 2030, biomass could supply 24% (= 91 EJ) of primary energy supply (total = 384 EJ) compared to today's 7% (= 22 EJ) out of a total of 338 EJ. The biomass supply could be derived equally from developing and industrialized countries.

The World Energy Council examined four 'Cases' of global energy supply up to the year 2020, spanning energy demand from a 'low' (ecologically driven) case of 475 EJ to a 'very high' case of 722 EJ, with a 'reference' case total of 563 EJ [4,21]. In the ecologically driven case, traditional biomass could contribute about 9% of total supply whereas modern biomass could supply an additional 5% of the total, equal to 24 EJ. New renewables combined (modern biomass, solar and wind, etc.) could supply 12% of the total. In the high-growth case, these contributions could be 8% and 5%, respectively, of a higher total supply.

In the Conventional Development Scenario studied in the Stockholm Environment Institute Polestar Project [22], a conservative view of bioenergy is taken. Bioenergy represents about 4–5% (36–45 EJ) of the global primary energy supply (over 900 EJ) in the year 2050. An optimistic scenario (GREENS) [23] proposes that biomass could supply 75% of the world's energy as soon as 2015.

Shell International Petroleum has developed two scenarios. In Sustained Growth (SG), abundant energy supply is provided at competitive prices, because productivity in supply keeps improving in an open market context. Global annual average energy per capita values increase from about 13 b.o.e. (75 GJ) today to 25 b.o.e. (143 GJ) by 2060 and 40 b.o.e. (229 GJ) in 2100. By 2020, new renewables, particularly biomass, photovoltaic and wind, become major players, representing about 10% (80 EJ) of the world's energy market. By 2050, renewables provide between 40 and 50% of the world's energy requirements. This model also envisages that, by 2060, about 200 EJ of primary energy could be obtained from 400 Mha of biomass plantations, with an average productivity of 25 t ha^{-1} [7].

Forecasting future energy demand and supply is notoriously difficult and imprecise, but what is evident from examining all these scenarios is that biomass could be a major contributor to future energy supplies, especially as a modern fuel, while still playing an important role as a traditional fuel or combined traditional and modern role. How much bioenergy will contribute in the next century will depend on many factors that are difficult to foresee at this stage.

3.1.4 ENERGY VALUE OF BIOMASS

If biomass is to be used as a source of energy on a large scale, there are a number of important specific attributes that need to be considered, e.g. growth rate, basic density, calorific value, ash and moisture content. High growth rate within a specific climatic and soil situation is a major factor because it can outweigh most other attributes. Most of the research in forest management, for example, has been directed towards increased productivity and/or characteristics that are more suitable for pulp and paper and not for energy [17].

Energy crops/forestry for power and heat purposes also require fuel specifications that are not yet fully met: low Cl to avoid corrosion; low K and Ca to prevent fouling caused by low ash melting temperatures; low N to reduce NO_x emissions; etc. A further area that needs more attention is the development of effective low-cost machinery for harvesting and processing energy crops.

1 Energy content, heating and calorific values. The energy content of biomass is a very important parameter, particularly from the standpoint of conversion to energy products. The heat liberated by the complete and rapid burning of a fuel per unit weight or volume is the heating or calorific value of the fuel. The net amount of energy available for heat is the total mass of the material in question, taking into account the amount of water it contains and the quantity of non-combustible material that will be left as ash after burning (the ash content). The energy available from biomass (which contains hydrogen) is expressed as either high or low energy content. The reason is that burning hydrogen produces superheated water vapor that escapes at the temperature of the stack gases. The low heating value (LHV) is the net heat liberated per kilogram of the fuel after the heat necessary to vaporize and superheat the steam formed from the hydrogen and from the fuel has been deducted. The high heating value (HHV) assumes that all the water vapor formed during combustion is condensed, whereas the LHV assumes that all products of combustion are in the gaseous state. Calorific values can easily be outweighed by variations in ash and moisture content. The heating value is a key factor when buying fuel because actual energy units are purchased.

2 Density. High-density woody species are preferred in terms of energy value, transport, handling and storage. The effectiveness of wood as a fuel depends on: (i) its density, which determines the rate of burning, combustibility and the heat produced per unit of wood volume; and (ii) its water content, which affects considerably how much heat is produced. Wood density also has a direct effect on initial green water content of the wood, the rate at which the wood dries out and the rate of burning at a given water content. The heavier the wood, when dry, the greater its calorific value. For example, some species, despite having a lower growth rate, have a much higher wood density and therefore are preferred for energy production to other species with a higher growth rate but lower density.

3 Energy balance. Energy forestry/crops must have a positive energy balance, or net energy ratio (NER). A major problem in calculating such balances is that each crop has different production and maintenance requirements, productivity (yields and total biomass), inputs, etc. It also depends on the end-use, including use of by-products. Thus the numerous values for the NER of a variety of energy forestry/crops that have been published reflect specific situations where they have been carried out. Woody biomass is usually considered to give the best energy and carbon benefits, but many herbaceous crops also have positive energy and carbon balances.

Estimates of energy balances by the Oak Ridge National Laboratory, USA, show a potential NER of various energy forestry/crops in the range of 10–15 times the energy inputs: e.g. 15.3 (in 1990) and 18.5 (in 2010) for hybrid poplar; and 10.3 (in 1990) and 11.7 (in 2010) for switch-grass. Results in the EU, after extensive research, have shown that the energy output ratio for the liquid transport fuel crops is in the range 1–6, and that of solid fuel crops is 14–30: e.g. 19–30 for SRWCs in Ireland and 15–20 for miscanthus in Greece [24,25]. These studies also show that there is a considerable potential for improving the NER and carbon balance through better production techniques (e.g. higher yields and more efficient harvesting, chipping, storing and drying methods).

4 Chemical characteristics of biomass. The chemical characteristics of biomass are very important and can vary considerably, given the many different types of species that can be used for either energy or chemicals. There is, generally, a lack of data describing the elemental characteristics of biomass fuels. Generally, biomass fuels have significantly different elemental characteristic compounds to those of fossil fuels, especially coal [26,27]. In trees, the composition of cellulose can vary from 40 to 50% of the dry weight, while the rest is composed of lignin and other cellulose-related compounds, such as hemicellulose. Many of the renewable carbon sources have a carbon content near to 45% (pure cellulose has a carbon content of 44.4%) once moisture and ash content have been taken into account. The basic elements of the cellulose fiber are carbon, oxygen, nitrogen and hydrogen, with a small content of ash. These percentages will vary slightly according to species.

The main characteristics of wood are expressed in terms of a Proximate Analysis (this indicates the exact chemical composition of a fuel without reference to its physical form) and the Ultimate Analysis (the determination of C and H in the material as found in the gaseous products of complete combustion). For the purpose of specifying and designing combustion equipment, the most common method used is the Ultimate Analysis.

5 Moisture and ash content. When biomass is burned, part of the energy released is used to turn the water that it contains into steam. It follows that the drier the biomass, the more energy there is available per unit weight or volume for heating. It is the moisture content that primarily determines the energy value. Thus, although wood has a higher energy value than the other forms of biomass at a given moisture content, it is possible for crop residues and dung, for example, to have a higher value than wood if they have lower moisture contents. Because the energy value of a unit weight of biomass is inversely proportional to the amount of water that it contains, it is necessary to calculate the moisture content. This can be measured in two ways: on a wet basis or a dry basis. The higher the ash content, the lower the energy value. For much woody biomass, the ash content is more or less constant at around 1% for all species, but for many herbaceous energy crops the ash content can be much higher, e.g. about 6% for *Cynara cardunculus*. It is therefore the moisture content rather than the species of wood that determines energy availability. Low ash content is important because this can affect the efficiency of the combustion equipment.

4 Chemicals from biomass

Almost every petroleum-derived chemical currently being produced could be produced from biomass. Biomass is already used as a source for various large-scale industrial chemicals, such as dimethyl sulfoxide, rayons, tall oil, tannins, specialty chemicals, etc. Biomass-derived chemicals can be divided into two main categories: those in which the plant has performed a major part in the synthesis, and those in which the chemical is synthesized from the more abundant biomass resources, such as wood and crop residue [28]. However, this area is discussed in some detail in other chapters and thus will only be mentioned briefly.

The biologically synthesized chemicals that are most easily used in the chemical industry are those that are either of similar characteristics to existing feedstock or are intermediate chemicals, and those that can readily be converted to industrial chemicals. The major difficulties with the biomass-based chemical industry are price and continuous feedstock supply for large-scale production of chemicals; for some petrochemicals, this is a capital-intensive industry that normally operates all year around. Biomass feedstocks pose additional difficulties with regard to supply, transport and storage, unless used on a small scale or for specialty chemicals. With the low cost of petrochemical feedstocks, the biomass-based chemical industry can compete only in specific circumstances and with specific chemical products.

Specifically, research has been done on natural rubber from guayule (*Parthenium argentatum*) and jojoba (*Simmondia chinensis*). Research has also been carried out involving the use of specific bacteria, molds or yeasts to synthesize the desired chemicals, e.g. the production of ammonia from algae. Research into the details of photosynthesis and molecular genetics could lead to the development of other plants or microorganisms that could synthesize specific predetermined chemicals.

Despite these difficulties, large amounts of chemicals are already obtained using lignocellulosic material to synthesize large volumes of chemical feedstock, which are converted in the chemical industry to a wide variety of more complex chemicals and materials. For example, about 95% of synthetic polymers can be derived from wood or other lignocellulosic materials.

Advances have been reported in the production of chemicals from various biomass sources: utilization of wheat straw to replace polypropylene and polyethylene; use of agricultural residues to substitute wood in pulping; use of MSW to produce lignocellulosic pulp; use of steam explosion of hardwood chips and agricultural residues to produce a fibrous mulch that can be used as a starting material for isolating polymers, etc.; and production of wood composite adhesives with fluidized bed pyrolysis oils to substitute 50–60 wt.% of the phenol and formaldehyde used in conventional phenolformaldehyde adhesive resins, etc.

5 Main sources of biomass

There are many biomass-based resources from which it is possible to obtain energy and chemicals. However, in reality, only a few could be considered as serious candidates, at least in the short to medium term. These resources can be grouped into herbaceous energy crops (HECs) and short-rotation woody crops (SRWCs).

5.1 *Herbaceous energy crops*

These are perennial crops, such as the various types of grasses (of different thicknesses and heights), that are capable of playing an important role in biomass energy supply, given their usually high productivity, acceptability by the farming community (especially considering that commercial technologies and machinery can be used), short growth cycles and diversity. The HECs comprise many plant varieties that can be harvested for their total above-ground cellulose material, although as yet only a handful can be regarded as serious contenders as biomass energy feedstocks. In America and Europe, two groups of perennial grasses and a few annual grasses are being investigated as potential energy crops. These include: (i) thin-stemmed perennials, e.g. switchgrass (*Panicum virgatum*) and reed canarygrass (*Phalaris arundinacea*); and (ii) thick-stemmed perennials, e.g. sugar-cane (*Saccharum officiale*) and miscanthus (*Miscanthus* sp.).

The productivity of herbaceous crops can vary significantly, but a major characteristic is their usually high yields: over 30 oven-dry tons (odt) ha^{-1} year^{-1} for sugar-cane; 13 odt ha^{-1} year^{-1} for switchgrass; and over 12 odt ha^{-1} year^{-1} for miscanthus. Experimental plot yields have recorded about 80 odt for energy cane, 23 odt for switchgrass and 28 odt for miscanthus. Such yields obviously depend on the agronomic and management inputs, but high yields can be expected over large areas based on the long experience of the sugar-cane industry in many parts of the world. Currently, the two main sources of fuel ethanol are sugar-cane in Brazil and corn in the USA, with an annual production of about 13 billion and 4.6 billion liters, respectively. Sugar-cane is one of the world's most important crops and the most promising source of energy for both liquid and electricity generation, at least in the short to medium term. Sugar-cane has a remarkable history of adaptability to many and differing ecosystems, resistance to adverse weather conditions and very high productivity.

Perennial HECs in particular have wide environmental advantages, including increasing soil organic matter, improving the soil structure and increasing the soil water- and nutrient-holding capacity, mainly due to their all-year-round ground cover and extensive rooting structure. These advantages are now being recognized by growers, environmentalists and energy users.

5.2 *Short-rotation woody crops*

Considerably less experience exists with SRWCs. In the USA and the EU, considerable research has been carried out to address some of the main issues; for example, since 1980 more than 100 woody species have been investigated in the USA to determine their suitability as energy carriers. So far, hybrid poplar and willow have proved to be the most promising. In the EU, results so far also show that willow and poplar have higher productivities; for example, in Sweden and Denmark they are being cropped and utilized commercially in rural heating systems and combined heat and power (CHP) plants. These facilities are also able to utilize energy grasses such as miscanthus and reed canarygrass, and in Finland, Sweden and Denmark large demonstration plots are being used for energy purposes [29,30].

The main challenge in the future is to develop efficient multi-fuel boilers/systems that allow the simultaneous utilization of various types of energy crops, including fossil fuels, and in particular co-firing with coal. In this type of system, one single crop does not have to be relied upon, because several different types of biomass and fossil fuels can be used together. This will represent a major step forward in the industrial use of biomass fuels.

6 Land use and availability

Many studies have been carried out on land availability and they give very wide-ranging results depending on the sources of data and assumptions used. Generally, it can be stated that there are large areas of degraded and abandoned lands worldwide, particularly in the tropics, which could benefit greatly from the establishment of environmentally sustainable biomass plantations. Bekkering [31] carried out an analysis of 117 tropical countries and found that 11 countries were suitable for expansion of the forest area up to 553 Mha. Trexler [32] found that, within 50 tropical countries, 67 Mha could be realistically converted to plantations over the next 60 years, more than 200 Mha could be regenerated and a further 63 Mha are available for agroforestry.

Hall [33] has estimated that, in Western Europe, if 10% of usable land (33 Mha) and 25% of recoverable residues were used, biomass could provide 9.0–13.5 EJ of energy, which represents about 17–30% of projected energy requirements in the year 2050. On a global scale it has been calculated that the land that *theoretically* could be needed to meet all the present energy consumption with biomass would be 950 Mha to replace all fossil fuels in the industrial countries and 305 Mha in developing countries, assuming a productivity of 12 odt ha^{-1} $year^{-1}$ and use of 25% potentially harvestable residues [34].

6.1 *Impact on food production*

'Food versus fuel' is an old, controversial and complex issue. As we have just shown, on a global scale there is land available. The doom and gloom about the declining food production is based on several misconceptions; for example, the decline in world cereal output has been largely unconnected to population growth but rather associated with the cut in grain production in the USA and the EU [35]. Other factors, such as food production and consumption, distribution patterns, hunger, lack of purchasing power, inequality, land and grains used for livestock, under-utilization of agricultural land, lack of appropriate investments, export of crops, land tenure, wars and political interference, should also be taken into consideration. Generally, the world's food production has been keeping ahead of population growth and diets are now more diverse. The most populous countries, such as India, provide a good example of changing trends in food production (which has kept ahead of population growth), energy, population and environment; for example, in 1995 India had over 30 Mt of surplus grain while at the same time halting deforestation, increasing tree cover and maintaining the same cultivated area [36].

It should be noted that both food and fuel are important requirements that need not compete, particularly when planning ensures ecological conservation and the sustainability of production methods. Forestry policies and programs such as agroforestry and integrated farming systems can in fact improve food security by providing food (from the tree directly and from animals in the habitat provided), fodder, energy and income for food purchase.

7 Costs of the biomass option

There have been important developments in recent years that have significantly reduced the cost of producing bioenergy. However, many biomass energy technologies have not yet reached a stage where market forces alone can make the adoption of these technologies possible. One of the principal barriers to the commercialization of all renewable energy technologies is that current energy markets mostly ignore the social and environmental costs and risks associated with conventional fuel use [37].

Existing infrastructure, tax regimes, subsidies, finance for R&D and the power of political interest groups all tend to work in favor of fossil fuels and nuclear power, at the expense of renewables such as biomass. Thus at present it is difficult for biomass to be competitive with such fuels, except in certain niche markets or where sufficient tax incentives exist. Some initial support, in terms of subsidies, taxes, fiscal incentives, etc., is needed to allow renewables to play a full part in the market place. An example is the UK's Non-Fossil Fuel Obligation (NFFO). The Electricity Act of 1989 enables the Secretary of State to require the electricity utilities to secure the availability of a certain amount of electricity from non-fossil fuels. The purpose is to secure additional generation capacity, e.g. 10% by 2010, from renewables in order to help them enter the commercial generating market. This policy has been quite successful in promoting and reducing the costs of renewable-based electricity, e.g. from a weighted average price of 4.3 p kWh^{-1} in 1995 to 3.46 p kWh^{-1} in 1997 ($1.62 = £1.00) [38].

To make biomass more competitive, attention needs to be focused on two key areas: reducing the cost of producing biomass fuels/feedstocks and reducing capital investment costs for converting biomass to useful energy carriers. High yields per hectare of land are essential for low biomass production costs. Agricultural crops for biomass energy, such as maize, sugar-cane and rapeseed, already achieve high yields as a result of long-standing R&D efforts. Trees and HECs, on the other hand, are still undergoing research as energy feedstocks, but are starting to be the focus for research into high-yielding species for energy production. At experimental stations, hybrid poplar cultivars in northwestern USA can produce up to 29.6 odt ha^{-1} $year^{-1}$ and switchgrass in southeastern USA up to 30.4 odt ha^{-1} $year^{-1}$ [39]. The US Department of Energy aims to produce ethanol from wood at a cost of $0.79 per gallon ($0.20 per liter) and $0.56 per gallon ($0.14 per liter) in the years 2005 and 2030, respectively [29].

Capital costs for biomass energy facilities tend to be high because of their novelty (except for ethanol) and relatively small scale. These factors raise prices compared to fossil-fueled plants, where economies of scale and experience are exploited. This underlines the need for demonstration plants for newer and more promising technologies, and their replication and progressive improvement. It has been predicted [30] that specific investment costs for biomass-integrated gasifier–gas turbine (BIG–GT) electricity generation plant would fall from 3000 $ kW^{-1} to 1300 $ kW^{-1} over the course of 10 replications.

Some authors [40] compared prices for plantation wood in Sweden ($2.39–3.39 GJ^{-1}) and Finland ($3.1 GJ^{-1}) with plantations in developing countries such as Brazil ($1.41–1.50 GJ^{-1}), India ($1.41–1.91 GJ^{-1}), Thailand ($1.69–1.91 GJ^{-1}) and the Philippines ($1 GJ^{-1}). Despite these low plantation costs in developing countries, wood can often be obtained even more cheaply by felling native forests, e.g. charcoal from plantation wood costs $3.03–3.15 GJ^{-1} to produce, compared to $1.43 GJ^{-1} from unauthorized deforestation and $2.05 GJ^{-1} from authorized deforestation.

Sale of heat in CHP facilities could generate revenue that would reduce the net cost of electricity production from biomass, and this could be an important measure to improve the competitiveness of biomass, given the economies of scale enjoyed by large fossil-fueled plants. BIG–GT has lower anticipated net electricity generation costs than the smaller coal integrated gasification combined cycle (IGCC) plant. In a UK study [41], a lower selling price for heat of 1.56 ¢ kWh^{-1} was assumed, while predicting that the net cost of electricity from wood residues would be

7.6 ¢ kWh^{-1} and from SRWC coppicing plantations would be 14.4 ¢ kWh^{-1}. Other authors [42] calculate that biomass feedstock prices would have to fall to \$1.5 GJ^{-1} for it to compete with coal.

7.1 *External costs*

Environmental and social costs are hard to estimate because human life and environmental amenity are difficult to assign a fixed value. All economic activity, and particularly energy production, also leads to external costs, which accrue to third parties other than the buyer and seller. These can be environmental (e.g. pollution damage to crops) or non-environmental (e.g. direct subsidies, national security of energy supply, R&D costs, goods and services publicly supplied). Renewable energy sources, which produce few or no external costs and have several positive external effects, are systematically put at a disadvantage. Internalizing external costs and benefits and reallocating subsidies in a more equitable manner must become a priority for all renewables to be in a better position to compete with fossil fuels. Some governments are trying to develop programs to account for external costs such as taxes on emissions and incentives for cleaner fuels, but few schemes have yet been put into practice and most have met with strong opposition.

Hohmeyer [43] has estimated external environmental effects of electricity production in Germany to be 0.011–0.061 DM (1982) kWh^{-1} for fossil fuels and 0.012–0.120 DM (1982) kWh^{-1} for nuclear power. When Hohmeyer includes other external costs and support, the total cost comes to 0.039–0.088 DM (1982) kWh^{-1} for fossil fuels and 0.097–0.208 DM (1982) kWh^{-1} for nuclear power (exchange rate \$1 = 2.43 DM at 1982 prices).

Ozdemiroglu [44] estimated the external costs of renewables compared to coal for electricity generation in Scotland, UK. Wind and hydro have lower external costs than biomass, but landfill gas and MSW combustion have higher external costs. Coal is estimated to have an external cost of 3.55–5.4 ¢ kWh^{-1} compared to energy crops at only 0.44–0.59 ¢ kWh^{-1}. Consideration of these external costs in planning investment in new generation capacity would thus reduce the cost of biomass relative to coal by 3.1–4.8 ¢ kWh^{-1}. Thus biomass for electricity at 7–10 ¢ kWh^{-1} would be competitive with the present electricity price in the range 4–5 ¢ kWh^{-1}. The energy security externality (ESE) associated with electricity generation in the UK is 0.005–1.65 p kWh^{-1} for oil, around 0.15 p kWh^{-1} for gas and about 0.31 p kWh^{-1} for coal [45].

In the USA, a total of 29 states take externalities into consideration in resource planning and/or acquisition, but only 22 do so solely in a qualitative manner and 5 states attempt to monetize the external costs, e.g. California, Massachusetts, Nevada, New York and Wisconsin. Focus has been almost exclusively on environmental externalities (not economic or social), particularly air emissions. There is a wide variation in the valuation of externalities between states, because the values adopted for carbon dioxide emissions range from \$1.21 t^{-1} in New York to \$25.24 t^{-1} in Massachusetts [46].

7.2 *Job creation*

Employment opportunities have been heralded as a major advantage of biomass because of the many multiplying effects that help to generate more economic activity and help strengthen the local economy, particularly in rural areas. However, estimation of the employment impact of biomass energy, or any change in investments in the economy, is complex and uncertain. The greatest uncertainty concerns the economic cost of investment in expanded energy supply from biomass

Table 9.4. Direct employment required in fossil, biomass and conservation systems for fuel production (person-years per million tons oil equivalent, t.o.e.) and for fuel production and power generation (person-years per TWh).

Sector	Jobs per million t.o.e.	Jobs per TWh
Natural gas	428	250
Petroleum	396	260
Offshore oil	450	265
Coal	925	370
Nuclear	100	75
Energy saving	2000	–
Bioenergy (solid fuels)	2500	1145
Wood energy	4500	1000
Bioenergy (net of displaced jobs)		28–406

From [47,48].

and the impact that this could have on displacing employment through, for example, reduced spending elsewhere in the economy. None the less, the evidence suggests that biomass for energy is a labor-intensive sector that is particularly favorable for rural regions.

In the EU, various attempts have been made to estimate potential employment from renewable energy resources. For example, the TERES II study using the SAFIRE market penetration model predicted a net employment of 0.5 million jobs created by the renewable industry by the year 2010 if the projections of the EU White Paper [15] are to be accomplished. It is interesting to note that sectoral studies of the wind, photovoltaic, biomass and solar industries give estimates of between 1.54 to over 1.67 million jobs.

Table 9.4 shows estimated job creation for power generation. It appears that the labor required to produce biomass fuels is approximately 4–10 times greater than that for fossil fuels, and total direct employment (including power generation) is 3–4 times greater than for fossil fuel systems. Compared with nuclear energy, biomass for electricity requires approximately 15 times as much labor [47,48].

However, two important points should be considered. Firstly, these figures only consider direct employment, while many indirect jobs would be created or destroyed by a change in the energy mix. The labor intensity in the sectors to which the energy system is linked is important. Indirect job creation or loss can be estimated using national input–output tables. Secondly, biomass energy is usually more expensive than fossil energy and some other renewables, such as hydro and wind. Therefore under present conditions a switch to biomass energy will increase the cost of energy to consumers, and thereby reduce household expenditure.

This effect of job displacement due to the higher cost of renewables, biomass in particular, could be significant. In a recent report for the UK Department of Trade and Industry's Energy Technology Support Unit (ETSU), it was estimated that in the year 2005 biomass for electricity would generate 2220 jobs TWh^{-1} (direct plus indirect). Of these jobs, 620 would be in the agricultural sector. However, the effect of the subsidy required via the NFFO would eliminate almost as many jobs [48,49].

In the USA, a study conducted by the Coalition of Northeast Governors Policy Research Center of residential and industrial wood energy use in 11 states in 1985 found that 13 000 jobs were supported directly in the fuel supply operations and a further 30 000 jobs were supported in related industries (including power generation). It was estimated that 35 000 jobs were displaced in other industries, giving a net employment benefit of 8000 jobs [14].

8 Management of biomass energy resources

One of the problems is that management practices have been determined until very recently by the need to produce food and animal feed in the case of crops, and by non-energy purposes, e.g. pulp and paper production, in the case of forest plantations. Thus comparatively little is still known about large-scale energy forest plantations. The most experience comes from Sweden where there is a planted area of about 16 000 ha of willows, from Brazil where about 3 million tons of charcoal are produced from eucalyptus plantations and limited experience from the USA and the EU where there are large energy crops/forestry R&D programs.

There is no question that we have the knowledge and the capacity to influence and to modify crops and trees to meet specific characteristics needed for energy purposes, e.g. fast growth, high density and low ash content. For example, genetic engineering techniques have been applied to alter wood chemistry to facilitate pulp production. There is no reason why such techniques cannot be applied to obtain more desirable energy species. Potential modifications of forest management systems for energy production may include matching of clones and species to sites, intermediate cuttings, precommercial thinnings, harvest cuttings, post-harvest treatment, stand rehabilitation, etc. Table 9.5 summarizes preliminary estimates of tropical plantation woody biomass productivity under different management conditions [17].

8.1 *General management practices*

The aims of any modern biomass energy system must be: (i) to maximize yields on a sustainable basis with minimum inputs; (ii) to optimize economic and social benefits to the local and wider communities; (iii) utilization and selection of appropriate plant material and processes;

Table 9.5. Preliminary estimates of tropical plantation woody biomass productivity potential under different management situations (oven-dry tons ha^{-1} $year^{-1}$).

	Semi-arid		Sub-humid	
Situation	Low	High	Low	High
No genetic improvement and no fertilizer or water added	2	5	5	10
Genetic improvement but no fertilizer or water added	4	10	10	22
Genetic improvement, fertilizer added but no water added	6	12	12	30
Genetic improvement, water added but no fertilizer added	8	18	11	25
No genetic improvement and no water added, but fertilizer added	3	7	8	15
Genetic improvement and fertilizer and water added	20	30	20	35

From [17]. Estimates refer to short-rotation plantation wood yields in semi-arid and sub-humid conditions, under various management options. The land type assumed is degraded forest land or non-arable agricultural land. Genetic improvement refers to a program of clonal selection and matching species and clones to sites, although biotechnology may have a role in future.

(iv) optimum use of land, water and fertilizer; and (v) to create an adequate infrastructure and strong R&D base.

Biomass production depends on the quality of the growing site, the planting density and the rotation age, e.g. wider spacing produces larger trees and planting more trees per hectare will not increase production once there is canopy closure. However, biomass production per unit area is more or less constant for trees of the same age on similar sites for a fairly wide range of planting densities. The availability of residue depends on many factors, e.g. whether there is whole-tree harvesting, thinning, pruning, etc. Clearly, with good management practices many more benefits can be obtained while reducing significantly any possible negative effects of energy forestry/ crop plantations, although much still needs to be learned. Some of these possibilities are discussed further.

Biological fertilization in energy forestry/crops offers many possibilities. It can take many forms, including: direct replacement of chemical nitrogen fertilizers; long-term amelioration of soil fertility; protection of soils from erosion and leaching losses; plant growth stimulation for optimal mobilization; and the use and recirculation of naturally available plant nutrient resources. In its widest sense these measures include: (i) direct cultivation of nitrogen-fixing woody species and interplanting with nitrogen-fixing trees or ley crops; (ii) soil amendments with various forms of organic matter (e.g. sewage sludge, wastewater, contaminated ground water, farmyard manure and green manure); (iii) stimulation or introduction of other rhizosphere microorganisms that interact positively with the energy forest/crop with respect to plant nutrient uptake; and (iv) biological fallow.

Few field data are available on erosion for comparison with arable crops. However, some authors [50] have reported average soil losses for US crops of 18.1 t ha^{-1} $year^{-1}$, and 2.4 t ha^{-1} $year^{-1}$ for conventional forests, compared to 2 t ha^{-1} $year^{-1}$ for SRWCs on a 5% slope, mainly during the first 2-year establishment period. Top-soil losses of Brazilian crops have been reported to be 12.4 t ha^{-1} $year^{-1}$ for sugar-cane, 38.1 t ha^{-1} $year^{-1}$ for beans, 26.7 t ha^{-1} $year^{-1}$ for peanuts and 33.9 t ha^{-1} $year^{-1}$ for cassava [13].

It is clear that whole-system, lifecycle approaches are required to estimate accurately the contribution of intensive forest management and bioenergy systems to local and global carbon balances. Results from various studies show that soil physical properties are influenced positively in energy plantations due to the less frequent use of heavy machinery and social disturbance compared to arable agriculture. Soil solution nitrate can also be reduced significantly in soils planted with fast-growing trees, as long as nitrogen fertilizers are applied in accordance with the nutrient demands of the trees [51].

Water availabilities are the most determinant factors in plantation productivity. There are complex interactions between soil water and root dynamics, and management options involved in fertilization, weed control, site preparation, etc., and thus much more work is needed on SRWCs and HECs.

A recent EU study has described in detail the achievements on energy crop production, processing and utilization (willow, poplar, eucalyptus, miscanthus, reed canarygrass and Cynara, for CHP; oilseed crops such as rapeseed and sunflowers for the production of biodiesel; and sugar-containing crops such as sugar beet and sweet sorghum for the production of bioethanol) [24]. A large number of crops were investigated for their potential use as energy crops in Europe, but only a few have gone beyond the R&D stage and have reached commercialized status mainly as a result of political and financial support given by some countries. The study demonstrated that both liquid and solid biofuels can be made economically feasible only with financial incentives, such as tax exemptions, for example in France and Austria for liquid biofuels, heavier taxes on fossil fuels in Denmark and Sweden of the order of 5–6 ECU GJ^{-1} for non-industrial users, NFFO contracts in the UK, grants for the farmers who cultivate energy crops grown on set-aside land and fiscal measures [24].

Despite large reductions in cost (e.g. in Sweden willow plantation costs were reduced from 1200 ECU to 600 ECU in only 5 years), there is still significant room for improving productivity and reducing costs. A number of major bottlenecks and gaps in knowledge were identified that require further R&D on energy crop production, including: breeding to develop better adapted plant material with low production inputs and a maximized fuel quality; low-cost establishment of perennial crops; efficient and environmentally sound weed treatment; energy crops, water needs and the implication for productivity and to ground water recharge; possibilities for utilization of wastewater, sludge and/or ashes in energy crops; influence of genotype, harvest time, fertilization, climate, etc. to maximize the crop quality for energy purposes; more effective dissemination of knowledge on new crops to farmers in order to speed up the successful introduction of new crops; more R&D on the relation between fuel requirements and selection of the most suitable species for energy; and greater efforts to stimulate the development of effective and low-cost machinery for harvesting and processing energy crops [24,25].

It is evident that there is a need to integrate further the agricultural practices and the energy sector. Also, energy crops for power and heat purposes require fuel specifications that are not yet fully met, particularly the need to obtain low Cl, K, Ca, N, low moisture content, etc. for energy forestry/crops.

9 Environmental aspects

A large number of studies, e.g. [14], have shown that biomass energy can have many environmental benefits if produced and used sustainably, and thus possible environmental impacts will be considered only briefly. The environmental concerns regarding biomass energy can be divided into two broad categories: traditional fuels and dedicated energy forestry/crops.

It is well known that traditional biomass energy is usually used very inefficiently and thus can result in serious negative environmental impacts. However, woodfuel in many parts of the world is obtained mainly from non-forest land and thus it has not been responsible for environmental degradation as was first thought in the 1970s. The main problem is its inefficient use, which largely relates to social, cultural and economic factors rather than the nature of the fuel itself. The reduction of potential negative impacts will depend on a combination of factors, including sociocultural changes (e.g. cooking practices), economic betterment and technological improvement (e.g. improved cooking stoves).

Major concerns that are often raised are the potential problems associated with large-scale use of energy/forest plantations. A large number of potential environmental and ecological impacts remain poorly understood owing to the lack of data on the long-term implications. Experience from Brazil with sugar-cane, and USA with corn, is of little direct relevance because this represents a prolongation of traditional agricultural and forestry practices. An exception may be Brazil's large eucalyptus plantations for charcoal production, although much of this experience relates to traditional forestry practices [52]. A particular focus of research has been the possible effects on crop land. Table 9.6 [53] summarizes the most important quantifiable land use impacts associated with the production of energy forestry/crops. Compared to agricultural crop production, energy forestry and crops have lower levels of erosion and fertilizer and pesticide application. These

Table 9.6. Comparison of agricultural intensity indices for energy crops and conventional crops.

	Reduction relative to corn–wheat–soybean average	
Index	Woody short-rotation	Herbaceous perennial
Erosion	12.5-fold	125-fold
Fertilizer	2.1-fold	1.1-fold
Herbicide	4.4-fold	6.8-fold
Insecticide	19-fold	9.4-fold
Fungicide	39-fold	3.9-fold

From [52].

trends are also consistent with other studies that show positive water quality benefits, better soil organic carbon levels when energy crops are grown on cropland (e.g. due to slower leaching of nutrients), more efficient use of nutrients and reduced NO_x.

Thus, although much needs to be learned of the possible impacts from energy plantations, experience thus far shows that these impacts can be significantly lower than for traditional agriculture. Much will depend on the management practices used. Many of the environmental and ecological impacts can be alleviated by using compensatory measures, e.g. soil acidification from biomass harvesting and leaching due to acid decomposition can be countered by liming and/or recycling of wood ash. Potential risks associated with wood ash application, such as pH or salt shocks for the soil microorganisms, can be overcome by using a granulated product, which acts somewhat like a slow-release fertilizer. In Sweden, Lunkqvist [54] considered bioenergy systems that included site-specific compensatory measures and concluded that these systems are likely to be sustainable, and Ericson [55] concludes that there is no reason to believe that energy forest plantations will have any significant environmentally and ecologically negative impacts from a nutritional point of view when proper management practices are applied.

Biomass energy, on the other hand, provides a number of strategies that can be used to tackle greenhouse gas emissions: (i) sustainable production and use of energy resources, which results in neutral CO_2 production; (ii) sequestration of CO_2, which creates carbon sinks; and (iii) direct substitution of fossil fuels, with the corresponding environmental and ecological benefits.

10 Future prospects

Although it is expected that market forces will be determining factors in the future development of bioenergy, past experience indicates that initial political and fiscal support is necessary for bioenergy to succeed, given the low price of conventional fuels, the hidden subsidies and the institutional barriers. For example, Austria provided political encouragement through favorable legislation, capital grants, cheaper finance and education. In Denmark, there has been political support at the highest level for green energy and sustainable development, and Finland allocated substantial funds to R&D. In Sweden, a catalyst to bioenergy was the decision to phase out nuclear energy and later to support CO_2 neutral renewable energies. In the UK, the main instrument has been the NFFO through a competitive bidding process. The USA has introduced a number of legislative, e.g. PURPA, and economic measures aimed at facilitating the introduction of alternative energies [29,56].

In developing countries the situation is more difficult. Conventional energy prices, e.g. for electricity, liquefied petroleum gas (LPG), kerosene and diesel, are often kept artificially low through subsidies to facilitate industrialization and for other social reasons, so little money is allocated to support bioenergy. Large countries such as Brazil, China and India have alternative energy programs of some scale. For example, Brazil has been subsidizing ethanol production, while in India artificially low priced energy supplies are hampering the development of bioenergy in most cases.

None the less things are changing rapidly, with the trend toward globalization and privatization of the energy sectors in many countries. Many state energy monopolies are gradually being exposed to greater competition and many are being privatized. This will have a major impact in the energy sector in general and the renewable energy sources in particular. It is important that there is a level playing field in energy production. Although subsidies may be politically acceptable for renewable energies until they reach some kind of maturity, ultimately it will be the market forces that play the main role.

It has been estimated [57] that a 0.2% levy on fossil fuel consumption in the Organization for Economic Cooperation and Development (OECD) countries would raise some $833 million annually. If half of the levy was spent on buying down the cost of the most promising electricity generation technologies (wind, solar thermal, photovoltaic and biomass energy), they could obtain full commercialization within 10 years or so. An example is UK's NFFO program, which has stimulated a decrease in the bid price for wind power from 14.4–17.6 ¢ kWh^{-1} in 1990 to 6.4 ¢ kWh^{-1} in 1994 and 3.53 ¢ kWh^{-1} in 1997, through a process of learning by doing [38]. The total subsidy in 1994 (NFFO — three projects) was about $27–35 million, or 48–64 cents per person per year over the 20 year contract period, or less than 0.2% of the UK consumers' electricity bill.

10.1 *Future energy trends*

All indications point to a more complex and varied future energy supply matrix in which renewable sources of energy will have a major and increasing share of the market. A more decentralized and diversified energy supply system would allow greater control at national, regional and local levels. Energy efficiency will increase, thereby allowing continued economic growth without necessarily increasing energy consumption per capita in industrialized countries. In developing countries, energy demand will continue to grow due to population growth and better living standards, but at a lower pace due to technological improvements. Fossil fuels will continue to be used on a large scale well into the next century, but with a diminished role on a percentage basis, while renewable energy sources will see their market share increase quite considerably.

10.2 *Bioenergy*

It is becoming clear that bioenergy could be a major source of energy in the next century. Modern biomass energy use will increase quite considerably, while traditional uses will experience a decline in relative terms. There is no question that in the next century biomass energy will be more efficiently used than is the case today due to technological developments. Also, there will be a wider combined and complementary use of traditional and modern biomass energy carriers to meet growing needs.

The use of bioenergy on a large industrial scale will require important advances in agrotechnologies to increase productivity and reduce costs, as well as in transformation technologies, e.g.

gasification. The most promising bioenergy market is for CHP and electricity, which can use already established technologies. In the medium term, emerging technologies such as BIG and BIG–GTs can open up new economic opportunities. A major gap in biomass energy data is that, although some research has been performed, it has usually been aimed at obtaining supply and consumption data along with conversion processes, with insufficient attention and resources allocated to more basic research, such as production and development of forestry/crops more in accordance with energy requirements, harvesting and integrated conversion processes. Additionally, little research has been done to study market flows, economics and the role of entrepreneurs in making bioenergies available to the end-user and whether these services are provided in a sustainable manner.

11 Conclusions

It is clear that there exists an enormous biomass energy potential. The main issue is, however, how to bring this resource to the consumer in a convenient, efficient manner and at a competitive cost. The most economically attractive forms of biomass energy are presently based on organic residues from agriculture, forestry or industry. Where these residues are locally available co-firing, CHP and district heating are already attractive. As plantation wood costs are reduced and capital costs fall, biomass to electricity will become increasingly competitive. Advanced gasification technologies for biomass to electricity have certain cost advantages over fossil-fueled plants, making them particularly promising. Liquid and gaseous fuels are generally less competitive at present.

Land availability and bioenergy production are intrinsically intertwined. Various studies have been carried out to determine how much land is available globally for non-agricultural purposes and all have shown that there are considerable amounts of land (ranging from 150 to 1200 Mha) that, if required, can be used for energy purposes without affecting food production. With proper support (R&D, infrastructure, financial, etc.) farmers have demonstrated repeatedly that they can produce far more food, and if more food is to be made available to those who presently have inadequate nutrition we need to make the necessary changes to the present food production and distribution system.

A range of technological advances is opening up new opportunities for bioenergy that were considered only a few years ago as long-term prospects. None the less, many biomass energy technologies have not yet reached a commercial stage and still require support. Exceptions are the use of residues from agriculture and forestry when they are readily available for generating heat, electricity and bio-gas, which can be sold at competitive prices. One of the principal barriers to the commercialization of all renewable energy technologies is that current energy markets mostly ignore the social and environmental costs and risks associated with conventional fuel use, the hidden subsidies, the long-term costs of depletion of finite resources and the costs associated with securing reliable supplies from foreign sources.

For bioenergy to succeed, particularly in its modern forms, some initial incentives (be they in the form of subsidies, financial incentives, carbon taxes, etc.) would be necessary to put bioenergy on more equal terms with the long-established fossil fuels. The experience from countries that have a significant modern bioenergy contribution clearly indicates that this is a necessary condition for competitiveness. In the longer term, market forces must be allowed to play their role where there is a 'level playing field'.

Future energy supplies could be more decentralized, with renewable energies playing an increasing role. Fossil fuels will continue to dominate well into the next century, but bioenergy in its various forms will also increase its market share, particularly with CHP generation and co-firing. Oil may still dominate the transportation system but biomass-based liquid fuels, with their environmental advantages, will increase their share.

12 References

1 Hall DO, Rao KK. *Photosynthesis. Studies in Biology*. Cambridge: Cambridge University Press, 1994.
2 Johansson TB, Kelly H, Reddy AKN, Williams RH. In Johansson TB, Kelly H, Reddy AKN, Williams RH (eds) *Renewables for Fuels and Electricity*. Washington, DC: Island Press, 1993.
3 WEC (World Energy Council). *Energy for Tomorrow's World*. New York: St. Martin's Press, 1994.
4 IPCC. *Climate Change 1995: Impacts, Adaptations and Mitigation of Climate Change: Scientific-Technical Analysis*, Intergovernmental Panel on Climate Change Working Group II Report. Cambridge: Cambridge University Press, 1996.
5 IEA *World Energy Outlook*. Intl. Energy Agency, Paris, 1998.
6 IIASA/WEC *Global Energy Perspectives*. In Nakicenovic N. *et al.* (eds). Cambridge: Cambridge University Press.
7 Shell International Petroleum. *The Evolution of the World's Energy System 1860–2060*. London: Shell International Petroleum Company, 1996.
8 Greenpeace. *Towards a Fossil Free Energy Future: the Next Energy Transition*. Boston Centre, Boston: Stockholm Environment Institute, 1993.
9 Dessus B, Devin B, Pharabod F. *La Hoille Blance* 1992; **1**: 1.
10 Lashof DA, Tirpak DA (eds). *Policy Options for Stabilizing Global Climate*, Report to Congress, Technical Appendices. Washington, DC: US Environmental Protection Agency, 1991.
11 Hall DO, House JI, Scrase JI. In Rothman H, Rosillo-Calle H, Bajay S (eds). *Industrial Uses of Biomass Energy*. London: Taylor and Francis, 1999.
12 Hall DO. *Energy Pol.* 1991; **19**: 711.
13 Moreira JR, Goldemberg J. *The Alcohol Program*, Report for the Ministry of Science and Technology (MCT). Brasilia: MCT, 1997.
14 Hall DO, Rosillo-Calle F, Scrase JI. *Biomass: an Environmentally Acceptable and Sustainable Energy Source for the Future*, Report prepared for the Division of Sustainable Development, DPCSD. New York: UN, 1996 (unpublished).
15 EU White Paper. *Energy for the Future: Renewable Sources of Energy* (White Paper for a Community Strategy and Action Plan). Brussels: European Commission, 1997.
16 Woods J, Hall DO. *Bioenergy for Development: Technical and Environmental Dimensions*, FAO Environmental and Energy Paper No. 13. Rome: FAO, 1994.
17 Ravindranath NH, Hall DO. *Energy Sustain. Dev.* 1996; **2**: 14.
18 FAO. *Regional Study on Wood Energy Today and Tomorrow in Asia*, FAO Field Document No. 50. Bangkok: FAO, 1997.
19 Hall DO, Scrase JI. *Biomass Bioenergy* 1998; **15**: 357.
20 Lazarus M. *Towards a Fossil Fuel Free Energy Future. The Next Energy Transition*. Boston, USA: Stockholm Environment Institute, 1993.
21 World Energy Council. *New Renewable Energy Resources — A Guide to the Future*. London: Kogan Page, 1995.
22 Raskin P, Margolis R. *Global Energy in the 21st Century: Patterns, Projections and Problems*, POLESTAR Series Report No. 3. Stockholm, Sweden: Stockholm Environment Institute, 1995.
23 Read P. *Responding to Global Warming*. London: Zed Books, 1994.
24 BTG. *EECO (European Energy Crops Overview Project)*, An EC-FAIR Concentrated Action (FIAR1-CT95-0512). The Netherlands: Biomass Technology Group BV, 1996.
25 Venedaal R, Jorgensen U, Foster CA. *Biomass Bioenergy* 1997; **13**: 147.
26 Nordin A. *Biomass Bioenergy* 1994; **6**: 339.
27 Kitani O, Hall CW (eds). *Biomass Handbook*. London: Gordon and Breach, 1989.

28 Office of Technology Assessment. *Energy from Biological Process, Vol. 2 — Technical and Environmental Analysis*. Washington, DC: Office of Technology Assessment, 1991; 226.

29 The President's Committee of Advisors on Science and Technology. *Federal Energy Research and Development for the Challenges of the 21st Century*, Report of the Energy Research and Development Panel. PCAST, 1997.

30 Elliott P. In Chartier P *et al.* (eds). *Biomass for Energy, Environment, Agriculture and Industry, Proceedings of the 8th EC Conference*, Vol. 1. Oxford: Pergamon Press, 1995; 129.

31 Bekkering TD. *Ambio* 1992; **21**: 414.

32 Trexler MC. *Mitigating Global Warming Through Forestry: a Partial Literature Review*, Report to GTZ. Bundestag, Bonn: European Commission, 1993.

33 Hall DO. *For. Ecol. Manage*. 1997; **91**: 17.

34 Hall DO, Rosillo-Calle F, Williams RH, Woods J. In Johansson TB *et al.* (eds). *Renewables for Fuels and Electricity*. Washington, DC: Island Press, 1993.

35 Dyson T. *People Planet* 1995; **4**(4): 13.

36 Hall DO. *Biomass Energy Options in Western Europe (OECD) to 2050*. Paris: IEA, 1995.

37 Hohmeyer O, Ottinger RL (eds). *Social Costs of Energy: Present Status and Future Trends*. Berlin: Springer-Verlag, 1994.

38 DTI. *Non-fossil Fuel Obligation for Generators of Electricity from Renewable Sources*, Fifth Round of Bidding-NFFO-5, Renewable Energy Bulletin No. 7. London: Department of Trade and Industry, 1997.

39 Ferrel JE. *Biologue* 1995; **13**(2): 33.

40 La Rovere EL, Nogueira AH. In Chartier P *et al.* (eds). *Biomass for Energy, Environment, Agriculture and Industry, Proceedings of the 8th EC Conference*, Vol. 3. Oxford: Pergamon Press, 1995; 2104.

41 Dagnall S, Dumbleton F. *A Study of Heat Markets and Incentives — Progress Report on Biomass*, ETSU BAC(95)PO4. London: Department of Trade and Industry, 1995.

42 Larson ED, Williams RH. In Goldemberg J, Johansson TB (eds). *Energy as an Instrument for Socio-economic Development*, UNDP/SEED/BPPS. New York: UNDP, 1995; 91.

43 Hohmeyer O. *Social Costs of Energy Production*. Berlin: Springer-Verlag, 1988.

44 Ozdemiroglu E. In *Symposium on Development and Utilization of Biomass Energy Resources in Developing Countries, UNIDO, Vienna, Austria, December 11–14, 1995*. London, UK: Economics for The Environment Consultancy (EFTEC), 1995.

45 Pearce DW, Lockwood B, Ozdemiroglu E, Steele P. *Non-Environmental Externalities of Electricity Fuel Cycles in the United Kingdom*, CSERGE. London: University College, 1995.

46 Sweezey BG, Porter KL, Feher JS. *Biomass Bioenergy* 1995; **8**: 207.

47 Grassi G. In Chartier P *et al.* (eds). *Proceedings of the 9th European Bioenergy Conference*, Vol. 1. Oxford: Pergamon Press, 1996; 419.

48 Scrase JI. *Biomass Energy and Employment in the European Union*, Biomass Users Network. London: Kings College London, 1997 (unpublished).

49 Weinrech VS. In *ETSU/Altener Conference on the Social and Economic Impact of Renewable Energy*. Harwell, UK: ETSU, 1996.

50 Pimentel D, Krummel J. *Biomass* 1987; **14**: 15.

51 Hansen EA. *Biomass Bioenergy* 1994; **5**: 431.

52 Rosillo-Calle F, Rezende MAA, Furtado P, Hall DO. *The Charcoal Dilemma: Finding a Sustainable Solution for Brazilian Industry*. London: Intermediate Technology Publications, 1996.

53 Ranney JW, Mann LK. *Biomass Bioenergy* 1994; **6**: 211.

54 Lunkquvist H. *Whole Tree Harvesting: Ecological Consequences*, Maritimes Region Information Report M-X-191. Canada: Canadian Forest Service, 1994.

55 Ericson T. *Biomass Bioenergy* 1994; **6**: 115.

56 Klass DL. *Energy Pol.* 1995; **12**: 1035.

57 Leach G. *SEI (Stockholm Environment Institute) Newsl.* 1995; **8**: 1, 6.

10 Biological Conversion of Biomass to High-quality Chemical Carriers

D.O. HALL and F. ROSILLO-CALLE
Life Sciences Division, King's College London, Campden Hill Road, London W8 7AH, UK

1 Overview

1.1 *Trends in research on high-quality biofuels*

The pace of technological advances is opening up many new opportunities for biomass-based, high-quality fuels, which were regarded only a few years ago as long-term prospects. The main advances include: continuous fermentation, e.g. the Melle–Boinot and Biostil processes; simultaneous saccharification and fermentation; anaerobic fermentation; use of bacteria; heat recovery in the distillation process; treatment of cellulosic materials by steam explosion; improved processes for obtaining ethanol from cellulosic material; better use of by-products, e.g. bagasse, which is used for efficient electricity co-generation; production of methanol and hydrogen from biomass; fuel cell vehicle technology; etc.

Other relevant advances in biofuel production and use include improved biomass technologies with immediate potential applications, such as: biomass-integrated gasifier–gas turbine systems for power generation; improved techniques for biomass harvesting, transportation and storage; gasification of crop residues; biocrude technology; and co-firing technology. Policy developments include: greater recognition of the potential of biofuels for reducing polluting emissions; inclusion in most energy scenarios of the potential contribution of biofuels to the transport sector; greater commercial and financial experience in dealing with biofuels; and greater research and development (R&D) effort aiming to find cheaper feedstocks.

1.2 *Bioconversion processes*

There have been significant economic and technical improvements in recent years in bioconversion, although currently the only large-scale commercial processes still continue to produce ethanol fuel via fermentation either from sugar-cane or maize. Ethanol is also produced from grain, citrus waste, etc. but on a smaller scale. An area of considerable interest is the hydrolysis of cellulose, where important advances have been reported that can lead to large-volume/low-cost ethanol production. This could have major implications for the transportation sector.

1.3 *Liquid biofuels for transportation*

Worldwide interest in expanding liquid biofuel production has subsided since the mid-1980s due to the low cost of oil. World production of ethanol fuel in 1997 was estimated to be about 18 billion liters annually, with Brazil responsible for about 13 billion and the USA for 4.6 billion liters. Ethanol fuel programs have brought important socioeconomic and environmental benefits but they face considerable pressure to cut costs to compete with oil and thus have an uncertain future without some kind of governmental support. However, the combination of technical improvements, cost reduction and environmental considerations can put ethanol fuels on a stronger footing in the future.

It is unlikely that in the short to medium term we will witness a large-scale introduction of biodiesel, because it will require large subsidies to be able to compete with diesel, which is inexpensive in many countries. In addition, the economics and environmental implications of the use of biofuels still need to be fully established. Currently, the most important sources of biodiesel are soybean and rapeseed. Large-scale production of biodiesel from rapeseed has become a reality in Europe, e.g. in the EU there is an annual capacity of over 100 000 t $year^{-1}$. Outside the Organization for Economic Cooperation and Development (OECD) countries biodiesel will probably be confined to special small-niche markets.

Small-scale production of bio-gas takes place in many countries around the world, but only three countries — China, India and Denmark — have national programs of any significance. Bio-gas can be used as an engine fuel. The most important technical and economic developments in bio-gas production of recent years have been in centralized industrial plants where it is possible to combine waste disposal, air pollution abatement, energy and fertilizer production, as in the case of Denmark.

The production of biomethanol and hydrogen, heralded as the fuels of the future by some experts, continues to be a medium- to long-term prospect. Currently, there are no commercial plants in operation and both hydrogen and methanol continue to be produced via fossil fuel sources.

1.4 *Co-generation in the sugar-cane-based ethanol industry*

The importance of co-generation to the biofuel industry rests on the fact that it may be a key factor in determining its viability, particularly in the case of ethanol from sugar-cane. The best alternative may be to have an integrated system in which the production of ethanol fuel and the co-generation of electricity, heat and other by-products become an integral part of the whole system. Advances in gasification technology thus could have a profound impact on the economics of liquid biofuel production.

1.5 *Environmental considerations*

Transportation is the world's fastest growing energy use sector and one of the major contributors to pollution; over 1800 million tons of carbon equivalent are produced annually, which is expected to increase significantly if new alternatives are not found. Of the various alternatives to reduce emissions, the environmentally sustainable biofuels could play a significant role. These benefits are already being felt in Brazil and the USA, particularly in some of the major cities where pollution has been cut dramatically. If biofuels were introduced worldwide, and blended with gasoline, the environmental benefits could be considerable. The most promising sources are ethanol from sugar-cane and cellulose.

2 Introduction

During the 1970s through to the mid-1980s, there was a considerable surge of interest in alternative energy resources caused by the increase of oil prices, and many programs and projects started up around the world. In the production of liquid fuels, Brazil, USA, three countries in Africa and, to a lesser extent, Europe were all of importance. The subsequent fall of oil prices in real terms resulted in many casualties. Nevertheless, considerable advances have been made in the production and use of biofuels and, in certain cases, they are reaching economic viability if hidden subsidies to the traditional energy sector are taken into account.

This chapter assesses the main achievements in bioconversion processes for biofuel production from sugar, starches and cellulosic biomass, the use of liquid biofuel in the transportation sector (ethanol, methanol and hydrogen together with their application in fuel cells), the production and utilization of bio-gas and the implications of co-generation in ethanol production from sugar-cane. Finally, some environmental implications of biofuels in the transport sector are examined.

3 Bioconversion processes — sugars, starches and cellulose

Bioethanol can be obtained from a large variety of sources, grouped as: (i) sugar-containing material, e.g. sugar-cane, sugar beet, etc., which can be converted directly to ethanol; (ii) starchy material, e.g. maize/corn, root crops, etc., which must first be hydrolyzed to fermentable sugars; and (iii) cellulosic material, e.g. woody crops, etc., which contain cellulose- and hemicellulose-derived sugars. Cellulosic material must be converted to sugar either by acid or enzymatic hydrolysis. Lignin, which comprises about 20–30% of wood, cannot be fermented. Woody crops are the most abundant source of biomass and thus if the technology can be developed to produce ethanol or methanol on a large scale and at a low price, this will have major implications for liquid fuel production and utilization in the transportation sector.

Currently, the main feedstocks for bioethanol are sugar-cane and corn. The advantages of ethanol include: (i) it can be obtained from a large variety of biomass sources, e.g. sugar-cane, sugar beet, corn, sweet sorghum, potatoes, grapes and others; (ii) its production is labor-intensive compared with other energy production methods, and can contribute to employment and the reduction of rural migration; (iii) the technology used for its production is well known and commercially proven and available, excluding cellulose material; (iv) the energy balance from its production is highly positive for most types of feedstock; (v) the by-products from the production of ethanol have, in many cases, an economic value; (vi) ethanol from biomass is, in principle, a renewable fuel and therefore the possibility exists to reduce the net carbon dioxide emission per energy output; (vii) ethanol is safe to operate, store and handle, and can be used with few adaptations in existing facilities built to operate with gasoline or other oil products; (viii) ethanol is a flexible fuel and can be used neat or blended with gasoline, methanol, ethers, higher alcohols and diesel fuels; (ix) as an octane booster, ethanol can substitute for lead in gasoline and can be used to reduce the content of aromatics in gasoline, which leads to energy savings in the refining process and a significant reduction in the toxicity of gasoline's exhaust emissions.

3.1 *Ethanol production from sugar-cane*

Sugar-cane is the world's largest source of ethanol fuel, mainly produced in Brazil, and thus most of the commercial experience relates to this source. The main advances in ethanol production have taken place in: continuous fermentation, e.g. the Melle–Boinot and Biostil processes; use of bacteria; heat recovery in the distillation process; and better use of by-products, e.g. bagasse, which is used for electricity co-generation and as animal feed.

The process for producing sugar-cane ethanol, which is used in most of the Brazilian distilleries, comprises the milling of whole and burnt stalks, batch fermentation with yeast recycling or continuous fermentation and distillation in stainless-steel columns under low pressure. Standard sugar-cane has an average sucrose content of 12.3% (glucose content, 0.5%) and produces 133.7 kg of glucose per ton of cane. After allowing for losses in the washing, milling and juice treatment stages, 119.4 kg t^{-1} cane can be recovered in the fermentation stage. Given the appropriate operating efficiencies, this should yield 70 l of dehydrated ethanol or 73 l of fuel ethanol (with 4% moisture content) per ton of sugar-cane [1]. Fermentation is a key step in the production of ethanol and thus, despite important advances, costs still need to be reduced. A technologically advanced, integrated fermentation–distillation system has been in use in some Brazilian mills. The main characteristics of the process are: continuous fermentation in only one tank, fermentation feeding with concentrated juice (a syrup with 60–65% soluble solids), a pure yeast culture system, the use of a yeast adapted to the high osmotic concentrations that result from stillage recycling, yeast recycling, recovery of ethanol from the exhausted stream, integrated fermentation–distillation, a divided still distillation column, concentration of the stillage from 6% soluble solids to 55%, automatic controls and adaptability to existing installations. Concentration of the stillage results in a large saving in transportation and application as a liquid fertilizer. Stillage has become an important source of fertilizer and also of bio-gas production in some Brazilian distilleries [2].

Distillation usually takes place in columns with bubble cap trays at low pressure (1.5–2 bar), heated with exhaust steam from either the milling turbines or the electricity generators. Although many old distilleries still use distillation techniques that are more than 50 years old, many innovations are being introduced to conserve energy, reduce investment and recycle stillage. The new distillation systems use only 1.5–2 kg of steam per liter of ethanol, as against 3–3.5 kg formerly; they also produce only 1 or 2 l of stillage volume per liter of ethanol, compared to 12–14 l for the old systems, and have distillation efficiencies of 99–99.8% compared to 98–99% for the old equipment.

The approximate amounts of water used for the manufacture of 1 m^3 of fuel ethanol (96% pure) are: 60 m^3 for cooling the fermenting mash, 80 m^3 for cooling of the distillery condensers and 115 m^3 for conventional washing of the sugar-cane. For each liter of ethanol produced, electricity consumption is 0.15–0.18 kWh. Steam used in the distillery includes high-pressure steam (21 bar or higher) to drive the mill turbines and the electricity generators; the low-pressure exhaust steam from these systems is used for such processes as juice heating, juice concentration and distillation. The steam consumption per liter of ethanol produced in a distillery is about 1.5–2.0 kg for each liter in the still column, 1.5 kg l^{-1} in the rectification column and more than 1.5 kg l^{-1} for juice heating, for a total of about 5–6 kg l^{-1} [3].

In a 120 000 l day^{-1} distillery, 1715 tons of cane are crushed each day, which yield about 430 tons of bagasse with 50% moisture. Each kilogram of this bagasse produces 2.2 kg of steam and each liter of fuel ethanol produced requires 6 kg of steam. Therefore, the mill consumes about 330 tons of bagasse and has a surplus of 100 tons. At present, the sugar mills and distilleries do not have a market for all this surplus bagasse. Alternatives being studied include pelletized briquettes, pressed bales, charcoal and charcoal briquettes and furfural (JR Moreira, personal communication).

3.2 *Ethanol from corn maize, Zea maize*

In the USA, ethanol is produced from grain, cheese whey, citrus wastes and forestry residues, but the predominant source is corn (maize). Unlike Brazil, where inputs are far lower and outputs higher, in the USA inputs are high and outputs are low by comparison. Unless the productivity of maize to ethanol can be increased significantly in the near future with the consequent increase in ethanol production and reduction costs, it is difficult to foresee a long-term future for corn-based ethanol in the USA unless subsidies continue to promote ethanol as a clean fuel. New and cheaper feedstocks will need to be found — the most promising alternative in the medium to long term is ethanol from cellulose.

3.3 *Ethanol from cellulose*

Despite the considerable potential to produce ethanol from cellulose, currently there are not any large-scale commercial plants in operation, although in the past wood has been used as a source of ethanol in the USA, Europe and the former USSR. Cellulosic ethanol processes are differentiated primarily on the basis of the methods used to achieve hydrolysis and fermentation. There are two processes to obtain ethanol from woody biomass: acid hydrolysis and enzymatic hydrolysis.

Although acid hydrolysis is more technologically mature, enzymatic processes have roughly equal projected costs today because considerable research is being carried out on enzymatic hydrolysis and it is expected to have an increasing cost advantage as the technology is improved. This is partly the reason why most R&D for large-scale production of ethanol from cellulosic biomass has concentrated on enzymatic hydrolysis. In addition, acid hydrolysis appears to be environmentally less favorable than enzymatic processes. However, acid hydrolysis processes have the advantage in that they are equally effective with softwoods and hardwoods, while enzymatic processes seem to be more effective with hardwoods and herbaceous materials than with softwoods. Ethanol production via enzymatic hydrolysis includes pretreatment, biological conversion, product recovery and waste treatment. For an overview of conversion technology, see [4,5].

Enzymatic hydrolysis can, in principle, give very high yields of fermentable sugars (in excess of 90%) from both cellulose and hemicellulose. However, commercial plants using acid hydrolysis (dilute sulfuric acid or concentrated hydrochloric acid) have only been able to achieve conversion rates of about 40–50% [6].

All naturally occurring, and most refined, cellulosic materials require pretreatment to become accessible to the enzymes that mediate hydrolysis. Typically, hydrolysis yields in the absence of pretreatment are <20% of theoretical yields, whereas yields after pretreatment often exceed 90% of theoretical yields. The limited effectiveness of current enzymatic processes on softwoods is thought to be due to the relative difficulty of pretreating these materials (see [7,8]). Pretreatment is one of the most costly steps in cellulosic ethanol production because it can greatly influence the performance, enzyme production and biological inhibitors [9].

An element common to essentially all proposed processes for producing ethanol from cellulosic biomass is microbial fermentation. A variety of microorganisms, generally bacteria, yeast or fungi, ferment carbohydrates to ethanol under oxygen-free conditions. Cells carrying out such fermentations do so to obtain energy and are thus dependent upon ethanol production for growth and long-term survival.

In the absence of cell production, the maximum possible yield of ethanol is 0.51 (mass ethanol/mass carbohydrate, corresponding to 51% of the carbohydrate converted to ethanol on a mass basis), the balance being carbon dioxide. Typically, about 5–12% of the carbohydrate is converted to cells, which results in most proposed ethanol production processes converting not more than 47% of the fermented carbohydrate to ethanol and lower hydrolysis/fermentation costs rather than longer reaction times.

Ethanol from cellulose has considerable potential, given the vast amount of cellulosic material available. There is a need to accelerate R&D on advanced enzymatic hydrolysis technology for producing ethanol from cellulosic material to make it competitive with gasoline. The main challenge is to find cost-effective ways to convert the cellulose and hemicellulose into component sugars via hydrolysis and then ferment those sugars to ethanol. There are good prospects worldwide for making this technology fully competitive with petroleum in transport by using advanced technology [5].

A new process for producing ethanol from cellulosic material, which appears to have considerable potential for reducing the costs of cellulosic ethanol, is under development by the National Renewable Energy Laboratory (NREL) of the US Department of Energy (US DOE). After the preparation of the cellulosic material, it is pretreated with sulfuric acid and then steam is used to expose the cellulose and convert xylan to xylose, 2% of the resulting mixture is separated for conversion to cellulase, which is then combined with the rest of the mixture fermented in SSF. The inclusion of xylose fermentation in this step increases the output of ethanol by 25% over previous processes. The NREL initial estimate ranges from US$0.78 to US$1.27 per gallon ($0.20–$0.33 per liter). However, other cost projections predict even lower production costs by the year 2010 or 2015 [10].

A further example of new development is the organism *Zymomonas mobilis*, which is a genetically engineered organism capable of fermenting five- and six-carbon sugars to ethanol. It allows lignocellulosic biomass to be fermented in less time and with greater yields than conventional methods, lowering significantly the cost of producing ethanol. This technology still has considerable potential for lowering the cost of ethanol by at least a further 10% [11].

The US DOE has a goal of producing around 9 billion liters of ethanol via enzymatic hydrolysis in the year 2005 and as much as 85 billion liters by the year 2030. The strategy is to launch the technology in high-value niche markets where it would be used as an oxygenate or octane-boosting additive to gasoline until about the year 2015, and thereafter also as a neat fuel. It should be pointed out that the estimated cost of producing ethanol from wood has been reduced from $4.6 per gallon ($1.20 per liter) in 1980 to about $1.20 per gallon ($0.33 per liter) in 1997. The cost goals of the US DOE are $0.79 per gallon ($0.20 per liter) and as low as $0.56 per gallon ($0.14 per liter) by the years 2005 and 2030, respectively [11].

However, despite the considerable research being conducted, particularly in the USA and the EU, much R&D effort is still needed to achieve the commercial reality of ethanol from cellulose. In particular: (i) to develop new microorganisms capable of converting cellulose and hemicellulose to ethanol, with or without preliminary hydrolysis to glucose and other simple sugars; (ii) to improve the efficiency of processes that separate lignin from cellulose and hemicellulose; and (iii) to design and demonstrate continuous conversion plant [6]. Costs also need to be reduced significantly, e.g. feedstock, power cycle and other infrastructure components, biological conversion, pretreatment and capital.

4 Liquid biofuels for transportation

Worldwide interest in liquid biofuels for transportation increased considerably in the 1970s and 1980s due to high oil prices. In the 1990s, this interest has subsided considerably, particularly in developing countries, because real oil prices have actually declined. However, in industrial countries, e.g. the USA and the EU, interest in liquid fuels has continued to grow, stemming from environmental, agricultural and social considerations. For example, in the EU, a committee proposed a target that up to 5% of the liquid fuel market could be supplied by bioethanol and biodiesel by the year 2010, requiring some 7 Mha [12]. In the USA, the Clean Air Act of 1990 provides various incentives for biofuels.

Thus, the combination of technological and environmental factors and high fuel efficiency may make the introduction of new transportation fuels possible at a faster rate than previously thought a decade ago, despite low oil prices. This will, however, depend on a mix of political, commercial and technical factors. High-quality liquid fuels will be reserved for road and air transportation, dominated by fossil fuels at least until the year 2030, but with an increasing share from biofuels. Experience so far shows that there must be clear long-term political guidelines and support if biofuels are to succeed.

4.1 *Ethanol fuel*

Brazil and the USA have pioneered large-scale ethanol fuel programs; on a smaller scale, various other sugar-cane-producing countries, e.g. Zimbabwe (about 40 Ml year^{-1}), Kenya (15–16 Ml year^{-1}) and Malawi (about 6 Ml year^{-1}) have ethanol fuel programs. In the late 1980s, Argentina had the second largest sugar-cane fuel ethanol program with a production in 1987 of 220 million liters and an installed capacity of 450 million liters, most of which is currently idle. In the EU, trials of bioethanol have been carried out in Germany, Italy and Sweden and a small amount has already been blended with gasoline in France.

Brazil's ProAlcool program (the world's largest) was set up in 1975 and is currently replacing about 200 000 barrels per day of imported oil. At its peak (late 1980s), almost 5 million automobiles ran on pure bioethanol and a further 9 million ran on a 20–22% blend of alcohol and gasoline. From late in the 1980s, the combination of high demand for ethanol, higher prices for sugar and uncertain government policy resulted in a shortage of ethanol, and as a result the fraction of new neat-ethanol cars dropped from 51% in 1989 to almost zero in 1997. From 1976, over 140 Mm^3 of gasoline equivalent have been replaced by ethanol. The estimated foreign exchange savings are about US$50 billion [13].

The success of the program was due to a combination of factors, including guaranteeing a price, providing financial incentives to ethanol producers and car owners and investments in R&D. However, ProAlcool has certain economic difficulties (mainly due to changes in the oil and sugarcane markets) that will have to be overcome or the industry's future may not be as secure without a stabilized supply and demand of alcohol.

The USA is the world's second largest fuel ethanol producer. In 1994, production was about 5.3 billion liters (1.4 billion US gallons) and a further 908 million liters (240 million gallons) of new capacity was reported to be under construction. Further expansion is planned, as ethanol is expected to enter into the octane market in the form of ethyl-*tert*-butyl ether (ETBE) and as 'neat' fuel. Ethanol is now produced in 21 states, and gasoline blends (gasohol) represent 10% of the USA's fuel sales and are used by over 100 million motorists [14].

The political birth of fuel ethanol in the USA was initiated with the introduction of the Energy Act of 1978, which provided some economic incentives. The ethanol industry has had its ups and downs in recent years. Provisions of the Clean Air Act of 1990 and the National Energy Policy Act of 1992 have created new market opportunities for ethanol, methanol and natural gas by phasing in requirements for fleet vehicles to operate on cleaner fuels. The Governors' Ethanol Coalition estimates that, by the year 2005, some 5 million vehicles will be running on non-petroleum fuels.

4.2 *Biodiesel*

Biodiesel is a fuel with similar properties to conventional diesel oil, which enables it to be used in conventional diesel engines. Biodiesel can be produced from any oil-bearing crops, e.g. rapeseed, *Jatropha curcus*, soybean, etc. Rape methyl ester (RME) produced from oilseed rape is the main source in Europe and Canada, while soya oil is used in the USA. An advantage of biodiesel is that there is considerable experience with rapeseed as a food crop and virtually the same agricultural practices can be applied. Also, the large-scale commercial production of biodiesel from rapeseed is possible, e.g. there is a production capacity of about 30 000 t year^{-1} in Austria (which accounts for 5% of the diesel market), approximately 20 000 t year^{-1} in France and a further 60 000 t year^{-1} plant capacity in Italy [15].

There has been increasing interest in biodiesel but, excluding a few countries such as Brazil, the EU and possibly the USA, the market will be small and localized because of high costs and the high demand for edible oils. In remoter rural areas of some developing countries where there is a high production potential, biodiesel could play a role in meeting local needs. For example, in Brazil with some 40 000 small isolated rural villages it makes sense to produce biodiesel to meet local needs, e.g. lighting, pumping water, etc. The higher cost may be justified on the grounds of high connection costs to the national grid and the transportation cost of diesel [16]. In the short term, biodiesel will probably be confined to OECD countries that, for environmental, agricultural and other reasons, can afford to pay high subsidies. The potential of biodiesel as a fuel depends on a number of factors, including establishing its costs and the environmental benefits. The CO_2 benefits of biodiesel depend on the inputs used and the utilization of by-products, which are variable, and thus they still need to be clearly established. Potential environmental benefits include negligible sulfur, reduced particulates and biodegradability within a few days (making it particularly suitable for use on inland waterways and other fragile environments).

4.3 *Bio-gas production and utilization*

Bio-gas is produced by the anaerobic fermentation of organic material. Production systems are relatively simple and can operate on small and large scales practically anywhere, with the gas being as versatile as natural gas and used for both combustion and engines. Anaerobic digestion can make a significant contribution to the disposal of domestic, industrial and agricultural wastes, as has been well demonstrated in Denmark. There are many different types of anaerobic systems that can provide treatment of wastewater, sludges and solid waste from a large variety of raw materials. For a detailed description, see [17].

The increasing commercial interest in this area is partly due to environmental considerations (coupled with energy supply), both in developed and developing countries. This has been helped by financial incentives, energy efficiency advances, dissemination of the technology and the training of personnel, particularly in China, with an annual production of approximately 1.2 billion m^3, and India, where in the mid-1980s there were about 1.2 million bio-gas plants operating where bio-gas technology has been considerably improved [18]. During the past few years, this technology has also been carefully examined in Denmark, which has been at the forefront of demonstrating large commercial bio-gas plants for treating manure and for heat and power generation. A significant change in bio-gas technology has been a shift away from energy efficiency towards more 'environmentally sound technology' that allows the combination of waste disposal with energy and fertilizer production. Bio-gas technology is reaching technical and economic maturity in many situations, but the economics of bio-gas production show considerable variations, depending on the specific circumstances. In June 1996, there were 18 centralized bio-gas plants in operation (plus several at the planning stage) with an annual input of 1.2 Mt of biomass (75% animal manure and 25% organic waste) and bio-gas production of 40–45 Mm^3, equivalent to about 25 Mm^3 of natural gas at about 2 PJ; the objective is 4 PJ in the year 2000 and 6 PJ in the year 2005 [19].

4.4 *Methanol fuel*

Any biomass with a moisture content of 50–60% or less can be converted to synthesis gas and then to methanol. In fact, any biomass material can be a potential feedstock, including municipal solid waste (MSW). Methanol can be blended up to 15% with gasoline without any engine modification, but higher proportions require modified engines. Until recently, biomass-derived methanol was not seen as a promising alternative to fossil fuels due to the high costs of biomass feedstock and economics of scale. However, important technical advances of recent years are changing this view. Methanol fuel is in fact being used in the USA, Germany and other countries, although only on an experimental basis.

Some experts have predicted that methanol could become a major fuel in the future and that biomass-derived methanol can make major contributions to energy requirements for road transportation if used in fuel cell vehicles [20]. Such systems offer attractive economics in the year 2010 time frame, and the potential for both very low emissions of local air pollutants and low net CO_2 emissions if the biomass is grown sustainably. The EU has an ongoing program for thermochemical conversion of biomass to produce methanol, with the aim of commercializing methanol from biomass by the year 2010. However, considerable R&D is still needed to determine which gasification process is the most suitable, and the potential environmental impacts also need to be better understood [21].

4.5 *Hydrogen*

Chapter 5 explains in detail the potential of hydrogen as an energy source. Recent interest in hydrogen stems from its potential as a high-quality energy carrier that yields low or zero emissions. Some experts predict that hydrogen may become an energy carrier as important as electricity. It is clear that if hydrogen can be produced on a large scale and at a competitive cost, then it could become a major transportation fuel [22]. There are various promising technologies, e.g. the low-temperature fuel cell that has wide market opportunities in transport and stationary combined heat and power (CHP) applications. However, much greater investment in R&D on hydrogen-using and -producing technologies is still needed in order to optimize hydrogen production as an alternative to fossil fuels and also for sequestering CO_2. This is a major task because hydrogen production on a large scale requires improvements in a chain of technologies involving production, conversion, storage, transmission, distribution and utilization. The adoption of hydrogen as a fuel thus depends upon each of these areas achieving sufficient technical development, so that the complete change becomes technically and economically competitive [23].

4.6 *Fuel cells*

Chapter 8 describes in detail the fundamentals of fuel cells. There is an increasing interest in fuel cells, particularly in the transportation sector where this technology looks very promising. The transport niche is at present occupied by the fuel cell (SPFC), specifically that made by Ballard Power Systems in Canada. It has the advantages of a high power density and solid electrolyte, low-temperature operation and acceptable efficiency. The SPFC operates at temperatures of about 80–100 °C, meaning that it is less dangerous in the event of an accident and also that it can be brought to operating temperature more quickly and easily. It has a solid electrolyte and is capable of high power densities (over 1 kW l^{-1}). The alkaline fuel cell is already in use in the space industry, but it requires fueling on pure hydrogen and oxygen. It also has a highly corrosive electrolyte, a problem it shares with the PAFC and the MCFC. Considerable R&D is still needed for fuel cells to be widely used in transportation. So far, the largest R&D program is to be found in Japan, where a variety of fuel cells are being developed and produced, followed by the USA, Canada and the EU [24].

5 Co-generation in the sugar-cane ethanol industry

Co-generation will be intrinsically linked in many ways to biofuel production in the future, e.g. ethanol production from sugar-cane and bio-gas production from advanced plants, as is the case in Denmark. Generation of bioelectricity could have a major impact both on the economics and the environmental aspects of biofuel production; in fact, it may determine its viability in many parts of the world and this is briefly discussed in this section.

In the past, most of the by-products of the sugar industry were treated as waste products, but today producers are becoming increasingly convinced that the sugar economy will be more stable and profitable as by-products become an integral part of the industry. More than 50 derivatives are being produced on a commercial scale and at various stages in sugar-producing countries, and a further 100 are technically and economically feasible. A number of possibilities have been tried, or are being investigated, with varying degrees of success: (i) co-generation of electricity from bagasse and tops and leaves; (ii) utilization of bagasse as building material, furniture and for charcoal production; (iii) production of animal feed; (iv) production of furfural and paperboard; (v) utilization of by-products, e.g. yeasts; and (vi) improved methods of stillage use and disposal, etc.

Bagasse is one of the most important and promising by-products of the sugar-cane industry. The main uses are the production of steam and electricity, pulp, paper and board, animal feed, etc. Tops and leaves are also increasingly becoming attractive for a variety of uses, such as energy, animal feed, etc. About 200 million tons of bagasse were produced worldwide in 1995, of which 95% was used as fuel in the sugar factories, saving approximately 40 Mt of oil. The rest was used for pulp and paper, board, furfural, animal feed, etc. [25].

Sugar producers have been using bagasse to raise steam for on-site processes for centuries, but very inefficiently. Because large amounts of bagasse were available, there was little incentive to improve energy efficiency, especially as disposing of the surplus bagasse was often a major problem. Recently, many sugar-cane distilleries have become energy self-sufficient and some are selling electricity to the grid.

The most important technical concepts being developed for biomass power generation include: direct firing of biomass; co-firing of biomass with coal; gasification; and pyrolysis of biomass to produce biocrude, either in liquid or gaseous forms. Much of the work on coal gasifier–gas turbine systems is directly relevant to biomass-integrated gasifier–gas turbines (BIG–GTs) and the gas turbine/steam turbine combined cycle (GTCC). The BIG–GTCC technology uses either low-pressure or pressurized gasification; although pressurized BIG–GTCC is more efficient, it is likely to be more capital intensive below a certain capacity range [27]. The sugar-cane industries that produce sugar and fuel ethanol and the pulp and paper/timber industries are particularly promising targets for near-term applications of BIG–GT technologies.

The BIG–GTCC technology has received considerable attention and is expected to be commercially available in the year 2000. A number of BIG–GTCC systems have been, or are being, built: at Varnamo, Sweden, 6 MWe + 9 MW combined-cycle district heating co-generation; 30 MWe BIG–GTCC demonstration plant in northeast Brazil at a total cost of $75 million; two other projects ongoing in the USA, a 3–5 MW air-blown pressurized, bubbling fluidized bed gasification of bagasse for gas turbine power generation in Hawaii and a 45 MW plant in Burlington, Vermont using wood chips. The EU also is putting considerable R&D into the gasification technology, e.g. 8–15 MWe by the late 1990s, 20–30 MWe by the year 2000 and 50–80 MWe by the year 2005. The plants must be biomass-based, be able to use high-yield energy crops and demonstrate advanced energy conversion systems and be environmentally friendly [27].

There are many studies suggesting that integral cane harvesting can be economical if tops and leaves could be used for energy purposes. The amount of tops and leaves (0.092 tons per ton of wet stems) and the amount of waste (0.188 tons per ton of wet stems) altogether comprise some 18 dt ha^{-1} $year^{-1}$ [28]. Assuming that at least a quarter is recoverable, an extra amount of 53.7 GJ ha^{-1} $year^{-1}$ as transport fuel would be obtained and, considering productivities of 90 t cane ha^{-1} $year^{-1}$ and 90 l of ethanol per ton, it is estimated that as much as 348 GJ ha^{-1} $year^{-1}$ could be obtained (194 + 100 + 54) instead of 111 GJ ha^{-1} $year^{-1}$ (JR Moreira, personnal communication).

Despite the obvious potential of co-generation it also has its problems, which need to be addressed, including: (i) legal restrictions, e.g. private capital may be restricted in the electricity generation industry; (ii) long lifecycle of existing electricity generation plants, which limits large-scale investment in co-generation in the short and medium term, as is the case in Brazil where most investment has been limited to replacing old equipment; (iii) short harvesting cycle of sugarcane (6–7 months), which acts as a barrier to year-round generation unless appropriate storage occurs; (iv) difficulties in connecting to the national or regional grids in the more remote rural areas; and (v) low tariffs, usually due to subsidies [29].

6 Some environmental considerations

Automobiles generate more air pollution than any other human activity, contribute very significantly to the emissions of greenhouse gases, notably CO_2, and are in the fastest growing energy consumption sector worldwide. Each year the transportation sector produces 1800 Mt of carbon equivalent emissions (MtCe), or 30% of the world carbon emissions, of which 465 MtCe is produced by the USA [30]. Because of the rapid growth of the transport sector, this has major environmental implications. By the year 2010, the current fleet of 550 million cars may have increased to 1.1 billion and CO_2 emissions will have risen by 65% by the year 2010 over the 1990 level due to the increased use of oil-derived liquid fuels [31].

There are various alternatives to reduce emissions from the transportation sector, e.g. increased fuel efficiency, introduction of new materials, better use of public transport, introduction of non-fossil fuels, etc. The use of alternative environmentally sustainable transportation fuels such as ethanol, methanol, biodiesel etc. offers a good potential for reducing emissions, with considerable experience in countries like Brazil and the USA. The use of these fuels blended with gasoline should be encouraged whenever possible.

In Brazil, the large-scale use of ethanol fuel has played a significant role in reducing the level of pollution. For example, the introduction of ethanol–gasoline blends had an immediate

Table 10.1. Light vehicle emissions from gasoline and ethanol in Brazil.

		Pollutants (g km^{-1})			
Year	Fuel	CO	HC	NO_x	Aldehydes
Pre-1980	Gasoline	54.0	4.7	1.2	0.05
1986	Gasohol	22.0	2.0	1.9	0.04
	Ethanol	16.0	1.6	1.8	0.11
1990	Gasohol	13.3	1.4	1.4	0.04
	Ethanol	10.8	1.3	1.2	0.11
1995	Gasohol	4.7	0.6	0.6	0.025
	Ethanol	4.6	0.7	0.7	0.042

From [13].

Table 10.2. Potential greenhouse gas emission reduction from cellulose ethanol, USA.

	Year		
	2005	2015	2025
Ethanol production (billion gallons)*	2	15	40
Ethanol costs (\$ gallon^{-1})	0.78	0.69	0.69
Ethanol subsidy (\$ gallon^{-1})	0.3	0.04	0
Emission reduction (MtCe)†	5	38	100

From [30].
* 1 gallon = 3.785 liters.
† Million tons of carbon equivalent.

impact on the air quality of the large cities, particularly São Paulo. Lead additives were gradually reduced as the amount of alcohol in the gasoline was increased and, finally, they were completely eliminated in 1991. The aromatic hydrocarbons, which are particularly toxic, were also eliminated simultaneously. In addition, many other pollutants were also drastically reduced (see Table 10.1) [13].

Various studies have looked at the potential of reducing pollution from cellulose-based ethanol, which offers the potential of reducing pollution by 80–95% in carbon emissions from vehicles. Difiglio [30] has estimated that, in the year 2025, the use of 40 billion gallons (154.4 billion liters) of cellulose ethanol in the USA would reduce emissions by 100 MtCe, and the total cumulative GHG reduction would be 930 MtCe (Table 10.2). Although the initial cost of reduced carbon emissions is high, the sector's long-term cumulative cost is estimated to be 40 times lower, some \$10 MtCe^{-1}.

The way forward in the sugar-cane ethanol-based industry is to have an integrated system in which ethanol and the co-generation of electricity and other by-products become an integral part of the whole process. Environmentally, co-generation offers considerable potential for pollution abatement. For example, it has been estimated that the net contribution of sugar-cane amounts

Table 10.3. Net CO_2 emissions (measured as tons of carbon equivalent, tCe) in sugar-cane production and utilization in Brazil, 1996.

	Net CO_2 emissions (10^6 tCe $year^{-1}$)
Fossil fuel use in the agroindustrial phase	+1.28
Methane emissions from sugar-cane burning	+0.06
N_2O emissions	+0.24
Ethanol substitute for gasoline	−9.13
Use of bagasse as substitute for fuel oil (food & chemical industries)	−5.20
Net contribution (carbon uptake)	−12.74

From [13].

to 12.7 MtCe $year^{-1}$ (Table 10.3). Over a time span of 100 years, 330 tC ha^{-1} can be avoided in carbon emissions, showing that in the long term ethanol is very effective for decreasing CO_2 emissions. For example, the production of 600 kWh of electricity per ton of sugar-cane would replace 6.3 GJ of fuel oil (assuming 33% conversion efficiency to electricity), which represents 504 GJ ha^{-1}, abating 11.9 tC ha^{-1} and thus increasing fourfold the present emission abatement capability to 15.2 tC ha^{-1} $year^{-1}$ [13].

The CO_2 emissions avoided in Brazil with the use of ethanol and bagasse correspond to nearly 18% of total emissions from fossil fuel use in the country. Although this is an important contribution to CO_2 reduction, the cost to Brazil is approximately US$200 per ton of carbon (about US$2 billion), which is a significant figure even when ethanol production is not aimed at a carbon abatement program [13].

It is clear that cost-effective or near-cost-effective technologies and alternatives exist for reducing emissions without curtailing economic growth. Biofuels, particularly ethanol from sugar-cane and in the future ethanol from cellulose, offer very good prospects for reducing pollution quite considerably. These fuels should be encouraged to be used as blends whenever possible. Although petroleum fuels will continue to dominate, evidence seems to indicate that, by the year 2020, about a quarter could be supplied by biomass-based liquid fuels.

7 Conclusion

Biofuels are an environmentally sustainable alternative to fossil fuels in the transportation sector. Considerable experience has already been gained from the experiences in Brazil and the USA. Despite continuing low oil prices, there is considerable interest in biofuels, particularly in the industrial countries for environmental, agricultural and other socioeconomic reasons. At present, the main sources of biofuel production are sugar-cane in Brazil, maize in the USA and bio-gas in Denmark. New technological developments are opening up many new possibilities such as ethanol and methanol from cellulose, hydrogen and fuel cells, which could, in the long term, provide the large-scale production and utilization of such fuels in many countries.

The transport sector is one of the fastest growing industries in the world and a major source of pollution. One of the most promising alternatives to reduce emissions without drastic changes and without curtailing growth is the large-scale use of biomass-based liquid and gaseous fuels. Such alternatives can be technically and economically viable in the medium to long term.

8 References

1 Moreira JR, Serra GE. In Ashok D (ed.) *Alternative Liquid Fuels*. ERC: Wiley Eastern, New Delhi 1990.
2 Cortez LAB, Freire WJ, Rosillo-Calle F. *An Assessment of Biodigestion of Vinasse in Brazil*, Inter. Sugar (in press).
3 COPERSUCAR. *Combate a Poluicao: Avaliacao do Programa COPERSUCAR*. Piracicaba, Brazil: Centro de Tecnologia Copersucar, 1979.
4 Wyman CE. *Biores. Technol.* 1994; **50**: 3.
5 Lynd LR. *Annu. Rev. Energy Environ.* 1996; **21**: 403.
6 ETSU (Energy Technology Support Unit). *An Appraisal of UK Energy Research, Development, Demonstration & Dissemination*, Vol. 7. 1994; 237.
7 Converse AO. In Saddler JN (ed.) *Bioconversion of Forest and Agricultural Residues*. Wallingford, CT: CAB International, 1993.
8 McMillan JD. In Himmel *et al.* (eds) *Enzymatic Conversion of Biomass for Fuel Production*, American Chemical Society Symp. Ser 5. Washington, DC: ACS, 1994; 411.
9 van Walsum P, Allen SG, Spencer MJ, Laser M, Antal M. *Adv. Biochem. Biotechnol.* 1996; **57/58**: 157.
10 US DOE. *Scenarios of US Carbon Reductions — Potential Impacts of Energy Technologies by 2010 and Beyond*. Washington, DC: US DOE, 1997.
11 PCAST (The President's Committee of Advisors on Science and Technology). *Federal Energy Research and Development for the Challenges of the 21st Century*. Washington, DC: Energy Research and Development Panel, 1997.
12 European Communities Economic and Social Committee. *Opinion on the Proposal for a Council Decision Concerning the Promotion of Renewable Energy Sources in the Community*, Document CES(92) 1314. Brussels: European Union, 1992.
13 Moreira JR, Godemberg J. *The Alcool Program*. Brazil: Ministry of Science & Technology, 1997.
14 GEC (Governors' Ethanol Coalition). *Ethanol Policy Statement*. Lincoln, Nebraska: GEC, Nebraska Energy Office, 1995.
15 ETSU. In *An Appraisal of UK Energy Research, Development, Demonstration & Dissemination*, Vol. 7. 1994; 535.
16 Moreira JR. In *Development and Utilization of Biomass Energy Resources in Developing Countries*, UNIDO Symposium, Vienna, Austria, 1995.
17 Ross CC, Drake TF, Walsh JL. *Handbook of Biogas Utilization*. Washington, DC: US DOE, 1996.
18 Ravindranath NH, Hall DO. *Biomass, Energy and the Environment: a Developing Country Perspective from India*. Oxford: Oxford University Press, 1995.
19 DEA (Danish Energy Agency). *Biomass for Energy — Danish Solutions*. Copenhagen: Danish Energy Agency, Ministry of Environment and Energy, 1996.
20 Williams RH, Larson ED, Katofsky RE. *Energy Sustain. Dev.* 1995; **1**: 18.
21 ETSU. In *An Appraisal of UK Energy Research, Development, Demonstration & Dissemination*, Vol. 7. 1994; 519.
22 Williams RH. In Ayres RU, *et al.* (eds) *Ecostructuring*. Tokyo: UN University Press, 1996.
23 ETSU. In *An Appraisal of UK Energy Research, Development, Demonstration & Dissemination*, Vol. 4. 1994; 91.
24 ETSU. In *An Appraisal of UK Energy Research, Development, Demonstration & Dissemination*, Vols 1 and 7. 1994; 567, 235.
25 *Statistical Data*. London: International Sugar Organization (ISO).
26 Marrison ED, Larson CI. In *Second Biomass Conference of the Americas*. Golden, CO: National Renewable Energy Laboratory, 1995; 1272.
27 Larson ED, Marrison CI. *Economic Scales for First-Generation Biomass-Gasifiers/Gas Turbine Combined Cycles Fuelled from Energy Plantations*, 41st American Society Mechanical Engineering Gas Turbine and Aeroengine Congress. Washington, DC: ASME, 1996.

28 Alexander AG. *The Energy Cane Alternative*, Sugar Series Vol. 6. Amsterdam, The Netherlands: Elsevier, 1985.
29 Rosillo-Calle F, Cortez LAB. *Biomass Bioenergy*, **14**; 115.
30 Difiglio C. *Energy Policy* 1997; **14/15**: 1173.
31 Waston RT, Zinyowera M, Moss RH and Dokken DJ (eds) *Climate Change 1995: Impacts, Adaptation and Mitigation of Climate Change: Scientific–Technical Analysis*. Cambridge: Cambridge University Press, 1995.

11 Thermal Biomass Conversion Technologies for Energy and Energy Carrier Production

A.V. BRIDGWATER
Bio-Energy Research Group, Chemical Engineering, Aston University, Aston Triangle, Birmingham B4 7ET, UK

1 Introduction to thermal conversion

This chapter reviews the status and prospects for biomass pyrolysis and gasification. Fast pyrolysis of biomass has been researched extensively for the production of liquid products that may be readily integrated into the fuel infrastructures of both industrialized and developing countries as well as for the production of high-value chemicals. The technology has now achieved some commercial success for the production of chemicals and is being actively promoted for liquid fuel products. The characteristics of fast pyrolysis processes and the most recent developments are described. The applications that have been researched to date include direct firing in engines, turbines and boilers, upgrading to high-quality hydrocarbon fuels and recovery and production of chemicals. Some cost data are included. The conclusions are that the technology is advancing rapidly and applications are being successfully developed for heat and power. Recovery of chemicals and the production of transport fuels still have some way to go before commercialization due to high costs in the case of transport fuels and poorly developed markets in the case of chemicals.

In the latter part of this chapter, the technology and costs involved in an integrated system for the production of electricity from wood by gasification are reviewed. All the elements required with respect to wood handling and preparation, gasification, gas quality and gas cleaning are examined and the selection criteria for delivery of a clean gas to a gas turbine or engine are established. Special emphasis has been placed on the technology status and key uncertainties that are considered to be crucial to the success or failure of a biomass-based integrated gasification/combined cycle (IG/CC) system.

The main conclusions are that biomass handling, storage, drying, comminution and screening are well established and present no uncertainties in operation and performance. The technology of biomass gasifiers is sufficiently advanced to justify a substantial demonstration plant to prove the total IGCC concept and obtain reliable performance data. There are still areas of uncertainty, but these are relatively minor and will not be resolved until and unless a large integrated plant is built. Gas cleaning has been developed successfully in laboratories to the point where large-scale demonstration and long-term operating experience are necessary. This area can be considered the least developed and most likely to create problems in a demonstration plant. Turbine and turbine fuel specifications are imperfectly defined although engines are known to be more tolerant of contaminated fuel gas.

2 Fast pyrolysis for liquids

2.1 *Introduction*

Pyrolysis is the thermal degradation of carbonaceous materials between 400 and 800 K either in the complete absence of an oxidizing agent or with such a limited supply that gasification does not occur to an appreciable extent. This latter case is sometimes termed partial gasification and is used

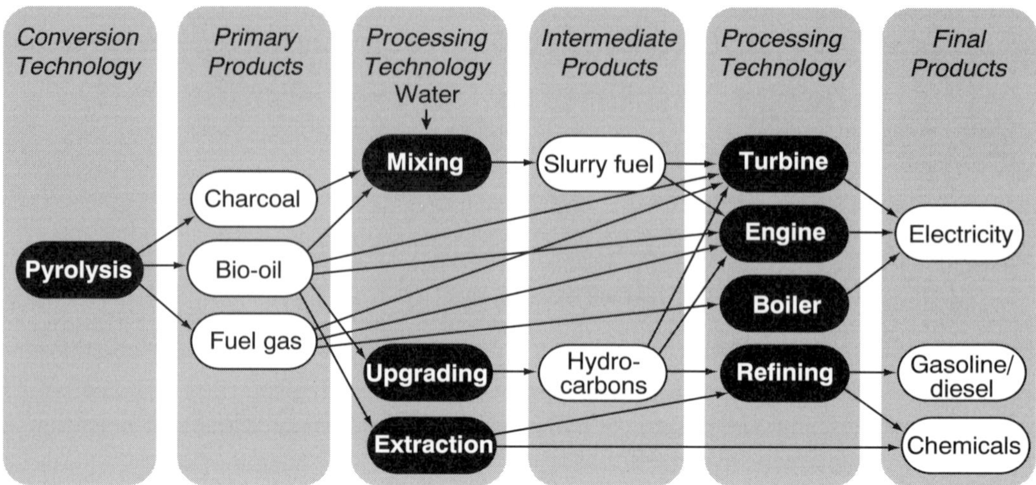

Figure 11.1. Pyrolysis conversion technologies and products.

to provide the thermal energy required for pyrolysis at the expense of char and liquid yields. The thermal energy for pyrolysis may be provided in a variety of ways, either directly or indirectly. Indirect methods include external heating, such as firing a rotary kiln with product gas, and direct methods include hot gas transfer or molten metals.

The products of pyrolysis are a gas, liquid and char. The relative proportions of each depend on the method of pyrolysis and the reaction parameters. Conventional, or slow, pyrolysis requires a slow reaction at low temperatures to maximize the solid char yield. Altering the pyrolysis conditions can substantially change the proportions of gas, liquid and solid product. Very high heating rates at moderate temperatures and rapid product quenching cause the primary vapor products of pyrolysis to condense before further reaction breaks down higher molecular weight species into gaseous products. These high reaction rates also minimize char formation and under some conditions no char is formed. At higher reaction temperatures, the major product is gas.

Figure 11.1 summarizes the main products and secondary processing options. Pyrolysis at these high heating rates is known as fast or flash pyrolysis. Fast pyrolysis at temperatures around 500 K can yield up to 80% mass yields of pyrolysis liquids or bio-oil. Fast pyrolysis at temperatures above around 600 K can be used to maximize gaseous products according to the temperature and residence time.

The modes of pyrolysis are summarized:

- Slow pyrolysis at low temperatures and long reaction times maximizes charcoal yields at about 30 wt.%, comprising about 50% of the energy. This is usually known as carbonization. The gas and liquid products are by-products that may be used in the process for energy or exported.
- Fast pyrolysis maximizes liquid yields at up to 80 wt.% at relatively low temperatures of typically 500 K, but less than 650 K, and at very high reaction rates and short residence times of typically less than 1 s.
- Fast pyrolysis at higher temperatures of above 650 K, at very high reaction rates and short residence times, maximizes gas yields at up to 80 wt.%.
- 'Conventional' pyrolysis at moderate temperatures of less than about 600 K and moderate reaction rates gives approximately equal proportions of gas, liquid and solid products. This is less efficient for liquid production than fast pyrolysis in giving a multiplicity of products that are difficult to handle and market. In addition, the liquid is significantly different chemically.

Table 11.1. Typical characteristics of pyrolysis modes.

Pyrolysis mode	Residence time	Heating rate	Reaction environment	Pressure (bar)	Temp. (K max.)	Major product
Carbonization	Hours/days	Very low	Combustion products	1	400	Solid
Conventional	5–30 min	Low	Primary & secondary products	1	600	Gas, liquid + solid
Fast (liquid)	<1 s	High	Primary products	1	<600	Liquid
Fast (gas)	<1 s	High	Primary products	1	>700	Gas
Vacuum	2–30 s	Medium	Vacuum	<0.1	400	Liquid
Hydropyrolysis	<10 s	High	H_2 & primary products	~20	<500	Liquid
Methanolysis	0.5–1.5 s	High	CH_4 & primary products	~3	1050	BTX* & C_2H_4

* BTX = benzene, toluene and xylene.

Other work has attempted to exploit the complex degradation mechanisms by carrying out pyrolysis in unusual environments. These variations are summarized in Table 11.1.

The problems of fast pyrolysis are well understood, and include:

- high heat transfer rates into the feedstock;
- careful temperature control to maximize the product quantity and quality;
- short vapor residence times to minimize secondary reactions;
- condensation and collection of the liquid fraction that tends to be produced as a 'fog' of dense white aerosol or vapor and micro-droplets of primary liquids, which makes collection very difficult.

Fast pyrolysis is now well established as a thermal biomass conversion technology with considerable potential, particularly for high yields of liquid fuels and chemicals. Increasing attention is being focused on liquid fuels due to their ease of handling, storage and transportation and their ready utilization in many heat and power applications. This is seen as a major advantage over gasification and combustion, which have to use the energy products immediately and cannot store or transport them. Fast pyrolysis, however, is less well developed than combustion, which is widely used for heat and power production, and gasification, which has now reached substantial demonstration scale in Europe and North America [1]. A variety of transport fuel and chemicals can be derived by secondary processing of these liquids as well as fuel gas and charcoal [2]. The essential features of a fast pyrolysis process are very high heating and heat transfer rates, moderate and carefully controlled temperature and rapid cooling or quenching of the pyrolysis vapors [3,4].

2.2 *Pyrolysis technologies*

A wide range of unusual reactor configurations have been devised. The types of pyrolysis reactors are summarized in Table 11.2. Commercial operation is currently only being achieved from a transport or circulating fluid-bed system, which probably offers good potential for scale-up. Fluid beds have also been researched extensively and, as well as being an ideal research and development (R&D) tool, have been scaled up to pilot plant size with plans in hand for demonstration by several systems and suitability for commercial operation. Substantial developments can be expected in performance and cost reduction in the coming years.

Table 11.2. Reactor configurations for biomass fast pyrolysis and proponents.

Reactor type	Organization
Ablative plate	Aston University
Ablative vortex	NREL
Ablative other	BBC + Castle Capital
Circulating fluid bed	CRES, ENEL + Pasquali
Entrained flow	Egemin, GTRI
Fluid bed	CPERI, INETI, IWC, NREL, RTI, Union Electrica Fenosa, Aston University, Hamburg University, Leeds University, Sassari University, Stuttgart University
Rotating cone (transported bed)	Twente University + BTG
Transported bed with solids recirculation	Ensyn, ENEL, VTT
Vacuum moving bed	Laval University + Pyrovac

2.3 *Fast pyrolysis system*

A fast pyrolysis system consists of an integrated series of operations starting with delivery of a roughly prepared feedstock: whole-tree chips from short-rotation coppice; wood waste from furniture manufacture; energy crops such as miscanthus or sorghum; or agricultural residues such as straw. The main steps are shown in Fig. 11.2, following which some key features are emphasized.

Reception and storage Low-capacity systems of up to around 3 t h^{-1} feed can consist of a concrete pad for tipping delivered feed and a front-end loader to move it between steps. Larger-sized plants will increasingly require more sophisticated reception, storage and handling systems analogous to those employed in pulp and paper mills. This will include weighbridge, tipping units, conveyors and bunker storage.

Feed drying This is usually essential unless a naturally dry material such as straw is available. Waste low-grade process heat would be employed, for example, in a rotary kiln. As moisture is generated in pyrolysis, bio-oil usually contains at least 15% water, which cannot readily be removed.

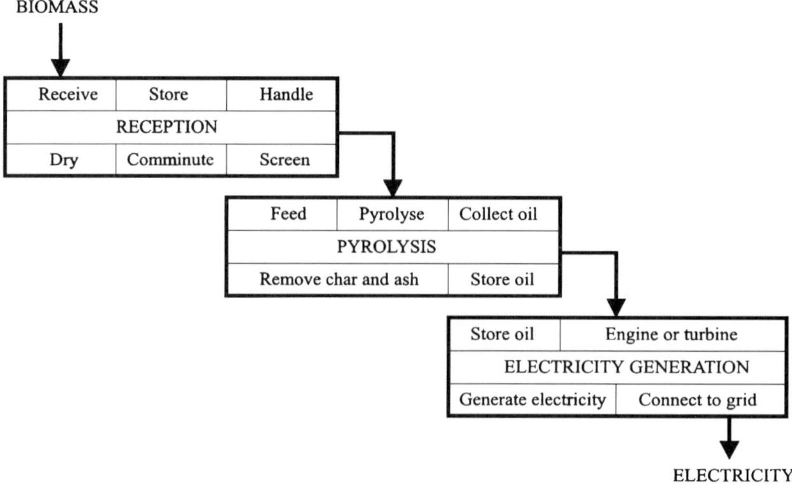

Figure 11.2. Steps in an integrated fast pyrolysis processing system for liquid fuels for electricity production.

Comminution Particles have to be very small to fulfill the requirements of rapid heating and to achieve high liquid yields. This is costly and reactors using larger particles have an advantage.
Reactor configuration A wide variety of configurations have been tested (see Table 11.2), showing considerable diversity and innovation in meeting the basic requirements of fast pyrolysis. The 'best' method is not yet established.
Char + ash separation Some char is inevitably carried over from cyclones and collects in the liquid. Almost all of the ash in the biomass is retained in the char so successful char removal gives successful ash removal. Char separation, however, is difficult and may not be necessary for all applications.
Liquid collection This has long been a major difficulty. Larger scale processing usually employs some type of quenching or contact with cooled liquid product, which is effective. Careful design is needed to avoid blockage from differential condensation of heavy ends. Light-end collection is important in reducing liquid viscosity.

3 Liquid product characteristics

3.1 *Introduction*

Of particular interest is the direct production of liquids because of their much higher energy density that reduces transport and handling costs, and because of their easy substitution for conventional fuels in many static applications such as boilers and for process heat.

Fast pyrolysis liquid approximates to biomass in elemental composition, with a slightly higher heating value of 22–23 MJ kg^{-1} on a dry basis and about 18 MJ kg^{-1} as produced with around 20 wt.% water from water of reaction as well as any water in the feed. The liquid is often referred to as 'oil', 'bio-oil' or 'bio-crude', although it is oleophobic. It is composed of a very complex mixture of oxygenated hydrocarbons. The complexity arises from the degradation of lignocellulosics, and the broad spectrum of phenolic and similar compounds that result from uncontrolled degradation.

Bio-oil has a number of special features and characteristics [3,5,6], as summarized in Table 11.3, which have caused some concern over direct utilization. However, the number of successful

Table 11.3. Fast pyrolysis liquid characteristics — typical data.

Moisture content	20%
pH	2.5
Specific gravity	1.20
Elemental analysis (moisture-free basis)	
C	56.4%
H	6.2%
N	0.2%
S	<0.01%
Ash	0.1%
O (by difference)	37.1%
C/H ratio	9.1
HHV (moisture-free basis)	22.5 MJ kg^{-1}
HHV as produced	18.0 MJ kg^{-1}
Viscosity (at 40 K)	51 cP
Pour point	–23 K

HHV, high heating value.

applications is increasing and considerable progress is being made in understanding the properties and controlling those that are less attractive. Density, viscosity, surface tension and heating value are known to be typical key properties for combustion applications in boilers, furnaces and engines, but other special characteristics such as char level, particle size and ash content will have a major effect. Oil quality standards need to be defined and test methods established, which is being carried out by an International Energy Agency (IEA) Bioenergy Activity [7].

These properties of bio-oil have been claimed to make it relatively unstable in both chemical and physical terms, causing problems in storage, utilization (such as atomization) and upgrading. However, a number of successful applications have been demonstrated and considerable progress is being made in understanding the phenomena and controlling the less attractive properties.

Phase separation is one of the major differences between primary (fast) pyrolysis liquids and secondary products. Primary liquids can absorb up to 50 wt.% of water before separation occurs, whereas secondary liquids have a maximum water loading of about 20%, depending on the technology employed. Water separation from slow pyrolysis processes is thus likely, leading to a potentially significant wastewater disposal problem that does not occur with fast pyrolysis products. Another difference is the viscosity and related properties such as the pour point, which are much lower for primary oils. Toxicity tests have suggested that primary products are significantly less hazardous than secondary liquid or tar products due to the lower temperature of formation and less vapor-phase reactions.

Feedstock can have a significant effect on liquid properties through the inherent inorganic materials naturally present in many biomass forms, including particularly cations of alkali metals. This is very important for chemical production when very low cation levels promote higher levoglucosan levels.

3.2 *Liquid properties*

Water The water content is important because it has several effects on the product, including: reduces the heating value; affects the pH; reduces the viscosity; influences both chemical and physical stability; reduces potential pollution problems by eliminating wastewater disposal requirments; and could affect subsequent upgrading, including catalytic processes. The interactions are poorly understood. The water is difficult to remove because evaporation or distillation at temperatures of around 100 K can cause significant and potentially deleterious physical and chemical changes in the liquid. Lower temperature drying is not successful due to the nature of the relationship between the water and the organic component in which the water seems to be chemically combined, such as water of hydration. Utilization and consideration of the liquid on a 'wet' basis is preferred.

Solids Particulate levels in the liquid may be high from char and ash carry-over. Separation of solids can be carried out in the hot vapor stream prior to condensation and/or in the liquid product. This aspect is receiving attention in the design and operation of pilot plant and commercial units as more information arises from application assessments.

Alkali The alkali metal content may be appreciable from ash that is carried over. This is a potential problem in gas turbine applications. Reduction is possible by hot gas filtration of the hot vapors before liquid collection.

Acidity The low pH arises from the organic acid content (acetic and formic acids), and the liquid is therefore corrosive. Mild steel is not suitable for handling or storage. Polypropylene successfully overcomes this problem, and more demanding applications such as pressurized systems or pumping may demand stainless steel.

Stability Polymerization or deterioration of the liquid can be caused by temperatures above around 100 K, which adversely affect properties such as viscosity, phase separation and deposition of a bitumen-like substance due to polymerization and other reactions resulting from the high oxygen content. Heating the liquid to reduce viscosity for pumping or atomization needs to be considered carefully, to avoid exposing the liquid to any temperature above 100 K, and thoroughly tested, although in-line steam heating appears to be successful. Exposure to air also causes deterioration, but at a slower rate than temperature increase. Exposure to oxygen must be minimized because oxygen is absorbed by the liquid and is thought to contribute to the polymerization phenomenon. Pyrolysis liquid has been stored in this way in a usable form for up to 2 years without problems.

Health Health hazards associated with pyrolysis liquids are poorly understood. The liquids may possibly be toxic and/or carcinogenic.

Miscibility Pyrolysis liquids are hydrophilic and oleophobic. They cannot be assimilated into a conventional fuel marketing infrastructure without upgrading to give a product that is compatible with conventional fuels. The alternative is either to dedicate a specific application, such as firing a boiler, or to create a discrete pyrolysis liquid storage, distribution and utilization system that is managed by experts who understand the special problems of this product. There will still be problems as described above, but these would be minimized in a dedicated system.

3.3 *Liquid product upgrading*

Extensive work has been carried out on catalytic upgrading that is proven but not well developed [2]. Most attention has been paid to either hydrotreating, to give paraffinic hydrocarbons, or zeolite cracking, to give aromatics. Hydrotreating is based on technology that is established in the petroleum industry and is, in principle, readily adaptable. Zeolite cracking has been demonstrated extensively for alcohol feeds. Neither technology is available commercially yet and robust mass balance and performance data have not been produced.

A slurry can be made with the char, which improves the heating value of the resultant fuel and improves the energy efficiency of the conversion process. Little work has been undertaken and the need for additives is unknown. A problem is the atomization requirements of the resultant slurry fuel, which may place restrictions on applications.

3.4 *Gas product from pyrolysis*

The gaseous product from pyrolysis based on indirect heating is usually a medium-heating-value fuel gas of around 14–18 MJ N^{-1} m^{-3}, depending on the feed and processing parameters. If direct heating by partial oxidation or partial gasification is carried out, the gas is a low-heating-value gas comparable to that obtained from air gasification processes, typically 4–6 MJ N^{-1} m^{-3}. It has a high level of hydrocarbons, including low-molecular-weight uncondensed pyrolysis liquids and saturated and unsaturated hydrocarbons from the complex thermal degradation processes. The heating value is enhanced by the sensible heat if the gas is used hot and kept hot, and also by the relatively high condensable liquid content.

3.5 *Solid product from pyrolysis*

When pyrolysis is optimized for charcoal production, yields of up to 30 or 40 wt.% on dry feed are obtained. This occurs in slow pyrolysis with reaction times of hours or even days with the traditional beehive kilns. Partial carbonization at lower temperatures gives higher solid yields because the product contains a high level of volatiles; this is also referred to as torrified wood.

At the very high heating rates encountered in fast and flash pyrolysis, very low char yields result and have been reported as approaching zero under some process conditions. This avoids the marketing and design problems of a multiple product process, because the associated by-product — gas in the case of liquid production, for example — can be used as process heat.

4 Fast pyrolysis technology

4.1 *Introduction*

Fast or flash pyrolysis is now well established as a thermal biomass conversion technology with considerable potential, particularly for high yields of liquid fuels and chemicals [8]. Increasing attention is being focused on liquid fuels due to their ease of handling, storage and transportation and their ready utilization in many heat and power applications. Fast pyrolysis is the main route for direct production of liquids from biomass and wastes by thermochemical conversion. A variety of liquid products can be derived by secondary processing, as well as fuel gas and charcoal and a number of potentially valuable chemicals.

4.2 *Engineering aspects of pyrolysis*

The problems of pyrolysis are fairly well understood, and include:

Feed drying This is usually essential unless a naturally dry material such as straw is available. Because moisture is generated in pyrolysis, bio-oil always contains at least about 15% water, which cannot be removed.

Particle size Particles have to be very small to fulfill the requirements of rapid heating and to achieve high liquid yields. This is costly and reactors using larger particles have an advantage. The trade-off between reactor type and feed characteristics has not been evaluated.

Pretreatment Acid washing to deionize or hydrolyze, or addition of inorganic ions, is practiced to enhance the production of specialty chemicals and liquid yields. Processing is by orthodox leaching-type methods.

Reactor configuration A wide variety of configurations have been tested showing considerable diversity and innovation in meeting the basic requirements of fast pyrolysis. The 'best' method is not yet established.

Heat supply The high heat transfer rate that is necessary imposes a major design constraint. Hot circulating solids are currently preferred for heat supply within the reactor. For ablative reactors, heat supply to the reactor is the major limitation on throughput.

Heat transfer This is effected either by gas–solid heat transfer or solid–solid heat transfer. The latter is more effective and occurs to some extent in most reactors.

Heating rates The low thermal conductivity of wood will prevent the very high claimed heating rates throughout the whole pyrolyzing particle, which therefore has to be sufficiently small.

Reaction temperature Total product yield is maximum at typically 500–520 °C for most forms of woody biomass. Definition of bio-oil quality is needed to optimize reactor parameters.

Vapor residence time This is widely considered crucial to the performance of fast pyrolysis, which is true for some chemicals but is less important for fuels. The time–temperature window of the product vapors is more important than either individually.

Secondary cracking Long vapor residence times and high temperatures cause secondary cracking of primary products, reducing the yields of specific products and overall liquids. This may not be disadvantageous.
Char separation Some char is inevitably carried over from cyclones and collects in the liquid. Separation is difficult.
Ash separation The alkali metals from biomass ash collect in the liquid and cannot be readily separated except by hot gas filtration, which is still not proven.
Liquid collection This has long been a major difficulty for researchers. Larger scale processing usually employs some type of quenching or contact with cooled liquid product, which is effective. Careful design is needed to avoid blockage from differential condensation of heavy ends. Light ends are important in reducing liquid viscosity.

4.3 *Liquid products from pyrolysis*

Of particular interest is the direct production of liquids because of their much higher energy density, which reduces transport and handling costs, and because of their easy substitution for conventional fuels in many static applications, such as boilers and for process heat.

Fast pyrolysis liquid approximates to biomass in elemental composition with a higher heating value of 22–23 MJ kg^{-1} on a dry basis and about 17 MJ kg^{-1} as produced when it has about 25 wt.% water from water of reaction as well as any water in the feed. The liquid is often referred to as 'oil,' 'bio-oil' or 'bio-crude', although it is oleophobic. It is composed of a very complex mixture of oxygenated hydrocarbons. These properties have been claimed to make it relatively unstable in both chemical and physical terms, causing problems in storage, utilization (such as atomization) and upgrading.

The complex interaction of time and temperature on fuel product quality has not been explored, at least partly because the characteristics of pyrolysis oil for different applications have not been defined. Bio-oil has a number of special features and characteristics that require consideration in any application, including production, storage, transport, upgrading and utilization. These are summarized in Table 11.4, along with the major problems that have been encountered or predicted and suggestions for amelioration or solution. However, the number of successful applications is increasing and considerable progress is being made in understanding the properties and controlling those that are less attractive.

Oil quality needs to be defined by potential users and may differ by application. Although there are set standards and methods of measurement for conventional fuels, analogous standards and methods have not yet been defined for biomass pyrolysis liquids. Density, viscosity, surface tension and heating value are known to be typical key properties for combustion applications in boilers, furnaces and engines, but other special characteristics such as char level, particle size and ash content will have a major effect.

4.4 *Applications of liquid products*

4.4.1 COMBUSTION

Liquid products are easier to handle in combustion than solids and this is important in retrofitting existing equipment. Existing oil-fired burners cannot be fueled directly with solid biomass, such as wood chips or chopped straw, without major reconstruction of the unit, which is not attractive

Table 11.4. Characteristics of bio-oil and methods for modification.

Characteristic	Effect	Solution
Alkali metals	Deposition of solids in combustion applications, including boilers, engines and turbines. In turbines the damage potential is considerable, particularly in high-performance machines with, for example, coated blades	Hot vapor filtration; processing or upgrading of oil; or modification of application
Health	Toxic, possibly carcinogenic	Proper precautions
High viscosity	High pressure drops in pipelines leading to higher cost equipment and/or possibilities of leakage or even pipe rupture	Careful low-temperature heating and/or addition of water, and/or addition of co-solvents such as methanol or ethanol
Inhomogeneity	Layering or partial separation of phases; filtration problems	Modify or change process; modify pyrolysis parameters; change feedstock; additives
Incompatibility with polymers	Swelling or destruction of sealing rings and gaskets	Careful materials selection, e.g. PTFE
Low pH	Corrosion of vessels and pipework	Careful materials selection. Stainless steel, copper and some olefin polymers are acceptable
Miscibility	Not miscible with fossil fuels	None, other than keep separate
Poor stability	Reacts with oxygen, polymerizes with viscosity increase, separation of phases or layers, deposits	Maintain low temperature, limited air/oxygen access, mix before use
Suspended char	Erosion, equipment blockage, combustion problems from slower rates of combustion. 'Sparklers' can occur in combustion, leading to potential deposits and high emissions	Hot vapor filtration; liquid filtration; modification of the char, e.g. by size reduction, so that its effect is reduced; or modification of the application
Temperature sensitivity	Liquid decomposition at hot surfaces, leading to blockage	Recognition of problem and appropriate cooling facilities; avoidance of contact with hot surfaces
Water content	Has complex effect on viscosity, heating value, stability, pH, homogeneity, etc.	Recognition of problem and optimization with respect to application

PTFE, polytetrafluoroethene.

in uncertain fuel markets. However, bio-oils require only relatively minor modifications of the equipment, such as the burners, or even no modification at all in some cases. Powdered coal-fired furnaces can accept charcoal as a partial fuel replacement relatively easily as long as the volatile content is compatible with the furnace design.

4.4.2 POWER GENERATION WITH LIQUIDS

A key advantage of production of liquids is that the fuel production can be decoupled from the power generation. Peak power provision is thus possible with a much smaller pyrolysis plant, or liquids can be readily transported to a central power plant of engines or turbine. There are additional benefits from potentially higher plant availability from the intermediate fuel storage.

Bio-oil has been fired successfully in several diesel test engines where it behaves very similar to diesel in terms of engine parameters, performance and emissions [9–11]. Continuous runs of 10 h have been achieved on raw bio-oil without dilution or processing [8]. A diesel pilot fuel is needed, typically 5% in larger engines, and no significant problems are foreseen in power generation up to 15 MWe per engine. A 2.5 MWe silo-fitted gas turbine has also been successfully operated recently [12]. This may be more suitable for larger scale applications.

4.4.3 LIQUID PRODUCT UPGRADING

Extensive work has been carried out, on catalytic upgrading of bio-oil to give transport fuels, that is proven but not well developed. Most attention has been paid to either hydrotreating to give paraffinic hydrocarbons or zeolite cracking to give aromatics [2]. Recent tests have been carried out successfully with bio-oil on coal liquid upgrading processes [13].

Hydrotreating is based on technology that is established in the petroleum industry and is, in principle, readily adaptable, although the high water content appears to severely damage the catalyst support [14]. Catalyst development is continuing [15]. Zeolite cracking has been demonstrated extensively for alcohol feeds for synthesis of hydrocarbons, particularly gasoline. It has operated successfully on pyrolysis vapors but with relatively low yields and still unresolved catalyst and operational problems. Neither technology is available commercially yet and robust mass balance and performance data have not been produced. Mild catalytic upgrading for product quality improvement is currently the preferred interest.

4.4.4 CHEMICALS

Several hundred chemical constituents have been identified to date, and increasing attention is being paid to recovery of individual compounds such as levoglucosan and hydroxyacetaldehyde [16] or families of chemicals such as polyphenols [17]. The potentially much higher value of specialty chemicals compared to fuels could make recovery of even small concentrations viable [18]. An integrated approach to chemicals and fuel production offers interesting possibilities for shorter term economic implementation. Chemicals reported to have been recovered include polyphenols for resins with formaldehyde, calcium and/or magnesium acetate for biodegradable de-icers and levoglucosan, hydroxyacetaldehyde and a range of flavorings and essences for the food industry.

5 Pyrolysis technologies

5.1 *Introduction*

A wider range of unusual combinations and configurations have been devised for fast pyrolysis compared to gasification. Types of pyrolysis reactors were shown in Table 11.2. Commercial operation is currently only being achieved from a transport or circulating fluid-bed system [19], which probably offers the best potential for scale-up. Fluid beds have also been researched extensively and are an ideal R&D tool. Union Electrica Fenosa in Spain have successfully scaled up a fluid-bed system based on the University of Waterloo Process [20]. The most widely researched type of fast pyrolysis is based on ablation, which offers the potential for using larger sized feed and has a higher specific reactor capacity. Substantial developments can be expected in performance and cost reduction in the coming years. The current status of the more substantial plants has been reviewed recently [6].

5.2 *Status*

Up to 1989, a conventional pyrolysis demonstration plant of 500 kg h^{-1} was operating in Italy for liquid and char production with approximately 25% yield of each. A 200 kg h^{-1} fast pyrolysis pilot plant based on the University of Waterloo (Canada) process has been constructed in Spain by Union Electrica Fenosa, which started up in late 1992 and is still operational after having undergone several refits and modifications. Egemin in Belgium have built and operated a 200 kg h^{-1} entrained downflow pilot plant to their own design, which started up in July 1991 and operated until late 1992. The largest plant currently operational is the Ensyn plant supplied to ENEL in the framework of an EC JOULE project at the ENEL Bastardo power plant in central Italy. This began operation in 1996 and was operating at least until 1998 on demand. The rotating cone technology developed at the University of Twente is being developed further by BTG and a 50 kg h^{-1} unit has been supplied to China. A 200 kg h^{-1} unit was being designed in early 1998 for construction later in that year. Other units are believed to be planned.

In North America, a number of commercial, demonstration and pilot plants for fast pyrolysis are operating at a scale of up to 1100 kg h^{-1}. There are other smaller scale activities on a research and development basis. The downturn in government funding in both Canada and the USA has had an adverse effect on the extent of research, development and demonstration (RD&D) in this area, although there is still a substantial body of expertise and activity.

5.3 *Pyrolysis processes*

A conceptual fluid-bed fast pyrolysis process is illustrated in Fig. 11.3. Prepared biomass is dried to less than 10% water and ground to –2 mm before feeding to a fluid-bed fast pyrolysis reactor. Products are passed through a cyclone to separate char and then the liquid is rapidly cooled and quenched, typically with an electrostatic precipitator to recover as much of the aerosol product as possible. The total residence time for the hot vapors is less than 1 s, which 'freezes' the thermally unstable liquid intermediates of pyrolysis. Most fast pyrolysis processes employ similar concepts

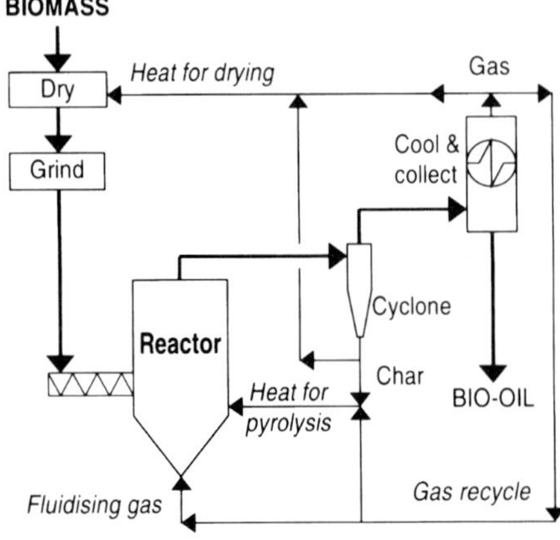

Figure 11.3. Conceptual fluid-bed fast pyrolysis system.

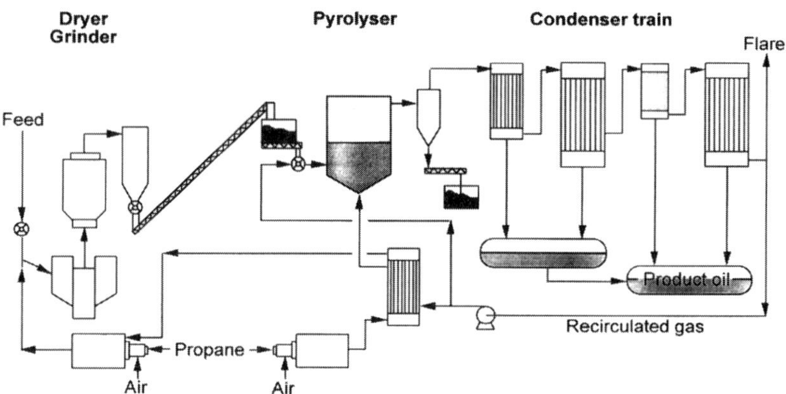

Figure 11.4. Union Electrica Fenosa 200 kg h^{-1} fluid-bed pilot plant.

to minimize vapor residence time. The liquid yield is up to 75 wt.% on a dry feed basis and has properties similar to those listed in Table 11.3.

To illustrate how these principles are applied in practice, Fig. 11.4 shows the Union Electrica Fenosa 200 kg h^{-1} fluid-bed pilot plant with wood preparation, feeding, fast pyrolysis and bio-oil collection in a condenser train.

6 Economics

The cost of bio-oil is mostly dependent on feed cost and capital cost. Some data are given in Fig. 11.5 [21]. Capital cost is mostly dependent on scale, which considerably outweighs additional feed collection and transport costs. The effect of learning on capital cost is also significant, with eventual costs being no more than half the current first-plant costs. There is evidence that this learning effect is already taking place.

Future systems may also benefit from other improvements in cost and performance as a result of learning. Probable improvements include:

- capital cost reductions (as already discussed);
- feed cost reductions;
- higher organic yields;

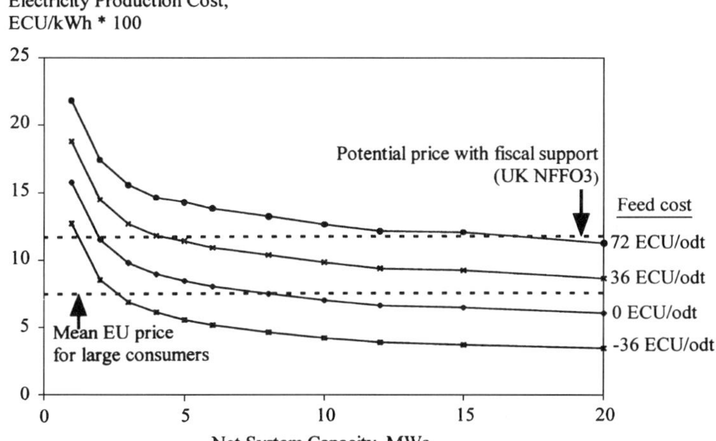

Figure 11.5. Cost of electricity from pyrolysis liquids at different feed costs.

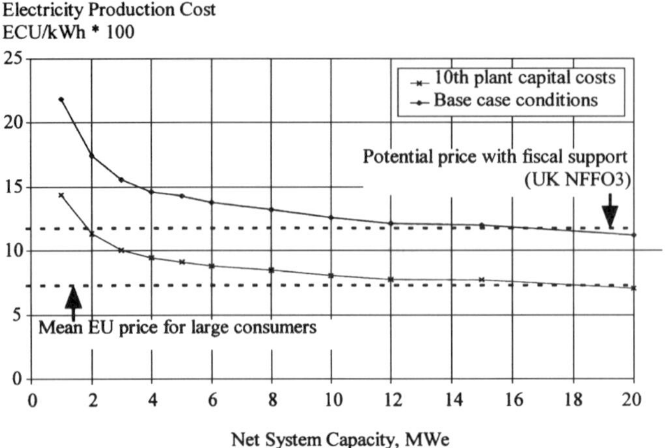

Figure 11.6. Potential electricity production costs in future systems.

- better reliability and hence higher availability;
- reduced labor requirements through optimized operator practices and automation.

It is extremely difficult to quantify such improvements, but conservative estimates are useful in showing how fast pyrolysis and engine systems could become viable in future. The following analysis presents potential production costs following moderate improvements in all of the above areas. The base case conditions are modified as follows:

- 10th plant capital costs are used;
- the feed cost is reduced to 36 ECU per oven-dry ton (odt) (a 50% reduction);
- the organic yield is increased to 65 wt.% dry feed (a 10% improvement);
- the system availability is increased to 95% (approximately 5% improvement);
- labor requirements are reduced by 25%.

The results are shown in Fig. 11.6, suggesting that there is considerable potential from continued RD&D to achieve viable operation.

7 Conclusions to thermal conversion

Fast pyrolysis offers considerable potential for the production of liquid fuels that may be used directly for direct firing and electricity production or upgraded to higher quality hydrocarbon fuels. High-value chemical specialties and intermediates are also possible, which will enhance the economic performance of fuel production systems. The performance and cost estimates for both crude pyrolysis liquids and refined hydrocarbon products show that, while crude liquids can be economic at low feedstock costs, refined hydrocarbons are still not competitive without credits for environmental and socioeconomic contributions. As feedstock cost is the major cost item, any process that can utilize waste materials with their low inherent cost will show economic viability in the short term. Substantial reactor and system developments can be expected in performance and cost reduction in the coming years.

8 Gasification

8.1 *Introduction*

Thermochemical gasification is the conversion, by partial oxidation at elevated temperature, of

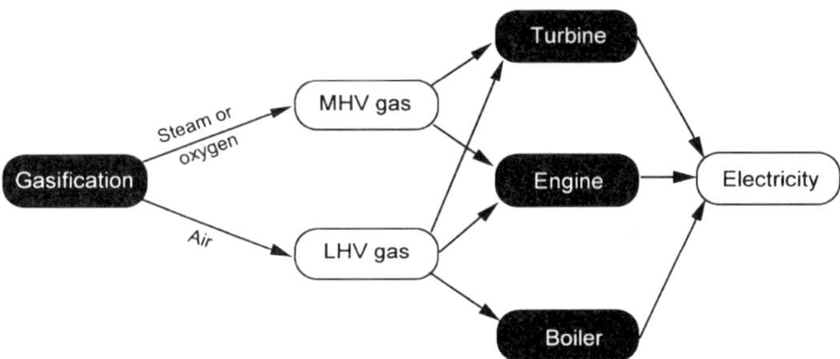

Figure 11.7. Gasification for electricity production.

a carbonaceous feedstock such as biomass or coal into a gaseous energy carrier. This contains carbon monoxide, carbon dioxide, hydrogen, methane, trace amounts of higher hydrocarbons such as ethane and ethene, water, nitrogen (if air is used as the oxidizing agent) and various contaminants such as small char particles, ash, tars and oils. The partial oxidation can be carried out using air, oxygen, steam or a mixture of these.

Air gasification produces a low-heating-value (LHV) gas (4–7 MJ N^{-1} m^{-3} higher heating value) suitable for boiler, engine and turbine operation but not for pipeline transportation owing to its low energy density. Oxygen gasification produces a medium-heating-value (MHV) gas (10–18 MJ N^{-1} m^{-3} higher heating value) suitable for limited pipeline distribution and as synthesis gas for conversion, for example, to methanol and gasoline. Such an MHV gas can also be produced by pyrolytic or steam gasification. Gasification with air is the more widely used technology because there is not the cost or hazard of oxygen production and usage, nor the complexity and cost of multiple reactors. The major routes and electricity generation options are summarized in Fig. 11.7.

8.2 *Principles*

Gasification occurs in a number of sequential steps:

- drying to evaporate moisture;
- pyrolysis to give gas, vaporized tars or oils and a solid char residue;
- gasification or partial oxidation of the solid char, pyrolysis tars and pyrolysis gases.

When a solid fuel is heated to 300–500 °C in the absence of an oxidizing agent, it pyrolyzes to solid char, condensable hydrocarbons or tar and gases. The relative yields of gas, liquid and char depend mostly on the rate of heating and the final temperature. Generally, in gasification, pyrolysis proceeds at a much quicker rate than gasification and the latter is thus the rate controlling step.

The gas, liquid and solid products of pyrolysis then react with the oxidizing agent — usually air — to give permanent gases of CO, CO_2 and H_2 and lesser quantities of hydrocarbon gases. Char gasification is the interactive combination of several gas–solid and gas–gas reactions in which solid carbon is oxidized to carbon monoxide and carbon dioxide, and hydrogen is generated through the water–gas shift reaction. The gas–solid reactions of char oxidation are the slowest and limit the overall rate of the gasification process. Many of the reactions are catalyzed by the alkali metals present in wood ash, but still do not reach equilibrium. The gas composition is influenced by many factors, such as feed composition, water content, reaction temperature and the extent of oxidation of the pyrolysis products.

Not all the liquid products from the pyrolysis step are converted completely due to the physical or geometrical limitations of the reactor and the chemical limitations of the reactions involved, and these give rise to contaminant tars in the final product gas. Owing to the higher temperatures involved in gasification compared to pyrolysis, these tars tend to be refractory and are difficult to remove by thermal, catalytic or physical processes. This aspect of tar cracking/removal in gas clean-up is one of the most important technical uncertainties in the implementation of gasification technologies, and is discussed below.

8.3 *Gasification technology*

8.3.1 INTRODUCTION

A range of reactor configurations have been developed, as shown in Table 11.5. Figures 11.8–11.10 show the configurations of the more common gasifier types. At the end of this section, Table 11.6 summarizes the key features of each reactor type, Table 11.7 summarizes the other relative advantages and disadvantages of the most common gasifier types and Tables 11.8 and 11.9 summarize performance data for most gasifier types. Each main type of gasifier configuration is described below, with significant features and limitations highlighted. These gasifiers have been reviewed [1,22].

8.3.2 DOWNDRAFT

The downdraft gasifier features a co-current flow of gases and solids through a descending packed

Table 11.5. Gasifier types.

Gasifier type	Mode of contact
Fixed bed	
Downdraft	Solid moves down, gas moves down
Updraft	Solid moves down, gas moves up
Co-current	Solid & gas move in same direction — downdraft
Counter-current	Solid & gas move in opposite directions — updraft
Cross-current	Solid moves down, gas moves at right angles
Variations	Stirred bed; two-stage gasifier
Fluid bed	
Single reactor	Low gas velocity; inert solid stays in reactor
Fast fluid bed	Inert solid is elutriated with product gas and recycled
Circulating bed	Inert solid is elutriated, separated and recirculated. This sometimes also refers to fast fluid bed or twin reactor systems
Entrained bed	Usually no inert solid; highest gas velocity of lean phase systems; can be run as a cyclonic reactor
Twin reactor	Steam gasification and/or pyrolysis occurs in the first reactor; char is combusted in the second reactor to heat the fluidizing medium for recirculation. Either can be any type of fluid bed, although the combustor is often a bubbling fluid bed
Moving bed	Mechanical transport of solid; usually lower temperature processes; includes multiple hearth, horizontal moving bed, sloping hearth, screw/auger kiln
Other	Rotary kiln: good gas–solid contact; careful design needed to avoid solid carry-over Cyclonic and vortex reactors: high particle velocities give high reaction rates

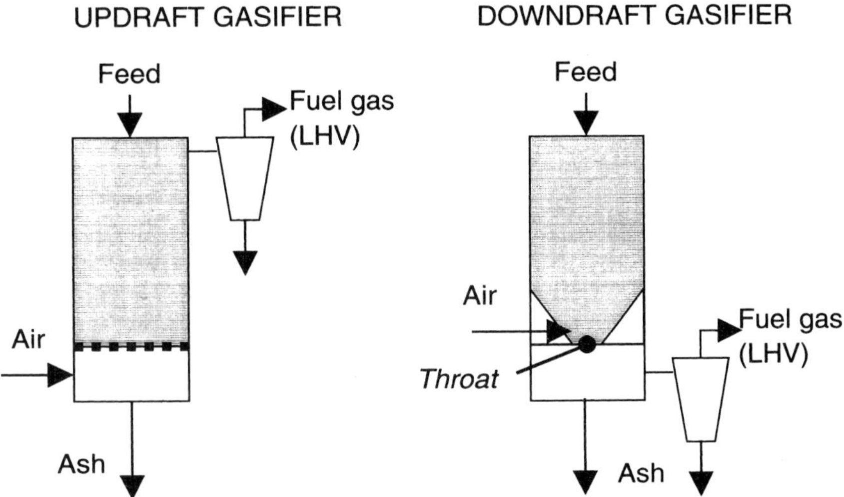

Figure 11.8. Fixed-bed gasifiers: updraft and downdraft.

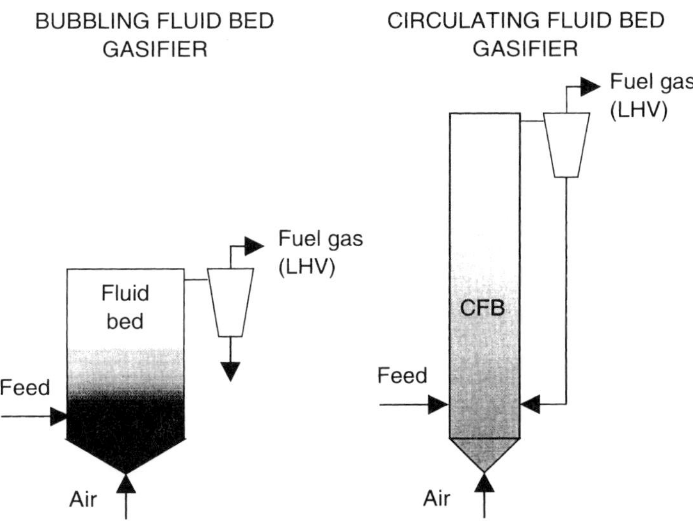

Figure 11.9. Bubbling and circulating fluid-bed gasifiers.

bed that is supported across a constriction known as a throat, where most of the gasification reactions occur. The reaction products are intimately mixed in the turbulent high-temperature region around the throat, which aids tar cracking. Some tar cracking also occurs below the throat on a residual charcoal bed where the gasification process is completed. This configuration results in a high conversion of pyrolysis intermediates and hence a relatively clean gas.

Downdraft gasification is simple, reliable and proven for certain fuels, such as relatively dry (up to about 30 wt.% moisture) blocks or lumps with a low ash content (below 1 wt.%) and containing a low proportion of fine and coarse particles (not smaller than about 1 cm and not bigger than about 30 cm in the longest dimension). Owing to the low content of tars in the gas, this configuration is generally favored for small-scale electricity generation with an internal combustion engine. The physical limitations of the diameter and particle size relationship mean

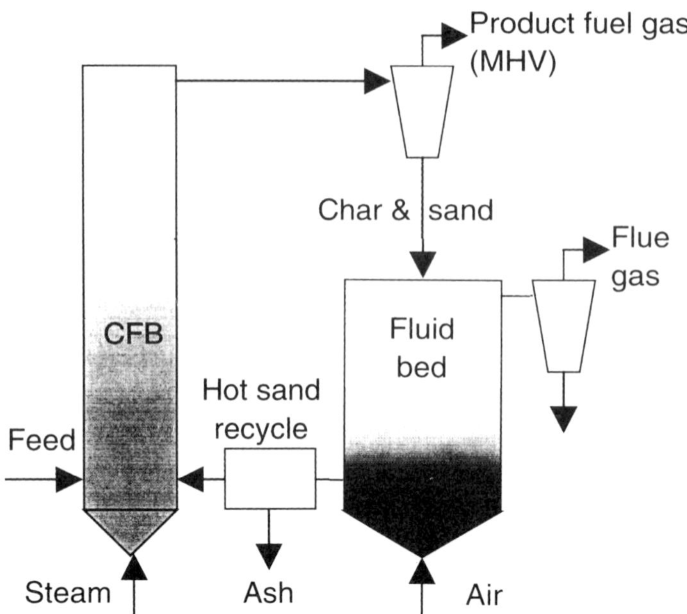

Figure 11.10. Twin fluid-bed gasifier.

that there is a practical upper limit to the capacity of this configuration of around 500 kg h^{-1} or 500 kWe.

A relatively new concept of a stratified or open-core downdraft gasifier has been developed in which there is no throat and the bed is supported on a grate. This was first devised by the Chinese for rice husk gasification and further developed by Syngas Inc. [23] from work carried out at the Solar Energy Research Institute (now the National Renewable Energy Laboratory — NREL) [24]. The concept is still widely practiced in China for rice husk gasification.

8.3.3 UPDRAFT

The updraft gasifier arrangement is shown in Fig. 11.8. The downward-moving biomass is first dried by the upflowing hot product gas. After drying, the solid fuel is pyrolyzed, giving char that continues to move down to be gasified and pyrolysis vapors that are carried upward by the upflowing hot product gas. The tars in the vapor either condense on the cool descending fuel or are carried out of the reactor with the product gas, contributing to its high tar content [25]. The extent of this tar 'bypassing' is believed to be up to 20% of the pyrolysis products [26]. The condensed tars are recycled back to the reaction zones where they are further cracked to gas and char. In the bottom gasification zone the solid char from pyrolysis and tar cracking is partially oxidized with the incoming air or oxygen. Steam may also be added to provide a higher level of hydrogen in the gas.

The product gas from an updraft gasifier thus has a significant proportion of tars and hydrocarbons, which contribute to its high heating value. The fuel gas requires substantial clean-up if further processing is to be performed.

The principal advantages of updraft gasifiers are their simple construction and high thermal efficiency: the sensible heat of the gas produced is recovered by direct heat exchange with the entering feed, which thus is dried, preheated and pyrolyzed prior to entering the gasification zone.

In principle, there is little scaling limitation, although no very large biomass gasifiers have been built.

8.3.4 FLUID BED

Fluid-bed gasifiers are a more recent development that takes advantage of the excellent mixing characteristics and high reaction rates of this method of gas–solid contacting. Although only recently applied to biomass, there are over 50 years of experience with peat. Fluid-bed reactors are the only gasifiers with isothermal bed operation. A typical operating temperature for biomass gasification is about 800–850 K. Most of the conversion of the feedstock to product gas takes place within the bed. However, some conversion to product gas continues in the freeboard section from reactions of entrained small particles and particularly from thermal tar cracking. In most cases, carbon conversion approaches 100%, unless excessive carry-over of fines takes place, which will occur with a top-feeding configuration.

The bubbling fluid-bed gasifier tends to produce a gas with a tar content between that of the updraft and downdraft gasifiers. Some pyrolysis products are swept out of the fluid bed by gasification products, but are then further converted by thermal cracking in the freeboard region (see Fig. 11.9).

Loss of fluidization due to bed sintering is also a commonly encountered problem, depending on the thermal characteristics of the ash, but the inherently lower operating temperature of a fluid bed and better temperature control provide an acceptable control measure. The problem is that alkali metals from the biomass ash form low-melting eutectics with the silica in the sand, resulting in agglomeration and bed sintering with eventual loss of fluidization. With biomass of high ash/inert content, it is better to use alumina or even a metallic sand such as chromite sand.

Carbon loss with entrained ash can, however, become significant and fluid beds are not economical for small-scale applications. Moreover, they incur higher operating (i.e. compression) costs.

Fluid-bed gasifiers also have the advantage that they can be readily scaled up with considerable confidence. Only the fuel distribution becomes problematic in large beds, although multiple feeding is an acceptable solution. Alternative configurations such as twin bed systems and circulating fluid beds are also available to suit almost every type of feedstock or thermochemical process. In catalytic thermochemical processes the bed material can be replaced by the catalyst, thereby avoiding costly impregnation techniques. Alternatively, a second catalytic reactor [2], as in the TPS system [22], or a thermal cracking reactor [27] can be added.

Fluid beds provide many features that are not present in the fixed-bed types, including high rates of heat and mass transfer and good mixing of the solid phase, which means that reaction rates are high and the temperature is more or less constant in the bed. A relatively small particle size compared to dense-phase gasifiers is desirable and this may require additional size reduction. The ash is elutriated and removed as fine particulates entrained in the off-gas.

8.3.5 CIRCULATING FLUID BED

The fluidizing velocity in the circulating fluid bed is high enough to entrain large amounts of solids with the product gas (see Fig. 11.9). These systems were developed so that the entrained material is recycled back to the fluid bed to improve the carbon conversion efficiency compared with the single fluid-bed design. A hot raw gas is produced, which, in most commercial applications to date, is used for close-coupled process heat or retrofitting to boilers to recover the sensible heat in the gas [22]. This configuration has been developed extensively for wood waste conversion in pulp and paper mills for firing lime and cement kilns [28,29] and steam generation for electricity.

8.3.6 ENTRAINED BED

In entrained-flow gasifiers no inert material is present but a finely reduced feedstock is required. Entrained bed gasifiers operate at much higher temperatures of about 1200–1500 K, depending on whether air or oxygen is employed, and hence the product gas has low concentrations of tars and condensable gases. However, this high-temperature operation creates problems of material selection and ash melting. Conversion in entrained beds effectively approaches 100%. There is little experience with biomass in such systems.

8.3.7 TWIN FLUID BED

Twin fluid-bed gasifiers are employed to give a gas of higher heating value from reaction with air than from a single air-blown gasifier (see Fig. 11.10). The gasifier in effect is a pyrolyzer, heated with hot sand from the second fluid bed, which is heated by burning the product char in air before recirculating it back to the first reactor. Steam is also usually added to encourage the shift reaction to generate hydrogen and to encourage carbon–steam reactions. Product quality is good from a heating value viewpoint but poor in terms of tar loading from the pyrolysis process.

8.3.8 COMPARISON OF PRESSURIZED AND ATMOSPHERIC OPERATION

The relative advantages and disadvantages of pressurized gasification systems have not been fully resolved.

Pressurized gasifiers have the following notable features:

- Feeding is more complex and very costly, and has a high inert gas requirement for purging.
- Capital costs of pressure equipment are much higher than atmospheric equipment, although sizes are much smaller [22].
- Gas is supplied to the turbine at pressure, removing the need for gas compression and also permitting relatively high tar contents in the gas. Hot gas clean-up also reduces energy losses and, in principle, is simpler and of lower cost than scrubbing systems.
- Overall system efficiency is higher due to retention of sensible heat and chemical energy of tars in the product gas.

Atmospheric gasifiers have the following notable features:

- For gas turbine applications, the product gas is required to be sufficiently clean for compression prior to the turbine. For engine applications, the gas quality requirements are less onerous and pressure is not required.
- Atmospheric systems have a potentially much lower capital cost at smaller capacities of below around 30 MWe [22].
- Gas compositions and heating values are not significantly different for either system.

8.3.9 SUMMARY

A summary of the relative advantages and disadvantages is given in Table 11.6.

Table 11.6. Gasifier characteristics.

Gasifier type	Characteristics
Downdraft	
Simple, reliable and proven for certain fuels	High residence time of solids
Relatively simple construction	Needs low-moisture fuels
Close specification on feedstock characteristics	High carbon conversion
Uniform-sized feedstock required	Low ash carry-over
Very limited scale-up potential	Fairly clean gas is produced
Possible ash fusion and clinker formation on grate	Low specific capacity
Updraft	
Product gas is very dirty with high levels of tars	Low exit gas temperature
Very simple and robust construction	High thermal efficiency
Good scale-up potential	High carbon conversion
Suitable for direct firing	Low ash carry-over
High residence time of solids	
Relatively simple construction	
Bubbling fluid bed	
Good temperature control & high reaction rates	Good scale-up potential
In-bed catalytic processing is possible	Low feedstock inventory
Greater tolerance to particle size range	Carbon loss with ash
Moderate tar levels in product gas	Good temperature control
Higher particulates in the product gas	High specific capacity
Good gas–solid contact and mixing	Can operate at partial load
Tolerates variations in fuel quality	
Easily started and stopped	
Circulating fluid bed	
Good temperature control & high reaction rates	High specific capacity
In-bed catalytic processing not possible	Very good scale-up potential
Greater tolerance to particle size range	High carbon conversion
Moderate tar levels in product gas	Good gas–solid contact
Relatively simple construction and operation	
Entrained flow	
Costly feed preparation needed for woody biomass	Slagging of ash
High temperatures give good gas quality	Very good scale-up potential
Only large-scale applications above about 10 t h^{-1}	Materials of construction
Carbon loss with ash	Low feedstock inventory
Produces tar-free gas and little methane	High conversion
Good gas–solid contact and mixing	High specific capacity
Twin fluid bed	
MHV gas produced with air, without oxygen	Catalyst can be added to bed
Complex and hence costly design	Scale-up complex
Moderate tar levels requiring cracking or cleaning	High specific capacity
Complexity requires capacities of greater than 5 t h^{-1}	Good gas–solid contact

Table 11.7. Typical gasifier characteristics (all air-blown).

Gasifier type	Reaction temp. (°C)	Exit gas temp. (°C)	Tars	Particulates	Turn-down	Scalability	Current max. ($t\ h^{-1}$)	Min. ($t\ h^{-1}$)	Max. (MWe*)
Fixed bed†									
Downdraft	1000	800	Very low	Moderate	Good	Poor	0.5	0.1	1
Updraft	1000	250	Very high	Moderate	Good	Good	10‡	1	20
Cross-current	900	900	Very high	High	Fair	Poor	1	0.1	2
Fluid bed									
Single reactor	850	800	Fair	High	Good	Good	10‡	1	20
Fast fluid bed	850	850	Low	Very high	Good	Very good	20‡	2	40
Circulating bed	850	850	Low	Very high	Good	Very good	20‡	2	40
Entrained bed	1000	1000	Low	Very high	Poor	Good	20‡	5	40
Twin reactor	800	700	High	High	Fair	Good	10‡	2	20
Moving bed									
Multiple hearth	700	600	High	Low	Poor	Good	5	1	10
Horizontal moving bed	700	600	High	Low	Fair	Fair	5	1	10
Sloping hearth	800	700	Low	Low	Poor	Fair	2	0.5	4
Screw/auger kiln	800	700	High	Low	Fair	Fair	2	0.5	4
Other									
Rotary kiln	800	800	High	High	Poor	Fair	10‡	2	20
Cyclone reactors	900	900	Low	Very high	Poor	Fair	5	1	10

* At 36% overall efficiency.
† Also referred to as moving bed because the biomass does move relatively slowly.
‡ Current maxima but capable of scaling up to larger capacities.

8.4 *Products of gasification*

The products of gasification will vary according to the reactor configuration and oxidant used. Ideally, there is complete conversion of all tars, hydrocarbons and char in the gasifier to give fuel gas. However, reactor design can give rise to incomplete oxidation, the extent of which is mostly determined by reactor geometry. Typical gasifier features are summarized in Table 11.7 and were compared in Table 11.6. Product characteristics are summarized in Table 11.8 and typical gas compositions are summarized in Table 11.9.

8.5 *Current status*

Table 11.10 summarizes all known recent and current activities on biomass gasification around the world that are either at a demonstration or commercial scale or have been developed to a point where they can, in principle, be demonstrated [22]. This is included to show the extent of activity that has taken place over the last 5 years. The table identifies those processes (in bold typeface) that have either been developed to a point where they can be potentially implemented at a substantial scale of 5 MWe or more or are being seriously considered. There are few technology types that are

Table 11.8. Gasification product gas characteristics.

	Capacity ($t\,h^{-1}$)	Capacity* (MWe)	Feed†	HHV ($MJ\,m^{-3}$)	Gas quality‡	Outlet temp. (K)
Downdraft, air	0.1–0.7	0.2–1.4	1	4–6	4	700–1000
Downdraft, oxygen	1–5	2–10	1	9–11	4	700–1100
Updraft, air	0.5–10	1–20	2	4–6	3	100–400
Updraft, oxygen	1–10	2–20	2	8–14	3	100–700
Single fluid bed, air	0.5–15	1–30	4	4–6	3	500–900
Single fluid bed, oxygen	2–10	4–20	4	8–14	3	700–1100
Single fluid bed, steam	1–10	2–20	4	12–18	3	700–900
Circulating fluid bed, air	2–20	4–40	3	5–6.5	2	700–1100
Circulating fluid bed, oxygen	2–20	4–40	3	10–13	3	800–1200
Twin fluid bed	1–10	2–20	4	13–20	3	750–1000
Cross-flow, air	0.1–0.5	0.2–1	2	4–6	1	600–900
Horizontal moving bed, air	0.5–5	1–10	5	4–6	2	300–800
Rotary kiln, air	1–10	2–20	5	4–6	2	600–1000
Multiple hearth	1–20	2–40	3	4–6	2	400–700
Secondary processing	–		–	–	5	1000–1200

* Conversion at 36% overall efficiency.
† Specificity: most specific = 1; least specific = 5.
‡ Relative assessment in terms of tars and particulates in raw gas: worst = 1; best = 5.

Table 11.9. Typical product gas compositions from different gasifiers.

	Gas composition (dry vol.%)					HHV ($MJ\,m^{-3}$)	Gas quality	
	H_2	CO	CO_2	CH_4	N_2		Tars	Dust
Fluid bed air-blown	9	14	20	7	50	5.4	Fair	Poor
Updraft air-blown	11	24	9	3	53	5.5	Poor	Good
Downdraft air-blown	17	21	13	1	48	5.7	Good	Fair
Downdraft oxygen-blown	32	48	15	2	3	10.4	Good	Good
Multi-solid fluid bed	15	47	15	23	0	16.1	Fair	Poor
Twin fluid bed	31	48	0	21	0	17.4	Fair	Poor
Pyrolysis (for comparison)	40	20	18	21	1	13.3	Poor	Good

not being exploited. There is only the pressurized indirectly heated fluid bed and the atmospheric entrained flow system that are not being developed.

9 Gas clean-up

9.1 *Introduction*

Gases formed by gasification will be contaminated by some or all of the constituents listed in Table 11.11. The level of contamination will vary, depending on the gasification process and the feedstock. Gas cleaning must be applied to prevent erosion, corrosion and environmental problems in downstream equipment.

Table 11.10. Recent and current gasification processes.

Company	Country	Gasifier type
Aerimpianti (TPS process)	**Italy**	**Circulating fluid bed**
Ahlström	**Finland**	**Atmospheric circulating fluid bed**
Arizona State University	USA	Twin fluid bed
Battelle Columbus	**USA**	**Twin fluid bed**
Battelle PNL	USA	Wet gasification
Bioflow (Ahlström/Sydkraft)	**Finland**	**Pressure circulating fluid bed**
Bioneer	**Finland**	**Updraft**
Ebara	Japan	Twin fluid bed
EFEU	Germany	Fixed bed with cracking
General Electric	USA	Updraft
Gotaverken	**Sweden**	**Circulating fluid bed**
Hitachi	Japan	Updraft
IGT	**USA**	**Pressure oxygen fluid bed**
JWP Energy Products (EPI)	USA	Fluid bed
LNETI	Portugal	Fluid bed
Lurgi GmbH	**Germany**	**Circulating fluid bed**
Manzano/Linz	Italy	Updraft
MTCI	**USA**	**Fluid bed**
NEI Fluidyne	New Zealand	Downdraft
Sofresid/Caliqua	**France**	**Updraft**
Southern California Edison	USA	Downdraft
Southern Electric Int.	USA	Fluid bed
Stein Industrie (ASCAB)	France	Pressure oxygen fluid bed
Tampella Power	Finland	Pressurized fluid bed
TNEE	France	Twin fluid bed
TPS (Studsvik)	**Sweden**	**Circulating fluid bed**
Tsukishima	Japan	Twin fluid bed
University of Sherbrooke	Canada	Fluid bed
Veba	Germany	Entrained flow
Ventec	UK	Downdraft
Voest Alpine	**Austria**	**Updraft**
Vølund	**Denmark**	**Updraft**
VUB	Belgium	Fluid bed
Wellman	**UK**	**Updraft**

Table 11.11. Fuel gas contaminants and their problems.

Contaminant	Examples	Problems
Particulates	Ash, char & fluid-bed material	Erosion
Alkali metals	Sodium & potassium compounds	Hot corrosion
Fuel-bound nitrogen	Mainly ammonia & HCN	NO_x formation
Tars	Refractive aromatics	Clog filters
		Difficult to burn
		Deposit internally
Sulfur, chlorine	HCl & H_2S	Corrosion
		Emissions

9.2 *Hot gas clean-up for particulates*

Gas streams from biomass gasification contain very small carbon-containing particles that are difficult to remove by cyclones. Tests using high-efficiency cyclones showed that particulate levels were not reduced to less than 5–30 g N^{-1} m^{-3} [30]. For this reason, barrier filtration methods, such as sintered metal or ceramic filters, are preferred. This is particularly important for pressurized systems, where the sensible heat of the gas needs to be retained as well as avoiding scrubbing systems for tar removal.

High-temperature ceramic or metal candle filters have been tested with gasification products from peat and coal. Many designs do not give a constant pressure drop, but this increases as the deposits build up. One solution is to layer the filters where removal efficiencies in excess of 99.8% have been reported. Tests on wood-derived gases have presented a further problem, with filter clogging by soot caused by thermal cracking of tars both in the gas phase and on the filter surface. This problem can be reduced by cooling the gas to below 500 °C and reducing gas face velocities across the filter surface. However, if temperatures fall below 400 °C there is still a potential problem of tar deposition. Recent developments employ ceramic candle filters with automatic pulsing to strip off the accumulated filter cake.

9.3 *Tar cracking*

9.3.1 INTRODUCTION

Tar concentration is mainly a function of gasification temperature, with reducing tar levels as temperature increases. The relationship between temperature and tar level is a function of the reactor type and processing conditions. The tars formed in pyrolysis are thermally cracked in most environments to refractory tars, soot and gases.

Tar levels and characteristics are also dependent on the feedstock. Tests have shown that tar production in wood gasification is much greater than in coal or peat gasification and that the tars tend to be heavier, more stable aromatics. These may partially react to give soot, which can block filters and appears to be a problem peculiar to biomass gasification. This implies that technology developed in coal gasification tar cleaning may not be directly transferable to biomass feeds.

There are two basic ways of destroying tars [2]:

- by catalytic cracking using, for example, dolomite or nickel (reviewed in [2]);
- by thermal cracking, e.g. by partial oxidation or direct contact.

9.3.2 CATALYTIC CRACKING

Pilot scale tests have shown that catalytic cracking of tars can be very effective. Tar conversion in excess of 99% has been achieved using dolomite, nickel-based and other catalysts at elevated temperatures of typically 800–900 °C. These tests have been performed using both fossil and renewable feeds.

Most reported work uses a second reactor. Some work has been carried out on incorporation of the catalyst in the primary reactor, which has often been less successful than the use of a second reactor (e.g. [31]), although this approach has been selected for the Biopower plant at Varnamo [22]. Elevated freeboard temperatures thermally crack tars and can reduce the load on the catalytic cracker.

Catalyst deactivation is generally not a problem with dolomite. An initial loss of activity is sometimes experienced as carbon compounds settle on the catalyst, but these compounds gasify as the bed temperature rises and the catalyst is reactivated. Metal catalysts tend to be more susceptible to contamination. Low hydrogen concentrations in the product gas will reduce the catalytic activity of metal-based systems. The low sulfur content of biomass gases can reduce the activity of metal sulfide catalysts by stripping out the sulfur.

9.3.3 THERMAL CRACKING

Tests on a fluid-bed peat gasifier at VTT have shown that tar levels can be reduced to levels found in downdraft systems by thermal cracking at 800–1000 K [5]. However, biomass-derived tars are more refractory and are harder to crack by thermal treatment alone. As indicated above, elevated freeboard temperatures in fluid-bed gasifiers provide some thermal tar cracking.

There are several ways of achieving thermal cracking:

- Increasing the residence time after initial gasification, such as in a fluid-bed reactor freeboard, but this is only partially effective.
- Direct contact with an independently heated hot surface, which requires a significant energy supply and thus reduces the overall efficiency. This is also only partially effective due to reliance on good mixing.
- Partial oxidation by addition of air or oxygen (e.g. [7]). This increases CO_2 levels, reduces efficiency and increases cost for oxygen use. It can be very effective, particularly at the high temperatures of 1300 °C or more achieved with oxygen gasification.

9.4 *Tar removal*

9.4.1 WATER SCRUBBING

Water scrubbing is widely assumed to be a proven technique for physical removal of particulates, tars and other contaminants. Unfortunately, most experience is not so reassuring and there are many reported problems, particularly in the poor removal efficiencies of tars, although surprisingly little hard data are available [32]. Tars require more physical capture and agglomeration or coalescence rather than simple cooling. Biomass-derived tars are known to be very difficult to coalesce and a complex treatment system is likely to be required even to attain tar removal levels of 90%.

A typical system will include a saturator to cool and saturate the gas for coalescence of particulates and tars in the next stage. A high-efficiency scrubber then follows to intimately contact the contaminants and reduce the pressure so that the water will condense onto the particulates and tar droplets, thus increasing their size and improving their susceptibility to agglomeration and coalescence. The final stage is to provide a high residence time tower to allow the system to equilibrate. Tar levels down to 20–40 mg N^{-1} m^{-3} and particulate levels down to 10–20 mg N^{-1} m^{-3} can be achieved with such a system. Soluble gases, such as ammonia, and soluble solids, such as sodium carbonate, are effectively removed.

These systems are fairly expensive and create a waste disposal problem by generating large quantities of contaminated water. The wastewater can usually be treated by conventional biological processes unless there is a high recycle ratio, when more concentrated solutions will be produced requiring special disposal.

Cooling the product will also reduce electrical efficiency, but it is essential for applications in engines to provide the highest energy density gas.

9.4.2 OIL SCRUBBING

A few attempts have been made to scrub with oil, which was thought more likely to capture the tars. The consequential problems outweighed any benefits.

9.4.3 ELECTROSTATIC PRECIPITATORS

This is an effective but costly way to remove tars. There is little experience on biomass-derived gasification products.

9.5 *Alkali metals*

Alkali metals exist in the vapor phase at high temperatures and will therefore pass through particulate removal devices unless the gas is cooled. The maximum temperature that is considered to be effective in condensing metals is around 600 K. Tests on alkali species have shown that their gaseous concentrations fall with temperature to the extent that concentrations are close to turbine specifications at temperatures below 500–600 K [10]. Thus it is possible that gas cooling to this level will cause alkali metals to condense onto entrained solids and be removed at the particulate removal stage. Alkali metals may also damage ceramic filters at high temperatures. A hot gas clean-up system will thus first have a cooler before the hot gas filter.

Alkali metals cause high-temperature corrosion of turbine blades, stripping off their protective oxide layer. For this reason, it is widely believed that their concentration must not exceed 0.1 ppm at entry to the turbine. There is no experience with modern coated blades in such an environment.

Alternatively, or additionally, water scrubbing can be used as described above.

9.6 *Fuel-bound nitrogen*

About 50–80% fuel-bound nitrogen is converted to ammonia and lesser quantities of other gaseous nitrogen compounds during gasification. These compounds will cause potential emission problems by forming NO_x during combustion. There are three ways of approaching the problem of NO_x emissions, any of which may be used singly or in combination:

- reduce the formation of NO_x by limiting fuel-bound nitrogen in the feedstock through careful selection of biomass types and/or blending;
- use low-NO_x combustion techniques;
- use selective catalytic reduction (SCR) at the exhaust of the engine or turbine.

Nitrogen-containing contaminants all exist in the vapor phase and will therefore pass through all particulate removal devices. Catalytic conversion methods can sometimes remove ammonia but this is dependent on the catalyst. Some catalysts are reported to increase ammonia content by releasing the nitrogen bound in the tars.

Water scrubbing is effective in removing these soluble impurities but results in loss of sensible heat and thus poorer efficiencies.

Selective catalytic reduction involves a reaction between ammonia and NO_x to form nitrogen and water. This is a well-established technology and is often specified in exhaust gases from engines and turbines. There is, however, a cost and efficiency penalty.

9.7 *Sulfur*

Sulfur is not generally considered to be a problem because biomass feeds have a very low sulfur content. However, the specification on turbines is typically 1 ppm or often much less, and even lower if co-contaminants are present, such as alkali metals. Some gas compositions have reported 0.01% sulfur, which represents 100 ppm. Sulfur removal may therefore be necessary for turbine applications, which can often be achieved with a conventional sulfur guard. Dolomite (CaO.MgO), often used for tar cracking, will also absorb significant proportions of sulfur.

Sulfur concentrations will be lower than those produced in the combustion of fossil fuels, and hence expensive sulfur removal trains will not be necessary. If a dolomite tar cracker is included in the process, this will reduce the sulfur levels considerably but possibly not to the low levels required. A sulfur guard, consisting of a hot fixed bed of zinc sulfide, is likely to be adequate for the concentrations expected. This would be relatively inexpensive to install but would create a waste disposal problem in terms of the zinc sulfide produced.

9.8 *Chlorine*

Chlorine is another potential contaminant that can arise from pesticides and herbicides as well as in waste materials. Levels of 1 ppm are often quoted, but this is a function of the temperature, chlorine species, co-contaminants and materials of construction. The behavior of chlorine and metals at elevated temperatures is well understood. Chlorine and compounds can be removed by absorption in active material either in the gasifier or in a secondary reactor, or by dissolution in a wet scrubbing system. Dolomite and related materials are less effective at removing chlorine than sulfur.

9.9 *Summary of clean-up methods*

A summary of the contaminants and the methods for clean-up is given in Table 11.12.

Table 11.12. Gas cleaning specifications.

Contaminants	Clean-up method
Ash	Filtration, scrubbing
Char	Filtration, scrubbing
Inerts	Filtration, scrubbing
NH_3	Scrubbing, SCR
HCl	Lime or dolomite, scrubbing, absorption
SO_2	Lime or dolomite, scrubbing, absorption
Tar	Tar cracking
	Tar removal
Na	Cooling, condensation, filtration, adsorption
K	Cooling, condensation, filtration, adsorption
Other metals	Cooling, condensation, filtration, adsorption

10 Power generation interfacing requirements

10.1 *Definition*

The product gases from gasification of biomass may be used in either gas turbines or engines for the generation of electricity. This section considers gas quality requirements and control techniques that are required to make the use of gas turbines or engines a feasible and viable proposition.

The gas quality requirement for a gas turbine is known to be very demanding but poorly specified and without any evidence that the specifications are necessary or justified.

10.2 *Gas quality requirements*

10.2.1 TURBINE OPERATION

Some typical conventionally stated turbine fuel gas specifications are summarized in Table 11.13. It must be emphasized that this list indicates some of the known problem areas and the specifications are not definitive but serve to show the extent to which the fuel gas may have to be cleaned according to various sources. Indeed, some of the specifications may turn out to be much more strict, and some contaminants will need to be possibly up to 10 times lower, particularly in difficult combinations.

Known major problems are alkali metals and sulfur. Sulfur is not normally associated with biomass, but trace levels of, for example, 0.1 wt.% can lead to levels of sulfur in the fuel gas of up to 100 ppm. This level is not acceptable (see Table 11.13) and will require reduction. Alkali metals are a major component of the ash of many biomass forms and their effect on turbines is well known, although the particular nature of the biomass-derived alkali metals and their association with other contaminants such as sulfur are not known.

Solid inerts such as char and fluid-bed material will clearly have a deleterious effect on any moving parts and will require almost total removal. Suggested limits are indicated in Table 11.13.

Table 11.13. Some notional turbine fuel gas specifications.

Contaminants	Tolerance examples
Minimum gas heating value (LHV)	4–6 MJ N^{-1} m^{-3}
Minimum gas hydrogen content	10–20%
Maximum alkali concentration	20–1000 ppb
Maximum delivery temperature	450–600 °C
Tars at delivery temperature	All in vapor form or none
Maximum particulate (ash, char, etc.) level, particle size:	
>20 μm	0.1 ppm wt.
10–20 μm	1.0 ppm wt.
4–10 μm	10.0 ppm wt.
NH_3	No limit
HCl	0.5 ppm
S (H_2S + SO_2, etc.)	1 ppm
N_2	No limit
Combinations	
Total metals	<1 ppm
Alkali metals + sulfur	<0.1 ppm

Tars are a potential problem if the gas has to be compressed as in an atmospheric gasifier because they will deposit in the compressor. Pressurized gasifiers overcome this problem by removing the need for a fuel gas compressor, as the gas can be filtered hot and burned hot with the tars remaining in the gas phase and combusted. Tars otherwise will require cracking and/or removal as discussed above.

Chlorine is a difficult contaminant because it interacts with most metals at the temperatures involved in gasification and combustion. Changing from a reducing (as in the gasifier) to an oxidizing (as in the combustor) environment exacerbates the potential problem. The reactions between chlorine and most metals are well known and the operating regimes are well understood.

Biomass often contains nitrogen, particularly from bark and some special biomass forms; NO_x generated from fuel-bound nitrogen may cause problems and gas cleaning should, therefore, reduce traces of HCN and NH_3 to a minimum. This is adequately dealt with in a water scrubbing system, but in a pressurized system with hot gas filtration, a post-combustion catalytic process (SCR) would be required.

There will be a trade-off between increasing the gas cleaning to a high standard and increasing the maintenance cost of the turbine. This interaction has not been studied and no data are available.

10.2.2 ENGINE OPERATION

Engines have the advantage of higher tolerance to contaminants than turbines (e.g. up to 30 ppm tars can be tolerated). If the gas is compressed in a turbocharger there will be similar but possibly less demanding quality requirements on the gas. There are no robust data on the gas quality specifications.

10.3 *Control requirements*

10.3.1 TURBINE OPERATION

If the gasifier operates at atmospheric pressure, the product gas will require compression prior to combustion as well as the air. This imposes severe gas quality requirements to avoid damage to the compressor. The air supply to the gasifier would probably be provided independently, although a bleed from the air compression loop could be used. This latter choice would, however, require extensive compressor modifications and impose severe control problems on the system analogous to those for a pressure gasifier.

A pressurized gasifier would either use compressed air from the compressed air loop on the turbine set or have an independent air compressor. The latter solves some of the potential control problems that arise from integration of the gasifier operation with the turbine, but at the expense of higher cost and lower system efficiency.

10.3.2 ENGINE OPERATION

Engines present a more tolerant control requirement through the use of conventional fuel mixing devices and orthodox engine management systems. They will tend to react positively and quickly to variations in gasifier output without adversely affecting the gasifier operation. There is extensive practical experience of such systems from small-scale gasifier operations as well as from landfill gas operations.

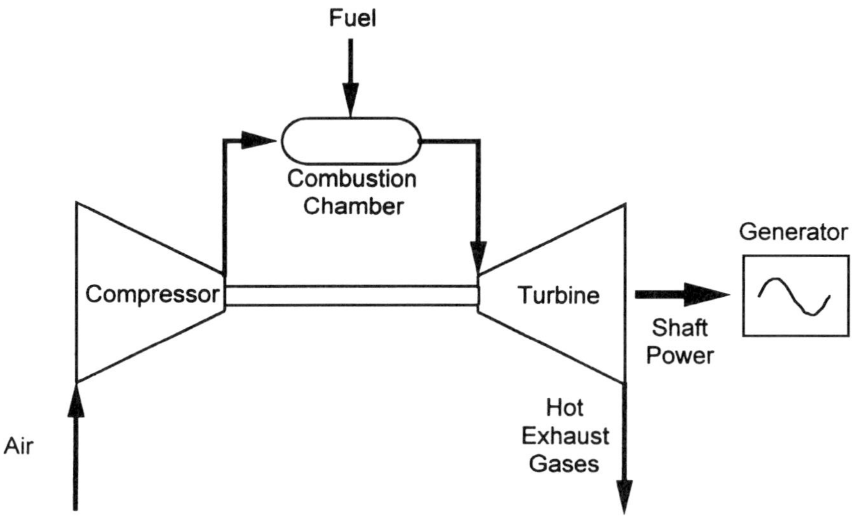

Figure 11.11. Simple gas turbine power cycle.

11 Power generation

11.1 *Definition*

The principle of power generation from a gas turbine is shown in Fig. 11.11. Air at ambient temperature and pressure is compressed in order to burn the compressed fuel gas. The hot exhaust is passed through a gas turbine where some of the heat energy is recovered as work energy, and the rest is discharged as waste heat. Some of the work energy is used to compress the air and the remainder is used to generate electricity.

The simple cycle shown is not very efficient because there is considerable energy wasted in the hot exhaust gases. Efficiency can, therefore, be increased by adding a heat recovery system after the gas turbine. These systems can either generate steam or preheat the air. The steam can power a steam turbine in a combined cycle mode, or the steam can be mixed with the combustion gases and fed through the gas turbine in a steam-injected gas turbine (STIG) cycle. The residual heat from the steam turbine or air preheater can be used as process steam or in district heating.

Gas turbines are proven in power generation when fueled by high-grade fossil fuels such as natural gas or liquid fuels such as diesel. Low-heating-value fuels such as gases formed in biomass gasification have not been demonstrated in gas turbines, although MHV producer gas from coal gasification has been used successfully at the Cool Water demonstration coal-integrated gasification/combined cycle (CIG/CC) 100 MWe facility in the USA [33].

11.2 *Biomass gasification to electricity system special requirements*

The special requirements of a biomass-based system are that if the gasifier operates at atmospheric pressure the product gas will require compression as well as the air. The air supply to the gasifier would probably be provided independently, although a bleed from the air compression loop could be used. This latter choice would require extensive compressor modifications and impose control problems on the system, thus a separate additional compressor is required.

A pressurized gasifier would either use compressed air from the compressed air loop on the turbine set or have an independent air compressor. The latter solves some of the potential control problems that arise from integration of the gasifier operation with the turbine, but at the expense of higher cost and lower efficiency.

Combined cycle operation will not only utilize the waste heat from the exhaust but also the heat recovered from the primary gas cooler before filtration. System optimization is thus a major requirement and requires careful consideration. The problem is exacerbated if a STIG is used or the gasifier requires steam.

11.3 *Fuel specifications and turbine/engine requirements*

Coal-integrated gasification capacities are far greater than the capacities proposed for biomass systems, typically ranging from 100 at a demonstration scale to 1000 MWe. The gas turbines used are bigger and more tolerant of gas contamination. Biomass-based systems are limited in size by the availability and collection costs of the resource. There are few advocates of biomass-based combined cycle power systems above 50–100 MWe, and few sites where biomass can be delivered in sufficient quantities: 40 dry ash free (d.a.f.) t h^{-1} at 45% efficiency for 100 MWe. Contaminant limits for systems based on biomass-integrated gasification (BIG) will have to be much stricter to ensure a long turbine life, as suggested in Table 11.13, although definitive limits have yet to be finalized.

11.4 *Fuel combustion*

11.4.1 FUEL CALORIFIC VALUE

The heating values of gasifier fuels are 4–10 times lower than those of conventional gas turbine fuels. This means that correspondingly more fuel will have to be burned to input an equivalent amount of heat energy. The additional throughput will mean that combustion chambers and burners will require modification, particularly to meet increasingly stringent environmental requirements. Contaminant limits may have to be tightened to account for the extra volume of fuel gas required.

11.4.2 PRODUCTION OF NO_x

A low flame temperature is predicted from combustion of LHV gas due to its dilution by nitrogen. This will reduce thermal NO_x production, which is not, therefore, expected to be an environmental problem. The composition of MHV gases, which are largely hydrogen and carbon monoxide with some methane, suggests that they will have a high flame temperature which could lead to NO_x problems. Medium-heating-value burners will have to be carefully designed to account for this or some form of dilution may be necessary, possibly by the nitrogen extracted in oxygen gasification, or by steam. Selective catalytic reduction processes are available for reducing NO_x in exhaust gas but there is an economic and energetic penalty.

Biomass often contains nitrogen, particularly from bark and some special biomass forms. The NO_x generated from fuel-bound nitrogen may cause problems and gas cleaning must therefore reduce traces of HCN and NH_3 to a minimum. Selective catalytic reduction may be required at the turbine or engine exhaust.

11.4.3 AUXILIARY FUEL CAPABILITY

It would be desirable for the turbine to have a dual fuel capability to supplement fuel gas and to ensure maximum availability. However, this will almost certainly require a new burner design, and also additional modifications to the combustion chambers.

11.4.4 INDIRECT FIRING

One way of avoiding the problems of gas cleaning is to burn the fuel in a separate combustor and heat the turbine gases indirectly via a high-temperature gas–gas heat exchanger (a Brayton cycle).

This approach is applied at the Free University of Brussels (VUB) combined heat and power (CHP) plant [34]. The main problems are the loss of efficiency caused by indirect heating of the turbine working gases and the temperature limits imposed by the heat exchanger materials, which limit the efficiency attainable. While indirect heating improves turbine reliability, there are costs incurred in flue gas treatment and the gas–gas heat exchanger.

11.5 *Energy recovery*

11.5.1 COMBINED CYCLE

Exhaust gases typically leave the gas turbine at temperatures of 500–600 °C for aero-derived turbines and 400–500 °C for industrial turbines, and so still have considerable thermal energy. This energy and other sources of high-temperature gas, such as the raw gas from the gasifier, can be recovered in a heat recovery steam generator (HRSG), which produces steam. This steam can be used in a steam turbine to generate extra electricity. The steam is then cooled and condensed in a cycle before passing once again through the steam generator. This combination of gas turbine and steam cycle is known as a combined cycle (CC). The waste heat from the steam turbine can be recovered as heat, e.g. for district heating.

Steam turbines and boilers are only normally economic in large-scale applications (>100 MWe), unless there are special circumstances such as low-cost feedstocks, and for this reason combined cycles are very sensitive to scale. Steam injection gas turbine cycles may be more appropriate to smaller generating systems, but there is no experience of such a system on biomass-derived gases.

11.5.2 STEAM INJECTION CYCLE

In steam injection cycles the HRSG raises steam, which is then mixed with the compressed air in the gas turbine cycle. Steam is injected at the combustion chamber and at points before entry to the gas turbine. This increases the gas turbine throughput and hence the power output. The exhaust gas stream will require a flue gas condenser. This system is commercially available in natural-gas-fired aero-derived gas turbines, although there is no experience in LHV applications. In such duties the steam may dilute the fuel to the point where combustion is unstable. There is also the possibility that turbine capacity may be exceeded.

This system uses the exhausted waste heat effectively without the need for large boilers or a steam turbine, which makes it insensitive to scale. It is favored because it is claimed to be the most economic option in small-scale facilities. There is, however, the problem of ensuring an adequate water supply.

Table 11.14. Turbine options.

Industrial	Lower efficiency More tolerant to contaminants Higher cost Higher availability Higher maintenance costs
Aero-derived	Higher efficiency Less tolerant to contaminants Lower cost Lower availability Lower maintenance costs

11.5.3 EVAPORATIVE CYCLES

Evaporative cycles are similar to STIG except that water is injected instead of steam. This is vaporized in the gas–gas heat exchanger. Performance is reported to be better than STIG or combined cycles, but the cycles are less established.

11.5.4 COMBINED HEAT AND POWER

In all the cycles noted, excess heat or steam can be used as process steam or in district heating. This will increase the overall system efficiency at the expense of additional capital cost.

11.6 *Status*

There are two basic machines for generating power: turbines and engines. There is no clear allocation of choice of machine and size of system, but the orthodox view is that engines are more suitable up to 5–10 MWe and turbines above 10–20 MWe for an atmospheric pressure gasifier and above 20–30 MWe for a pressure gasifier. Engine gen-sets are, however, available up to 50 MWe, and gas turbines have been successfully used at 3 MWe. Turbine options are listed in Table 11.14.

Turbines become more attractive at larger sizes, particularly for IG/CC and similarly advanced cycles when higher efficiencies can be achieved and economies of scale become more noticeable.

Engines have the advantages of robustness, high efficiency at low sizes, higher tolerance to contaminants than turbines (e.g. up to 30 ppm tars can be tolerated), easier maintenance and a wide acceptability. However, operation in CC mode is rarely justified because only a small increment in efficiency can be gained. There is poor economy of scale because capacity is more a function of number of cylinders than size of cylinders, and constant specific capital costs that are independent of size are typical.

12 Environment

12.1 *Introduction*

The operation of a gasification plant can result in occupational health and safety hazards unless adequate and effective preventative measures are taken and continuously enforced.

Table 11.15. Environmental aspects of gasification systems.

Process activity	Fuel preparation	Feeding system	Gasifier	Gas cleaning	Gas utilization
Environmental concerns					
Dust	*	*		*	
Noise	*	*	*	*	*
Odor	*		*	*	
Wastewater				*	*
Tar				*	*
Fly ash				*	
Exhaust gases					*
Hazards					
Fire	*	*	*	*	*
Dust explosion	*	*	*		
Mechanical hazard	*	*	*		*
Gas poisoning		*	*	*	*
Skin burns			*	*	*
Gas explosion			*	*	*
Gas leaks			*	*	*

A gasification system consists of:

- fuel storage, handling and feeding system;
- the gasifier, gas cooling and cleaning equipment;
- utilization of the gas.

Each part of the plant creates specific occupational health and safety hazards. Table 11.15 describes the main environmental concerns and major hazards associated with the operation of a gasification system.

This section examines the sources of environmental concerns, describes the measures that have to be taken to limit environmental impact and only considers in-plant environmental factors concerning gas, liquid and solid emissions and wastes. The major hazards are then discussed and safety guidelines are provided for appropriate operation of gasifier systems.

12.2 *Environmental aspects of gasification operations*

12.2.1 DUST

Dust is generated during feedstock preparation, storage and handling, feeding and fly ash removal by particulate collection equipment. The handling of solid materials is a notorious source of airborne particles, especially when the solids are dry and friable. Dust generation creates several problems, including:

- Airborne or entrained dust may form explosive mixtures with air, in which a primary explosion can render the dust airborne causing secondary explosions that can be devastating.
- The inhalation of dust is a potential source of lung damage.
- Eye and skin irritation may occur.
- Layers of combustible dust may cause smells or they may smoulder and ignite.

• Dust settlement on all exposed horizontal surfaces leads to safety problems for personnel in routine operations, as well as increased maintenance and aesthetic detraction.
• Increased friction and wear of mechanical equipment caused by dust deposition increases costs and reduces reliability, both increasing the potential for accidents.
Preventative measures include:
• minimization of solids handling and avoidance of rough handling to minimize attrition of fuel particles and suspension of dust;
• complete enclosure of all solids handling, particularly conveying equipment at the discharge points;
• installation of suction hoods and gas cleaning equipment to control localized dust sources, e.g. mills and screens;
• maintenance of an underpressure in enclosed environments to prevent the spreading of dust into adjacent premises, again with suitable gas cleaning equipment.

Solid particles such as cinders, fly ash, filter dust, charcoal, fluid-bed inerts and catalyst fines also arise in the product gas from gasification. Because such sources are localized they are, in principle, easier to control. In the discharge of fine material such as fly ash, they may need to be wetted to prevent re-entrainment during handling and disposal. Carbon formed by secondary cracking or incomplete gasification can also form explosive mixtures with air, but it is usually contained in appropriate vessels. Charcoal from biomass can be pyrophoric and needs to be cooled adequately prior to discharge and storage if arising in significant amounts. Some types of gasifiers may produce hot particles as a consequence of malfunction or equipment faults. These may ignite flammable materials and cause a fire.

From an occupational health viewpoint, dust particles may be classified by:
Size Particles larger than 5 μm (0.005 mm) are arrested by wet hairs in the nostrils. Those smaller than 0.2 μm (0.0002 mm) do not settle in the lungs and are breathed out again. Thus, the intermediate size range is the most dangerous.
Shape and composition Some materials are known to cause lung damage, for example asbestos (asbestosis) and silica (silicosis). The latter may arise from fluid-bed materials.

Dust originating from fly ash removal may be toxic due to adsorption of chemicals onto the dust particles. Several compounds with carcinogenic properties, such as benzo[*a*]anthracene and benzo[*a*]pyrene, are adsorbed onto the dust particles. They are dangerous to human health either after inhalation and skin contact and/or after accumulation in the food chain. Dust particles may also adsorb non-polar organic compounds in up to 40 wt.%, which may be higher for soot and carbon black with their very high specific surface areas. The dispersion of gasifier dust may lead to air and food contamination.

12.2.2 WASTEWATER AND CONDENSABLES

Wastewater and condensates may be produced during gas cooling and wet gas cleaning. The condensate is known to contain, for example, acetic acid, phenols and many other organic compounds. There is a risk of water pollution and adverse health effects from the tars and soluble organics.

The condensate and wastewater consist mainly of water and can be divided into an aqueous, i.e. water-soluble, and a non-water-soluble fraction consisting of tars and oils. Separation, however, is not always simple because wood tar tends to emulsify in the aqueous phase. The insoluble fraction consists mainly of tars, fly ash, phenolic compounds and light oils.

The tars in particular, as well as the condensates, are toxic and require careful evaluation of their occupational and safety aspects. Little research has been carried out to determine the mutagenic and carcinogenic effects of biomass tar, but research on coal tar has confirmed the above reservations. It is safe to assume that some of the tar components may be carcinogenic. High-temperature gasifier operations can increase this problem because the mutagenic and carcinogenic effects are related to the presence of polycyclic aromatics and their relative concentration increases as process temperature increases. Direct contact between skin and tars or condensates should be avoided by appropriate clothing and training.

Tars present an insignificant fire hazard because their flash points are comparatively high. Tar disposal has not been examined. It may be recycled to the gasifier or incinerated. Other disposal options are unlikely to be acceptable.

Because tars are such a potential problem in wastewaters and in their own right, every effort should be made to reduce their environmental impact by:

- cracking tars during or after gasification;
- applying hot gas clean-up and so avoiding wet gas scrubbing;
- reducing gasification temperatures to limit the production of refractive polycyclic tars.

Pressurized gasifiers are assumed to operate with hot gas clean-up with no wastewater and no tar production. Atmospheric pressure gasifiers are more likely to include a wet scrubbing system, particularly if an engine is specified for power generation.

Wastewater treatment is usually assumed to be relatively simple and of low cost, although there is remarkably little information on treatment methods or costs [32]. The design of the wastewater treatment plant would be expected to rely partly on chemical treatment, such as solvent extraction of phenolics with incineration of recovered organics, and partly on orthodox biological treatment of the dilute, low biochemical oxygen demand (BOD) phase.

12.2.3 FLY ASH AND CHAR

Fly ash and char present similar problems to those caused by dust, as described above. There is an additional risk of fire that dictates that fly ash and char should be stored moist. Disposal of this wetted mixture presents its own environmental problems. The solids need to be separated from the water in a water treatment facility. Extracted water will be contaminated and may require further treatment before discharge using orthodox water treatment technology. There are no known special problems. The solid fraction should be considered an industrial waste and discharged accordingly to licensed landfill sites.

12.2.4 ODOR PROBLEMS

Odors may arise because of:

- the degradation of organic matter (e.g. in refuse or sewage gasification);
- the occurrence of even minute gas leaks;
- the handling and storage of tar, wastewater, fly ash and other by-products.

Wood tar has a strong, characteristic and persistent odor, even in minute concentrations. The smell of coal tar is somewhat aromatic due to the presence of naphthalene, anthracene and phenanthrene. Tar derived from lignocellulosic feedstocks is more pungent.

When sulfur- or nitrogen-containing feedstocks are used, the product gas also contains odorous gases, such as H_2S, COS or NH_3. The tar and wastewater may be contaminated by even more strongly smelling organic sulfur and nitrogen compounds, although this is unlikely to be a problem with biomass-derived products owing to the very low levels of sulfur and nitrogen.

12.2.5 NOISE

Noise is produced whenever a mechanical part or an engine or motor is in operation. Particular plant areas where noise levels are likely to be significant are:

- reception, storage and handling equipment;
- the feeding system;
- the compressors;
- the gas turbine or engine.

The effects on humans of prolonged exposure to noise are well documented. Adequate measures must be taken to minimize noise, e.g. by using sound and vibration absorbent materials between supports, or acoustic enclosure. Operators are also required to be provided with ear protection plugs.

12.3 *Hazards of gasifier operation*

12.3.1 COMBUSTIBLE GASES AND VAPORS

A flammable gas is combustible only within a certain range of concentrations, bounded by the lower explosion limit (LEL) and the upper explosion limit (UEL): below the LEL, the mixture is too lean to sustain combustion; above the UEL, the reaction stops because of a deficiency in oxygen. In both cases, the generation of heat becomes too slow to give rise to the characteristic acceleration in reaction rates that marks the start of an explosion. The range between the LEL and UEL values depends on the reactivity of the flammable compound or mixture, and widens when the flammable gas or the combustion air is preheated or under pressure. Some data are given in Table 11.16.

Explosive mixtures could arise when:

- Air leaks into the gasifier plant as a consequence of a reduction in operating pressure. Reduced pressure may arise due to rapid cooling, condensation of vapor such as water, chimney effects, the suction of an induced draft fan or of an engine.
- Fuel gas leaks out of the gasifier plant into a confined space, thus building up a substantial concentration in an enclosed space. A source of ignition is necessary for an explosion, so explosion-

Table 11.16. Some LEL and UEL values and self-ignition temperatures in air.

	LEL (vol.%)	UEL (vol.%)	Self-ignition temperature (K)
Hydrogen	4.0	76	400
Carbon monoxide	12.5	74	*
Methane	4.6	14.2	540
Ethane	3.0	12.5	515
Ethene	3.1	32.0	490
Propane	2.2	9.5	450

* This temperature is highly dependent on the presence of traces of moisture.

proof or flame-proof or spark-proof motors would be specified in any such areas. In addition, such an atmosphere is also likely to present a lethal toxicity hazard from carbon monoxide, so suitable detectors should be fitted.

When a flammable mixture of gas and air is formed, an explosion may occur when the mixture is ignited. Ignition may occur from static electricity, sparking equipment such as motors or contact with a hot surface. In view of the wide explosion limits of the main components of product gas — hydrogen and carbon monoxide — the accidental formation of explosive gas mixtures should be prevented. Mixtures of product gas with oxygen-enriched air or pure oxygen have a higher UEL compared to mixtures with air. The LEL does not change significantly. This means that oxygen gasifiers present an even higher explosion risk, e.g. when oxygen breaks through the fuel layer or there is a perturbation in the fuel supply.

12.3.2 COMBUSTIBLE DUSTS

Combustible solids such as wood, flour and coal dust, with their very small particles, can also form explosive mixtures with air within certain concentration limits. These boundaries usually range from an LEL of 20–50 g m^{-3} and a UEL of 2–6 g m^{-3}. Numerous carbohydrate materials, including starch, sugar and wood flour, have given rise to extremely destructive explosions.

12.3.3 FIRE RISKS

The main fire risks in gasifier systems are associated with:

- fuel storage;
- combustible dusts formed in fuel comminution;
- fuel drying (in forced draft conditions a fire is likely to expand quickly);
- ignition procedure (especially for moving bed gasifiers);
- the product gas.

There are also the usual risks associated with any construction involving a thermal unit. Local rules and guidelines should be followed with construction and material selection of buildings. Adequate means for firefighting should be provided and the gasifier operators should be well acquainted with their existence, location and operating instructions. Such fires can be avoided with proper procedures and proper layout of the plant.

12.3.4 CARBON MONOXIDE POISONING

Carbon monoxide is a major constituent of product gas and is by far the most common cause of gas poisoning. It is particularly noxious due to the absence of color or smell. The accepted threshold limit value (TLV) is 50 ppm CO (0.005 vol.%), although concentrations and exposure are closely linked. There is extensive documentation available on the effects, treatment and controls that are laid down by the relevant statutory authorities.

Because carbon monoxide is an odorless, colorless gas, it may only be detected through instrumentation, and personal detectors are necessary when working in confined spaces. All operating personnel should be aware of the hazards presented by the gas. The best way of avoiding the risk of carbon dioxide poisoning is to build the gas generator in the open with the minimum of containment and with adequate ventilation, particularly where gases may collect.

12.3.5 OTHER TOXIC COMPOUNDS

It is well established that extremely toxic dioxins and furans (i.e. polychlorinated) are formed during most combustion and gasification processes when some chlorine is present. Under normal operating conditions the concentration of these compounds in wood-fired units is extremely small, although pesticide-treated wood and waste materials can provide the source of chlorine necessary for their formation. A close-coupled IG/CC system will normally provide satisfactory operating conditions for thermal destruction of these compounds and it is generally believed that these compounds do not present any problem.

12.3.6 OTHER HAZARDS

These include:
- skin burns;
- mechanical hazards;
- electrical hazards.

Some of the surfaces of the gasifier, the cyclones, the gas lines, the engine and its exhaust may get hot during operation and thus create a hazard to personnel for skin burns. For permanent stationary installations such surfaces should be insulated to protect the operators and also to reduce heat losses. Covers or rails should be installed to keep personnel at a safe distance. All equipment with moving parts, such as blowers, fans, screw conveyors, front-end loaders, pretreatment and feeding equipment, etc., present a hazard from moving parts and should be suitably protected. All electric appliances provide the potential for electric shocks and suitable precautions need to be taken using standard procedures and equipment specifications.

Any work on elevated equipment involves the hazard of possible falls. Similarly, there is some hazard from falling objects, tripping on hot equipment or slipping on oil-stained floors. Unauthorized, inexperienced or untrained personnel should be prevented from entering the plant to reduce the risk of injury through improper use of equipment or facilities.

For all 'normal' hazards that may arise in a process plant there are well documented and statutory requirements.

12.4 *Conclusions*

The gasification system has to be designed to meet all local environmental and safety requirements. If it is operated correctly and no accidents occur, then the environmental impact will normally be acceptable. Safety design is a major consideration and there are a range of precautions that have to be included and provided for that will be defined internationally, nationally and locally. In addition, an important aspect of safety and environmental management is good training. The general factors are outlined above but there may be additional specific requirements such as:
- dust from feed handling;
- dust explosions;
- gas explosions;
- carbon monoxide poisoning;
- tars and wastewater management;
- solids disposal;
- noise.

All of these factors can be adequately managed through good design and operation practice.

13 Status and technical uncertainties

13.1 *Typical system*

From the review and evaluation of the technologies, processes and systems contained in this chapter, a typical system consists of four main areas:

- Feed pretreatment.
- Gasifier.
- Gas clean-up.
- Turbine or engine.

13.2 *Pretreatment*

The pretreatment steps are relatively well established with a high level of reliability from experiences gained in the pulp and paper industry. The need to minimize capital cost with the lower quality specifications for a fuel product than paper feed permit a less demanding design.

13.2.1 STORAGE AND HANDLING

Wood as forestry waste or short-rotation coppice would normally be delivered in bulk as whole-tree chips. Wood waste, if acceptable as a feedstock, might be delivered in plank form requiring comminution. There are no perceived problems in handling or storing wood. This is common practice in pulp and paper mills throughout the world and in many smaller biomass combustion systems that operate in many countries.

13.2.2 DRYING

The drying requirement depends on the gasifier feed specification. Wet wood at typically 50% moisture wet basis is generally considered too wet, giving rise to a much dirtier gas, condensation problems and lower efficiencies. Drying to 15–25% is considered acceptable in energy and cost terms.

Drying may be carried out in the field and in the storage pile, but this is slow, unreliable, causes loss of material from biological degradation and can cause fires. Rotary kilns are widely specified as dryers using waste heat and/or combustion of biomass feed, perhaps as screenings or fines, again depending on the gasifier feed specification. Fluid-bed, silo and steam dryers have all been used successfully for biomass. None is very efficient, however, and the energy and economic costs are high, but these are outweighed by the higher downstream gas cleaning requirements consequent on not drying. The operation is well established with extensive experience to draw on.

13.2.3 PARTICLE SIZE CONTROL

Different gasifiers have different feed requirements. Fluid beds and circulating fluid beds are the most tolerant to particle size range, while fixed-bed gasifiers require regularly sized and relatively large particles with a minimum of fines. Entrained-flow gasifiers require a smaller particle size, which can increase technical and economic problems due to the difficulty in comminuting or grinding wood to small sizes. Comminution and screening are well-established operations in the pulp and paper industry and there are minimal uncertainties.

13.3 *Gasification*

Biomass gasification has been practiced for over 100 years, but with little commercial impact due to competition from other fuel sources and other energy forms. In the last 20 years, there has been a renewed interest worldwide, with many instances of substantial demonstration and commercial-scale plants. In particular, the last few years have seen a major resurgence of interest in large-scale biomass gasification processes, mostly due to environmental and political pressures required for CO_2 mitigation measures. Very few processes have proved economically viable, although the technology has progressed steadily. There is sufficient expertise and knowledge now available to have a very high level of confidence in modern gasification processes. The comments below refer to potential problem areas where concerns have been expressed or where special attention should be directed.

The recent attention for environmental reasons has created interest in major organizations who have the resources to thoroughly develop and market suitable technologies that meet the environmental and political requirements. The result has been consolidation of interest at an industrial level and substantial speculative investment in these technologies of the future.

13.3.1 FEEDING

Biomass has a number of peculiar properties that must be considered in designing feeding systems that relate to its grain structure. In devising handling and feeding systems where gas-tight seals are required, provision must be made for particles to fall away or be swept aside because blockage will result in major physical deformation. This is well known but poorly understood. Pressurized gasifiers are a special and extreme example of this problem where the feeding system can cost more than the gasifier.

Another problem, particularly with pressurized gasifiers, is the inert gas requirement, which can be considerable from purging feeders due to the high voidage of most bulk biomass. A commercial plant might consider recycling carbon dioxide from gas combustion, for example, rather than purchasing inert gas in bulk.

13.3.2 GASIFICATION

Recent large-scale demonstration and commercial biomass gasification plants have focused on fluid beds and circulating fluid beds rather than fixed beds [1]. This is due to a variety of reasons, including scalability, feed specification tolerance and controllability, which all favor fluid-bed and circulating fluid bed gasifiers. A 10 MWe IG/CC system, for example, might require one circulating fluid bed gasifier but four fixed-bed gasifiers, although the overall costs may be similar.

The advantages and disadvantages of the different types of gasifier have been described in Table 11.6 and systems that are currently available or are being developed are summarized later. It can be concluded that there are no perceived obstacles to successful demonstration of an advanced biomass gasifier.

13.3.3 ASH REMOVAL

The quantity of ash requiring removal and disposal from a biomass gasifier is relatively small, at typically 1–2% of the dry feed weight. Removal from the gasifier will vary according to the type of system. Fixed beds will usually have a rotating grate with screw or mechanical discharge from the base of the reactor. Fluid beds may have an overflow arrangement or extraction from the bed as a 'bleed', whereas circulating fluid beds will take a side-stream off from an appropriate place in the circuit. Each process will have its own proprietary system. Reliability is a function of the experience gained by the developer and the mode of ash removal.

Secondary and tertiary ash removal will arise from cyclones, hot gas filters and water washing systems. Apart from hot gas filters, where little operating experience has been obtained, these systems are well understood and reliable.

13.3.4 HEAT RECOVERY

The product gas will usually be hot, ranging from around 800 °C up to 1100 °C. It will need to be cooled before a hot gas filter to around 500–600 °C, or even lower if water washing is the first gas cleaning step. This provides the opportunity to recover heat as steam for combined cycle operation when up to 10% of the total energy content of the feed might be recovered. Particular care is needed to avoid tar deposition or fouling of the heat exchanger surfaces with ash, char or any other contaminants. Primary raw gas cleaning is thus very important.

The specification of the entire heat recovery and gas cleaning train requires careful evaluation and optimization, and generalizations other than identification of problems are not possible. The most convincing evidence of plausible design is the data from extensive operation with a quantitative appreciation of deviations from ideality.

13.4 *Gas cleaning*

This section of the overall system has received the least exposure to large-scale and long-term operation and so is the least certain aspect of the system.

13.4.1 HOT GAS CLEAN-UP

Much has been written and assumed about the effectiveness and performance of hot gas filters that cannot be substantiated for biomass-based systems. This is probably the least developed aspect of the entire system. There is no long-term or large-scale operating experience. Claims for effectiveness and the consequences of failure of such devices will require careful evaluation.

13.4.2 TAR CRACKING AND TAR REMOVAL

There is no clear view as to whether tar cracking or tar removal is preferable, although the current trend is to prefer cracking to reduce potential tar deposition problems and minimize washing requirements. Pressurized systems rely on high temperatures for tar cracking or catalytic tar cracking with hot gas filtration to give as hot a gas as possible to the turbine combustor to maximize efficiency; also the temperature requirements for alkali metal control and the filter materials need to be considered. The energy efficiency of the system is maximized in this way but combines several unproven or inadequately proven concepts. The high operating pressure also avoids the need for a fuel gas compressor, thus avoiding the need for a cool gas to compress before combustion.

Atmospheric pressure systems have fewer constraints from the complexity of high-pressure systems but may require the fuel gas to be compressed before combustion, in which case more cooling is required. At least one system advocates the use of both cracking and removal to ensure that a sufficiently clean gas is delivered to the turbine.

For engine applications, a clean cold gas is required, but not to the purity requirements of a turbine. Water washing may be adequate, depending on the gasifier.

13.4.3 HEAT RECOVERY

Heat may be recovered from the hot raw gas at several stages. Recovery from the raw gas is described above. Some further heat may be recovered after the hot gas filter but this is lower grade, although for larger plants a more sophisticated water–steam cycle may be justified on a countercurrent principle. Low-temperature heat is only worth recovering if there is a good local market for the heat. Careful optimization of the total heat system is necessary to obtain the high efficiencies promised in combined cycle plants.

13.4.4 WATER TREATMENT

Water treatment will be required if there is wet washing or if there are any condensates from the process. Although it is believed that most organics can be processed satisfactorily in conventional biological filters, there is a paucity of data to support this claim [11]. However, if there is a water scarcity and intensive internal recycling is used to give more concentrated wastewater, then there is a potential problem with phenols and related compounds. These may require incineration or other disposal.

13.4.5 GAS SPECIFICATION

Table 11.13 provided a list of contaminants and a possible specification for a turbine. Engine requirements would be much more relaxed. In neither case are there robust data available. Current experiences with ongoing projects will provide better guidelines but until a turbine is installed, and its performance carefully monitored, there will only be informed speculation about specifications. Even then, there will be a trade-off between more stringent and costly gas cleaning and higher maintenance costs of a turbine, which will require careful optimization.

13.5 *Power generation*

13.5.1 GAS COMPRESSION

Established turbosets are designed on the assumption that the compressor throughput matches the turbine throughput, with small allowances made for fuel addition in the combustion chambers and air bleeds from the compressor for blade cooling. The low energy density of LHV gases means that the volume of fuel injected will be substantial. In cases where the gasifying medium is taken from the turbine compressor, this is not expected to be a problem. However, where the gasifying medium is provided independently, there will be a large difference between compressor and turbine throughputs due to the addition of fuel gas, which will require redesign of the turboset to match the compressor with the turbine.

There is a potential problem when using the turbine compressor to supply the gasifier, whereby loss of output from the gasifier will mean loss of power at the turbine and thus loss of power to the compressor supplying the gasifier. Such a situation could rapidly shut down the system. Advanced control systems, possibly with auxiliary firing, would be needed to reduce this risk. Alternatively, additional and/or oversized compressors can be used.

Compressors are a well-established technology. However, there are uncertainties in this application in the potential mismatch between turbine and compressor and in the design of an adequate control system.

13.5.2 GAS TURBINE

Gas turbine manufacturers have produced turbines fueled by LHV gases, such as those produced for steel smelting operations. These tend to be large industrial units. Design work and trials are under way with aero-derived turbines but there are no reports of commercial applications with LHV gases. Whilst turbine manufacturers are confident that the combustion of LHV gases is technically feasible, there remains some uncertainty until this is proven.

Some auxiliary fuel capability would be beneficial at start-up and at times when the gasifier is unavailable or product gas output is limited. This would require redesign of fuel burners, which is again assumed to be technically feasible but at the expense of additional design cost and uncertainty with operation of a multi-fuel combustor.

The importance and uncertainty of gas clean-up have been noted above. It should be remembered that turbines of the scale likely to be used for biomass applications are far smaller than the 100+ MWe units usual in natural-gas-fired or coal-gas-fired applications due to the disperse nature of the biomass. This reduced capacity means that turbine components are more susceptible to damage from fuel contaminants because the protective layers are necessarily thinner. This is exacerbated by the high volume of fuel that is required. Turbine reliability is therefore closely related to the effectiveness and reliability of the gas clean-up system.

Complete combustion of the fuel may be difficult to achieve, which would result in high hydrocarbon emissions. This problem can be solved by increasing the size of the combustion chambers but this will add to the design cost of the turbine. Again, the solution is assumed to be technically feasible but unproven.

It is generally agreed that thermal NO_x emissions are unlikely to be a problem due to low flame temperatures. However, fuel-bound nitrogen could cause substantial NO_x emissions unless action is taken to remove the nitrogen compounds from the fuel or to remove NO_x from the flue gas.

13.5.3 ENGINES

The use of LHV gases such as landfill or digester gas in gas engines is well developed and machines exist that can be used in this application. Their disadvantage lies in the low-quality waste heat that is generated as a result of engine cooling and from exhaust gases. This makes application in combined cycle mode unlikely. Operation in co-generation mode is more accepted. Engines also have the advantage that they can be run on a variety of fuels or fuel combinations with relatively minor adjustment.

Gas cleaning is very important, although engines are considered to be less sensitive to contaminants than gas turbines.

The main uncertainty with engines is an economic one. Electrical efficiencies of engines at the rating required are higher than turbines in simple cycle mode. When turbines are operated in combined cycle or STIG modes, then they become much more competitive in terms of electrical efficiency, although the systems naturally become more complex and expensive. Also, there is a far greater potential for waste heat in a gas turbine cycle than in engine cycles. Thus, there is some uncertainty about the choice between turbines and engines at the likely scale of a biomass-based unit of 5–30 MWe.

13.5.4 STEAM GENERATION

Steam generation is an established technology and there are few problems associated with it other than the design of the system to make best use of the available waste heat. However, there is still a risk of corrosion if removal of sulfur and chlorine compounds from the product gas is inadequate.

The use of steam in combined cycles is established in large-scale generating systems. Application at the scale required is technically feasible, although it may not be economically viable.

The use of steam in STIG cycles is commercially accepted for high-grade fossil fuel applications but has not been tested when firing with LHV fuels. In the latter case, steam injection could make combustion unstable, and may also increase the turbine throughput to above its maximum capacity. Hence STIG, whilst offering substantial electrical efficiency improvements at a lower cost than combined cycles, must be viewed with some uncertainty.

Steam production from waste heat from engines is also established, although only low-pressure steam could be generated.

13.5.5 HEAT RECOVERY

The use of exhaust heat to preheat compressed air in a gas turbine cycle will require a high-temperature ceramic gas–gas heat exchanger. The application of ceramics in heat exchangers is an emerging technology and therefore this option is uncertain. Again, the importance of gas cleaning must be emphasized to prevent fouling at the heat exchanger surfaces.

14 Costs

14.1 *Gasifiers*

Gasification plant capital costs have been collated and normalized to early 1993 basis in Western Europe [22]. The results are summarized in Fig. 11.12, which includes conversion to an equivalent power output assuming 36% overall efficiency. There is a clear distinction between pressurized and atmospheric systems, which is accounted for by the significantly higher equipment and construction costs. There is also a considerable spread of data, particularly for atmospheric units. This is explained in part by variations in scope — what costs are included and also what units are included, such as the inclusion or exclusion of feed handling and pretreatment — and in part by cost variations in different locations resulting from varying labor rates and regional legislative variations.

14.2 *Cost analysis*

In order to identify the significant cost elements in a total system, an approximate breakdown of the main stages in a gasification process of 10 MWe is given in Table 11.17. It must be emphasized

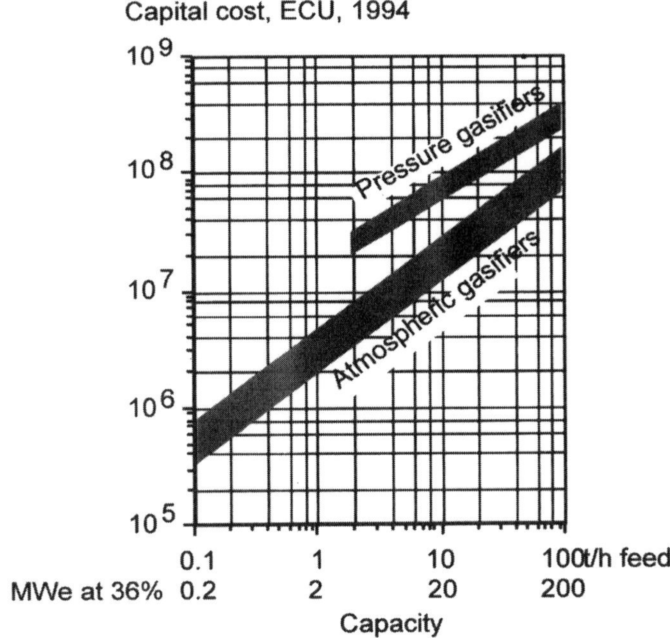

Figure 11.12. Installed plant costs (ECU 1994) for gasification systems.

Table 11.17. Approximate IG/CC plant cost analysis at 10 MWe.

Item	MECU	%
Reception, storage and handling	2.0	13
Comminution and screening	1.0	7
Drying	2.0	13
Gasification	6.5	43
Heat recovery	1.0	7
Tar cracking and/or removal	2.5	17
Subtotal	15.0	100
Power generation, gas and steam turbines	14.0	
Total	31.0	

that these are only indicative costs to establish order-of-magnitude relative costs between the main components.

14.3 *Learning effects*

It is well known that after the first plant of a new technology has been built, subsequent plants will cost less due to learning effects — the knowledge and experience gained in building and operating this first plant will improve the design and operation of subsequent plants. This learning effect has been widely applied to chemical production costs, for example, when learning effects of 15–20% are common [35]. This is defined as the cost reduction when production is doubled. A similar effect is found for capital costs of novel process plant and application of the 20% learning effect

Table 11.18. Integrated gasification combined cycle (IG/CC) system capital costs, 1993 basis (exchange rate = 1.2 ECU US$$^{-1}$).

System	MWe	Cost (ECU kWe^{-1})	Cost (US$ kWe^{-1})	Basis	Status
IG/CC	27	2300	2750	Brazil GEF	1st plant
IG/CC	27	1250	1500	Brazil	10th plant
IG/CC	5	3000	3600	Atmospheric gasifier	1st plant
IG/CC/CHP	6	5000	6000	Pressurized gasifier	1st plant
IG/CC/CHP	16	2670	3200	Atmospheric gasifier	1st plant
IG/CC/CHP	55	3500	4200	Pressurized gasifier	1st plant

CHP = combined heat and power.

results in a potential cost reduction of 50% by the time the tenth plant is built. There is only empirical evidence of this effect but it is widely known and is now being increasingly applied to examination of the replication potential of energy from biomass projects.

14.4 *System costs*

A major IG/CC project of 27 MWe is being evaluated for Brazil. The initial capital cost estimates were around US$2750 kWe^{-1} in 1992. It is hoped that careful design and R&D will bring this down to around $1500–1600 kWe^{-1} for the tenth plant due to the learning effect described above. Other cost data have been analyzed and representative figures for first plant are included in Table 11.18. Figure 11.12 clearly shows the higher cost of pressurized systems.

15 Conclusions to gasification

The process components involved in an integrated biomass to electricity system have all been individually tested at pilot scale or larger, but long-term operation and integration can only be achieved in a substantial demonstration plant. Several processes are in hand and more are planned for implementation in the near term. The key conclusions from this study are summarized below:

- The stage of development of biomass gasifiers is sufficiently advanced to justify a substantial demonstration plant to prove the total IG/CC concept and obtain reliable performance data. There are still areas of uncertainty, but these are relatively minor and will not be resolved until and unless a large integrated plant is built.
- Gas cleaning has been developed in laboratories to the point where large-scale demonstration and long-term operating experience are necessary. This area can be considered the least developed and most likely to create problems in a demonstration plant.
- Biomass handling, storage, drying, comminution and screening are well established in the pulp and paper industry as well as combustion systems worldwide and present no uncertainties in operation and performance. There is a need to optimize the cost and performance of the front end of the plant in relation to the gasifier performance and requirements, and the availability of heat and power energy from the clean-up and power generation stages.
- Turbine fuel specifications are imperfectly defined. There are a variety of requirements promoted by manufacturers that have not been substantiated in tests. There will be a trade-off between

higher levels of gas cleaning and higher maintenance costs that can only be resolved by large-scale and long-term operation.

• Engine fuel specifications are imperfectly defined. The level of uncertainty is lower than for gas turbines. Engines exist that can be, and have been, readily adapted to run on biomass-derived gases. However, fuel specifications are currently produced on an *ad hoc* basis.

16 References

1 Beenackers AACM, Maniatis K. In Kaltschmitt, MK, Bridgwater AV (eds) *Biomass Gasification and Pyrolysis*. Newbury, UK: CPL Press, 1997; 24.
2 Bridgwater AV. *Appl. Catal. A* 1994; **116**: 5.
3 Bridgwater AV. In *Proceedings of Conference on Bio-oil Production and Utilisation, Estes Park, CO, USA*. (NREL CP-430-7215, Golden, Colorado, USA, 1995).
4 Diebold JP, Bridgwater AV. In Bridgwater AV, Boocock DGB (eds) *Developments in Thermochemical Biomass Conversion*. London: Blackie Academic and Professional, 1997; 411.
5 Bridgwater AV. In Chartier P, Beenackers AACM, Grassi G (eds) *Proc. Biomass for Energy, Environment, Agriculture and Industry*. Oxford: Pergamon Press, 1995; 1591.
6 Bridgwater AV. In Kaltschmitt MK, Bridgwater AV (eds) *Biomass Gasification and Pyrolysis*. Newbury, UK: CPL Press, 1997; 53.
7 Diebold JP, Oasmaa A, Piskorz J, Bridgwater AV. In Bridgwater AV, Boocock DGB (eds) *Developments in Thermochemical Biomass Conversion*. London: Blackie Academic and Professional, 1997: 433.
8 Bridgwater AV. In Hall DO, Grassi G, Scheer H (eds) In *7th EC Conf. on Biomass for Energy and Industry*. London: Ponte Press, 1994; 166.
9 Solantausta Y, Nylund N-O, Westerholm M, Koljonen T, Oasmaa A. In *Proc. Power Production from Biomass*. Espoo, Finland: VTT, 1992.
10 Leech J. In Kaltschmitt MK, Bridgwater AV (eds) *Biomass Gasification and Pyrolysis*. Newbury, UK: CPL Press, 1997; 495.
11 Leech J, Bridgwater AV, Zarauzo A, Maggi R. *Final Report to EC*, Contract RENA-CT94-0070, 1998 (EC DGXII, Brussels, Belgium).
12 Andrews RG, Patniak PC. In *Bio-oil Production and Utilization. Proceedings of the 2nd EU/Canada Workshop on Thermal Biomass Processing*. Newbury, UK: CPL Press, 1996; 236.
13 Kaiser M. In Kaltschmitt MK, Bridgwater AV (eds) *Biomass Gasification and Pyrolysis*. Newbury, UK: CPL Press, 1997; 399.
14 Baldauf W, Balfanz U. In Kaltschmitt MK, Bridgwater AV (eds) *Biomass Gasification and Pyrolysis*. Newbury, UK: CPL Press, 1997; 392.
15 Maggi R, Elliott DC. In Bridgwater AV, Boocock DGB (eds) *Developments in Thermochemical Biomass Conversion*. London: Blackie Academic and Professional, 1997; 575.
16 Scott DS, Piskorz J, Radlein D. In Klass DL (ed.) *Energy from Biomass and Wastes XVI*, Paper 19. Chicago: IGT, 1992.
17 Chum HL, Diebold JP, Black SK, Scahill JW, Johnson DK, Evans R. In Klass DL (ed.) *J. Proc. Energy from Biomass and Wastes XV*. Chicago: IGT, 1991; 531.
18 Radlein D, Piskorz J. In Kaltschmitt MK, Bridgwater AV (eds) *Biomass Gasification and Pyrolysis*. Newbury, UK: CPL Press, 1997; 471.
19 Graham RG, Freel BA, Huffman DR. In Bridgwater AV (ed.) *Advances in Thermochemical Biomass Conversion*. London: Blackie Academic and Professional, 1994; 1275.
20 Cuevas A. *Proceedings of EC JOULE Contractors Meeting*, Athens, June 1993. EC DGXII, Brussels, Belgium.
21 Toft AJ, Bridgwater AV. *FAIR Contract 0216 Final Report*, November 1997. EC DGII, Brussels, Belgium.
22 Bridgwater AV, Evans GD. *ETSU Report B/T1/00207/REP*. Harwell, UK: ETSU, 1993.
23 Graboski MS, Brogan TR. In Klass D (ed.) *Energy from Biomass and Wastes XI*. Chicago: IGT, 1988.
24 Reed TB. *Biofuels and Municipal Waste Technology Research Program Summary: FY 1986*. US DOE/CH/10093-6, DE87001140, July 1987. NREL: Golden, CO, USA.

25 Bridgwater AV, Maniatis K, Masson HA. In Ferrero G-L, Maniatis K, Buekens A, Bridgwater AV (eds) *Pyrolysis and Gasification*, CEC, EUR 12736. 1990; 41–65; London: Elsevier Applied Science, 1989.
26 Belleville P, Capart R. In Bridgwater AV (ed.) *Thermochemical Processing of Biomass*. London: Butterworth, 1984; 217; Shand RN, Bridgwater AV. In Bridgwater AV (ed.) *Thermochemical Processing of Biomass*. London: Butterworth, 1984; 229.
27 LeMasle J-M. *Steine Industrie, Final Report EC Contract JOUB 0052*, 1991. EC: Brussels, Belgium, 1991.
28 Lyytinen H. *Tappi J.* 1987; **69**: 77–80.
29 Herbert PK, Loeffler JC. In *Proc. Conf. on Thermal Power Generation and the Environment*. Hamburg: Unipede and IEA, 1993.
30 Kurkela E, Ståhlberg P, Laatikainen J, Simell P. *Bioresource Technol.* 1993; **46**: 37.
31 Corella J, Herguido J, Gonzalez-Saiz J, Alday FJ, Rodriguez-Trujillo JL. In Bridgwater AV, Kuester JL (eds) *Research in Thermochemical Biomass Conversion*. London: Elsevier Applied Science Publishers, 1988; 411.
32 Mackie KL. *IEA Bioenergy Agreement Environmental Systems Group Interim Report*. New Zealand: FRI, 1993.
33 Corman JC. *General Electric Company Report No. ET14928-13*. Schenectady, USA: GEC, 1986.
34 De Ruyck J, Maniatis K, Distlemans M, Baron G. In Bridgwater AV (ed.) *Advances in Thermochemical Biomass Conversion*. London: Blackie Academic and Professional, 1994; 411.
35 Elliot P, Booth R. *Brazilian biomass power demonstration project*. Shell special project brief, September 1993. Shell UK, 1993.

12 Thermochemical Energy Conversion

V.N. PARMON

Boreskov Institute of Catalysis, Siberian Branch of the Russian Academy of Sciences, Prospekt Akademika Lavrentieva 5, Novosibirsk 630090, Russia

1 Introduction

The impact of chemistry on future energetics will, no doubt, largely increase when the base of energetics shifts more and more to new kinds of non-exhaustible energy resources.

For the distant future, when the cheap fossil combustibles are either exhausted or too expensive for general use, two non-exhaustible sources of primary energy will remain of most interest: nuclear energy liberated by either fission or fusion nuclear processes; and solar light. The direct use of both of these primary energies is impossible because they have to be converted initially into a form suitable for energy consumers. Chemical, and especially catalytic, processes are capable of converting any of the above-mentioned kinds of energy into energy of chemical bonds, thereby storing it in the convenient form of chemical fuels.

From the viewpoint of chemists, the natural and simplest idea for conversion and storage of these and other kinds of energy is their preliminary transformation into heat and then direct absorption of this heat by highly reversible energy-absorbing chemical processes capable of shifting their equilibrium when the temperature of the system changes. Thus, chemistry presents a unique chance to create very simple systems for conversion and storage of energy by using the ability of some of the simple chemical transformations to be reversible. The use of this ability for energy transformations is called 'thermochemical' energy conversion.

The earliest mention of the possibilities of thermochemical conversion of energy with a large list of most promising reversible chemical reactions was made nearly 40 years ago in a patent [1] and the idea was later developed for numerous applications (for reviews, see [2,3]). To date, thermochemical energy conversion may be considered as the simplest and most developed application of chemistry to the newest kinds of energetics.

2 Principles of thermochemical energy conversion: thermochemical processes for closed-loop systems

The idea of storing heat energy via its conversion to chemical fuel energy is very simple and is based on the thermodynamic consideration of the behavior of reversible endothermic reactions that proceed with a large increase of entropy, ΔS. At ambient temperatures, many of these reactions are characterized by a positive value of the Gibbs potential change ΔG, i.e. they are endoergonic and do not occur.

However, let us consider the behavior of a system capable of providing the simplest reversible chemical reaction:

$$A \rightleftarrows B$$

where A denotes the 'energy-poor' reagents and B denotes the 'energy-enriched' products. The change in the Gibbs free energy (ΔG) on accomplishing this reaction can be expressed as:

$$\Delta G = \Delta H - T\Delta S$$

where ΔH is the enthalpy of the reaction. At the initial temperature $\Delta G > 0$ but with a temperature rise the term $T\Delta S$ increases and the sign of $\Delta G = \Delta H - T\Delta S$ may change. Because ΔH and ΔS are of the same sign, there is a temperature:

$$T^* = \Delta H^\circ/\Delta S^\circ \approx \Delta H^\circ_{298}/\Delta S^\circ_{298}$$

at which $\Delta G^\circ = 0$. Thus, one should expect a dramatic shift of the reaction equilibrium when passing T^*. In other words, T^* is the temperature of the 'shift' of the equilibrium, temperatures above T^* providing conditions for the shift of equilibrium towards energy-enriched reaction products, i.e. towards the endothermic direction with $\Delta H > 0$. Using this shift one can store or release the corresponding enthalpy ΔH of the reaction by absorbing heat from the surroundings.

Heat energy can thereby be converted into energy of chemical bonds with the efficiency (η_G) coinciding with that of the Carnot cycle if we consider the storage of free Gibbs energy (see [4]):

$$\eta_G = \frac{\text{Amount of energy stored as Gibbs energy}}{\text{Amount of heat consumed}} \approx \frac{T - T_0}{T}$$

where T is the temperature of the reaction and T_0 is the temperature of storage of the reaction components. Note that for converting heat into the enthalpy of the reaction products there are no Carnot restrictions. Thus, the efficiency η_H of such energy conversion can be much higher than η_G and is limited mostly by the heat capacity properties of the reagent mixtures as well as the heat losses of the system as a whole [5].

If, after a prompt cooling ('quenching') of the reaction mixture, the back exothermic reaction proves to be kinetically slowed down, the energy-enriched products of the reaction will acquire the properties of a chemical fuel, i.e. have the ability to store the accumulated heat energy and release it when necessary in the presence of an appropriate catalyst. It should be noted that the energy-enriched mixture can be easily 'quenched', if it is obtained also via a catalytic process, simply by isolating the mixture from the catalyst.

The list of reactions in Table 12.1 denotes the most simple and widely discovered reversible catalytic reactions for the conversion of energy from nuclear and solar plants and chemical heat pumps. Most of these reactions proceed only in the presence of suitable catalysts and are well known as endothermic chemical processes commercialized on a large industrial scale. For example, the reactions of methane reforming (reactions 4 and 5 in Table 12.1) rely on the large-scale production of primary syn-gas for further conversion into other valuable products, including hydrogen.

For the practical application of reversible chemical reactions for thermochemical storage of energy, a very important parameter appears to be the 'specific power loading' (SPL) of the particular chemical reactor used for the energy conversion. The value of the SPL can be expressed as heat power converted into energy of chemical bonds per unit volume of the reactor:

$$\text{SPL} = \frac{\text{Amount of heat transformed to chemical energy}}{(\text{Volume of reactor}) \times \text{time}}$$

Clearly:

$$\text{SPL} \approx \Delta H^\circ \times (\text{The rate of the energy-storing chemical reaction})$$

As shown in [5] for the catalytic thermochemical processes, the magnitude of the SPL is determined primarily by the performance and activity of the catalysts used rather than by the amount of

Table 12.1. Reversible catalytic reactions proposed for conversion of energy at nuclear and solar plants and in chemical heat pumps, and their thermodynamic parameters.

No.	Reaction*	ΔG°_{298} (kJ mol-1)	ΔH°_{298} (kJ mol-1)	ΔS°_{298} (J mol-1 K-1)	T^*† (K)
1	$SO_3(g) \rightarrow SO_2(g) + \frac{1}{2}O_2$	70.1	98.8	94.5	1030
2	$C(s) + H_2O(l) \rightarrow H_2 + CO$	100.4	176.0	253.7	980
3	$C(s) + CO_2 \rightarrow 2CO$	120.1	173.5	177.2	980
4	$CH_4 + H_2O(l) \rightarrow 3H_2 + CO$	151.2	251.2	334.7	960
5	$CH_4 + CO_2 \rightarrow 2H_2 + 2CO$	171.4	248.2	258.3	960
6	$C_6H_{12}(g) \rightarrow C_6H_6(g) + 3H_2$	98.3	207.1	363.7	570
7	$C_6H_{11}CH_3(g) \rightarrow C_6H_5CH_3(g) + 3H_2$	95.3	205.4	369.6	560
8	$NH_3(g) \rightarrow \frac{1}{2}N_2 + \frac{3}{2}H_2$	16.8	46.2	99.5	470
9	$CH_3OH(g) \rightarrow 2H_2 + CO$	25.2	91.1	219.7	415
10	$N_2O_4(g) \rightarrow 2NO_2$	1.29	13.9	≈41.2	330‡
11	$CH_3OH(l) + H_2O(l) \rightarrow 3H_2 + CO_2$	8.8	133.0	410.3	320
12	$CH_3CHOHCH_3(g) \rightarrow CH_3COCH_3(g) + H_2$	23.5	59.2	119.7	495

* (g) Gaseous; (l) liquid; (s) solid.
† T^* is the temperature at which $\Delta G^\circ = 0$.
‡ Reaction is non-catalytic.

Table 12.2. Value of specific power loading (SPL) calculated for some endothermic catalytic reactions at a total pressure of 1 atm.

Reaction	Catalyst	Temp. (K)	Reagent mixture conversion	SPL (kW dm-3)
$CH_4 + H_2O \rightarrow CO + 3H_2$	TH-2	1020	0.7	70
($H_2O:CH_4 = 2.0$)	GIAP-8	1020	0.7	50
	K-3	1020	0.7	45
	NC-2	1020	0.7	30
$SO_3 \rightarrow SO_2 + \frac{1}{2}O_2$	V_2O_5	1070	0.7	100
	Fe_2O_3/Al_2O_3	1120	0.6	30
	Fe_2O_3/SiO_2	1120	0.6	15
$NH_3 \rightarrow \frac{1}{2}N_2 + \frac{3}{2}H_2$	Rh	720	0.7	230
	Pt	970	0.5	100
	Ru	970	0.5	20
$C_6H_{12} \rightarrow C_6H_6 + 3H_2$	Pt/Al_2O_3	670	0.7	115
	Pt	620	0.7	120
$CH_3OH \rightarrow CO + 2H_2$	CuO–ZnO	870	0.5	2
	Ru	620	0.6	7
	Co	620	0.5	6
	Ni	620	0.4	5

From [5].

heat (ΔH°) stored in the reaction. Therefore, the SPL seems to be more important than the value of ΔH, i.e. the heat absorbed and transformed by a given amount of the reaction mixture.

As a result, the best reactions for thermochemical conversion of heat energy (and vice versa) appear to be those for which the most active catalysts are known (see Table 12.2 and [5]). For some known catalysts the calculated and principally allowed values of SPL can reach a level above 100–150 kW dm^{-3}, which is comparable with the density of the huge heat fluxes released in the active zones of modern nuclear reactors.

These values of SPL are too huge to imagine in ordinary life. Indeed, they correspond to nearly 1000 1 kW electric boilers installed and operating simultaneously in a 10 l bucket. Such thermochemical processes appear to be the most efficient based on the known methods of absorbing and converting energy *per se*.

However, in conventional chemical industries that use the same endothermic catalytic reactions, typical average values of SPL do not usually exceed 3–10 kW dm^{-3}. These low values are the consequence of severe restrictions in the rate of heat transfer through the walls of common catalytic reactors. Theoretically predicted SPL values of over 100 kW dm^{-3} can be reached in order to intensify the heat transfer into the catalyst bed. As seen below, advanced thermochemical energy-converting systems that operate by converting light or ionizing radiation are able to reach the highest possible values of SPL.

3 Thermochemical conversion of nuclear energy to fuel energy

In addition to the power industry based on fossil fuels, nuclear energetics has become the second significant area of the power industry and tends to achieve momentum despite the shock caused by the nuclear disaster in Chernobyl.

The contribution of chemistry to nuclear energetics is extremely large, beginning with the discovery and isolation of radioactive chemical elements. When considering the less known but nevertheless common contribution of chemistry to nuclear energetics, one should mention also the catalytic technologies for the elimination of explosive stoichiometric mixtures of H_2 and O_2 formed in the primary circuits of nuclear reactors with water cooling. The formation of this explosive mixture is caused by radiation-stimulated water dissociation inside the active zone of the reactor, which provides a real danger to the safety of nuclear reactors. The explosive mixture has to be eliminated just inside the cooling circuit by catalyzing the interaction between the hydrogen and oxygen components of the mixture under sufficiently mild conditions to preclude explosion. The particular demands of the catalysts developed for this purpose are high catalytic activity at low temperatures under conditions of high water content of the reaction medium [6].

In the future, chemistry is expected to be of crucial importance not only for ensuring the safety of both existing fission and future fusion reactors but also for the preparation of future nuclear fuels themselves [7].

The contribution of thermochemical energy conversion to nuclear energetics can perhaps eliminate certain drawbacks of today's nuclear energetics. It is well known that conventional nuclear power plants produce electricity and/or middle- or low-potential heat (500–800 or < 500 K). But not all the needs of consumers can be satisfied by these electricity and heat potentials. It should be noted also that the amount of energy consumed can vary significantly from day to night, and in other respects, and nuclear reactors are not very suitable for fast changes in the regime of power generation. Therefore, in order to smooth the nuclear power consumption and provide consumers remote from nuclear power plants with middle- and high-potential ($T > 800$ K) technological heat, various processes aimed at intermediate accumulation of energy produced inside a nuclear reactor are being developed. Of special interest for this purpose are the reversible endothermic chemical processes mentioned in Table 12.1. These allow the accumulation and storage of energy in the form of chemical fuels and the release of this energy when required by accomplishing the back

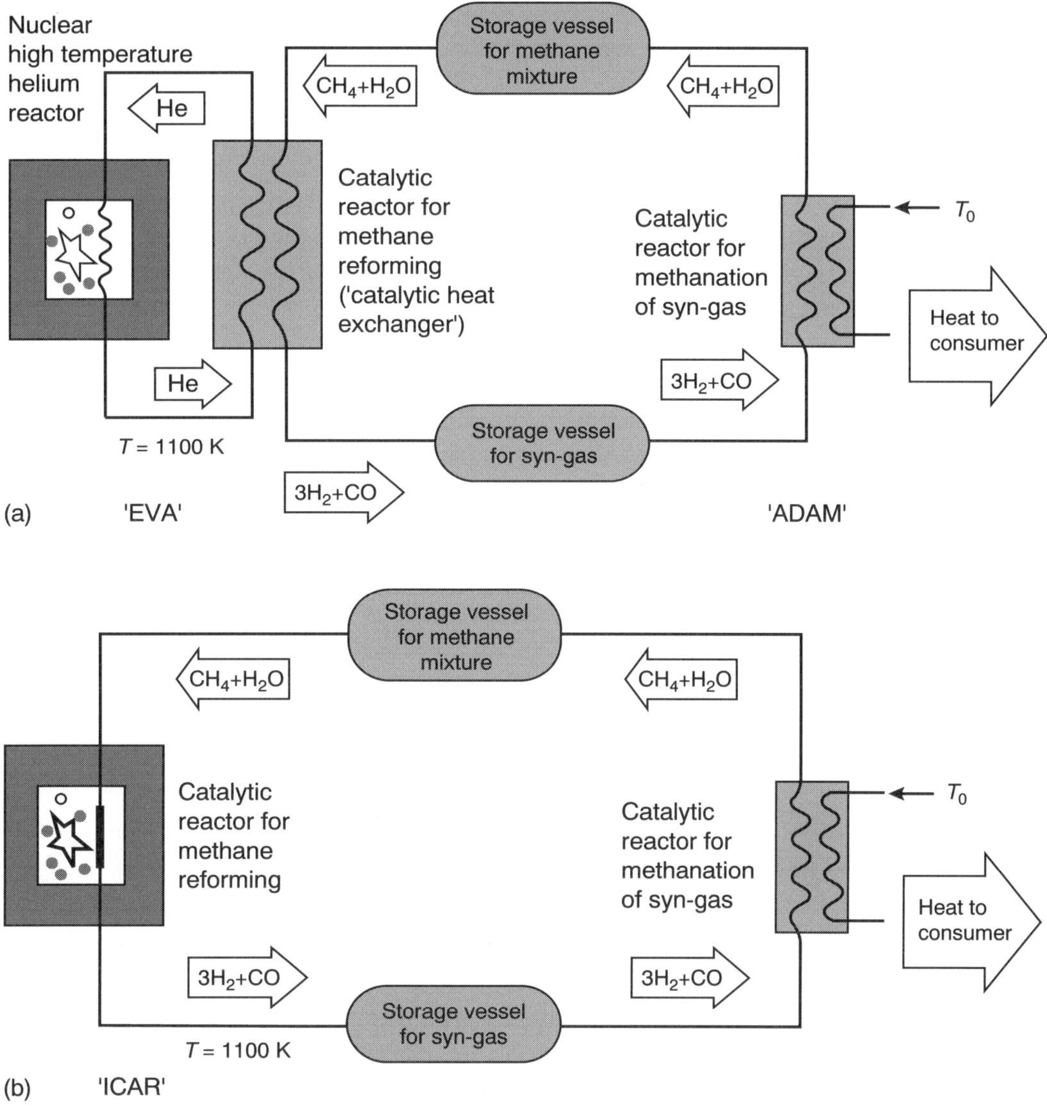

Figure 12.1. Principal schemes of the EVA–ADAM (a) and ICAR (b) cycles of thermocatalytic conversion of nuclear energy. The reversible reaction of the steam reforming of methane is used as the energy-converting process in both cases.

exothermic reaction. Most of these processes are catalytically well developed and have close analogs in the conventional chemical industry (see Table 12.1 and [1–3,8–12]).

With regard to nuclear fission energetics the so-called EVA–ADAM scheme, based on highly reversible chemical reactions with the participation of methane, is well known from the 1970s for indirect accumulation of released nuclear energy in the form of syn-gas, its storage and transportation over large distances and subsequent methanation in order to obtain high-potential heat for the final energy consumer (see Fig. 12.1a and, e.g., [8,9] as well as earlier issues of these sets of books).

The energy-storing part of the EVA–ADAM system is based on the highly endothermic catalytic reactions of methane reforming with steam and/or carbon dioxide to energy-enriched synthesis gas ('EVA' process, reactions 4 and 5 in Table 12.1). In the scheme, these reactions occur due to heat provided by hot helium flow preheated to 1120–1300 K in a specially designed high-temperature nuclear reactor. The syn-gas obtained serves as a good chemical accumulator of energy (see Chapter 6). For example, this gas can be transported by pipeline over long distances as a chemical fuel and then converted to the initial methane-containing mixture via catalytic methanation ('ADAM' process); during the methanation process the stored energy is released as high-potential heat with a temperature of *c*. 1000 K. Other versions of the 'EVA' process application have also been discussed. For example, syn-gas produced using energy from nuclear reactors can be used as a raw material for the production of liquid fuels or valuable chemical products such as methanol [8,9]. This approach to the use of nuclear energy allows substantial amounts of valuable organic raw materials to be saved, which at the moment are burnt to drive endothermic industrial chemical processes such as methane reforming.

Of practical interest are also thermochemical cycles based on the reversible process of SO_3 decomposition (see reaction 1 in Table 12.1 and [8–13]). An advantage of this reaction is the utilization of a highly reversible inorganic substrate for reversible conversions, which does not create problems with selectivity, catalyst deactivation, etc. However, the main disadvantage of this scheme is the corrosivity and toxicity of sulfur oxides.

Quite large (up to 1 MW heat power) pilot plants for testing the EVA–ADAM process were constructed and tested in the Julich nuclear center in Germany up until the early 1980s, these plants being expected to be joined up to a real nuclear reactor. Simultaneously, in the former Soviet Union, large-scale experiments were provided for the development of another thermochemical cycle for nuclear energy conversion, based on a highly reversible non-catalytic reaction (reaction 10 of Table 12.1):

$$N_2O_4 \rightleftarrows 2NO_2$$

(see [14,15]). A delay in accomplishing both of these plans originated from the non-expected Chernobyl accident and the switch of the West European economy to natural gas as the main chemical energy carrier for many fields of large-scale energy industry.

Even so, several fundamental problems still remain to be solved in the development of EVA–ADAM-like thermochemical technologies for nuclear power conversion. The first important problem appeared to be the design and construction of a reliable high-temperature nuclear reactor with an operating temperature of no less than 1120 K and a high power density of energy release in the active zone. The known projects of high-temperature helium nuclear reactors considered a power density of no more than 10 MW m^{-3}, which is more than one order of magnitude less than for conventional water-cooled nuclear fission reactors. Efficient high-temperature helium reactors are thus still to be constructed, with the Chernobyl accident halting many experiments in that direction.

In order to overcome this drawback of conventional EVA–ADAM systems with a separate high-temperature nuclear reactor, recently a new and much more elegant and efficient ICAR (immediate catalytic accumulation of ionized radiation energy) process was suggested for direct nuclear-to-chemical energy conversion (see Fig. 12.1b and [16,17]). The peculiarity of the ICAR technology consists of placing the energy-converting catalysts immediately into the active energy-releasing zone of a nuclear reactor (Fig. 12.2), or even using special multifunctional physicochemical structures to serve simultaneously as a nuclear fuel and a catalyst for the energy-storing chemical reaction (e.g. porous uranium oxide impregnated with catalytically active Ni or Ru; see Figs 12.3 & 12.4 [18]). In both versions of the ICAR process the heat needed for driving the endothermic process is evolved just inside the catalyst granules during dissipation inside them of the energy of fast ionizing particles emitted by the nuclear fission reaction. Thus, no special heat

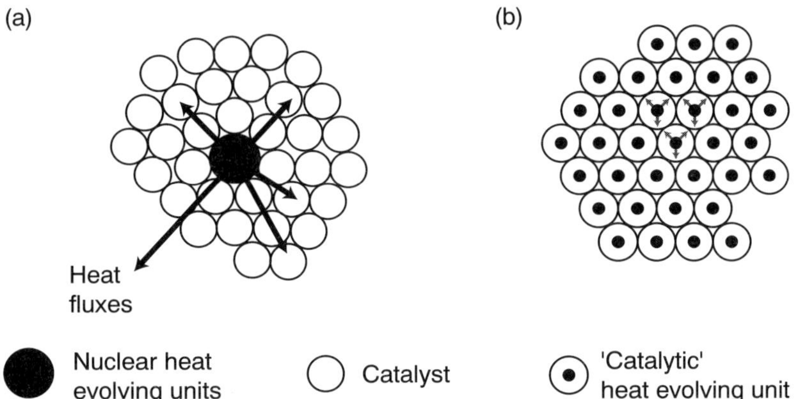

Figure 12.2. Two versions of loading the active zone of the nuclear reactor with energy-converting catalysts: (a) the nuclear heat-evolving units located between the catalyst bed; (b) the heat-evolving units incorporated in the catalyst for energy transformation.

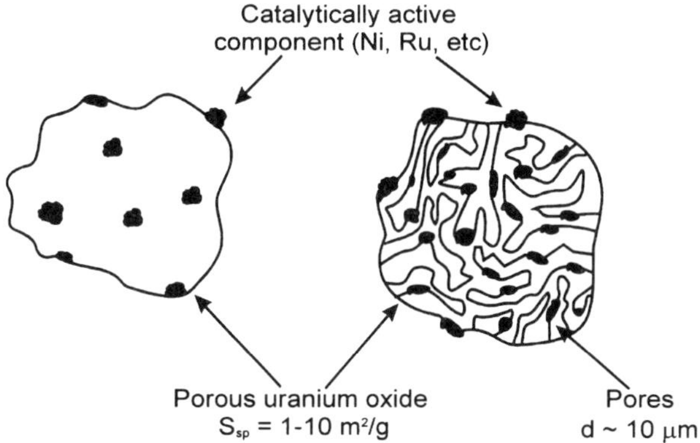

Figure 12.3. Catalytically activated nuclear fuel based on uranium oxide (from [18].) On the right, a UO_x granule covered with a thin layer of Al_2O_3 or TiO_2 is schematically represented; this additional coverage of UO_x with an inert layer diminishes pollution of the chemically reactive medium with radioactive residues of the nuclear fission reaction.

exchangers are needed to convert heat evolved at dissipation of energy of rapid and energy-rich particles created by nuclear reactions. Consequently, the SPL of nuclear energy conversion in the ICAR-type systems allows achievement of the highest possible values, which are restricted only by the intrinsic activity and thermal stability of the catalysts used.

Thus, the ICAR process seems to be considered as a direct radiation/catalytic conversion of the ionization radiation energy to chemical fuel. This conversion was successfully tested on the laboratory scale [17,18]. Special estimates have been made showing that radioactive contamination of the chemically reactive medium in this scheme cannot be considered dangerous or catalytically deactivating during long-term runs under severe radiation.

The ICAR process has exhibited high efficiency of the energy conversion combined with the ability to maintain a useful SPL of up to 100–200 kW dm^{-3} for the nuclear-to-chemical energy conversion (see Fig. 12.4). These SPL values correspond to those for energy release in modern fission nuclear power plants.

It should be noted that both thermocatalytic and radiation/catalytic methods of energy conversion may, perhaps, be applied to energetics of the future based on nuclear fusion reactors with

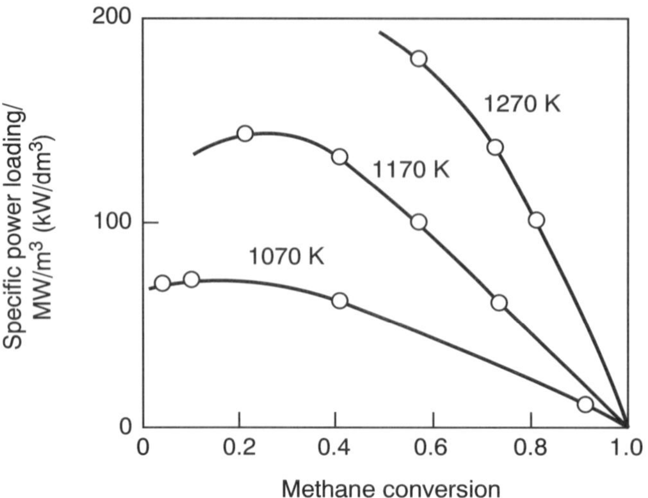

Figure 12.4. Catalytic properties of 0.8% Ru/UO_x catalyst in the reaction $CH_4 + H_2O \rightarrow 3H_2 + CO$. (From [18].)

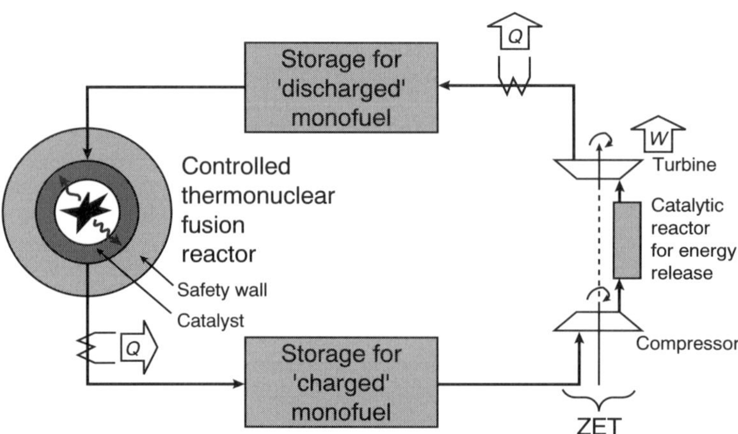

Figure 12.5. Scheme of the ICAR process coupled with a zero-emission turbine (ZET) (see Section 5) for adjustment to a nuclear fusion reactor (W, work generated by ZET; Q, flows of heat equilibrating the temperature of the chemically reactive 'working body' with the ambient).

blankets rather than conventional nuclear energetics based on nuclear fission. Clearly, ICAR technology could be based on fusion thermonuclear energetics, because today it seems to be the sole known technology that allows efficient use of the enormous fluxes of ionizing radiation released by future thermonuclear fusion reactors (see Fig. 12.5 and [16]). Also, one can expect an extended application of ICAR technology to the construction of middle- and low-power energy sources based on isotopic or small-scale nuclear reactors which accumulation of energy for its supply to short-term intensive energy consumers.

Another advantage of the ICAR process is its safety in an emergency situation when a rapid cooling of the active zone of a nuclear reactor is needed, e.g. when providing an endothermic reaction for steam reforming of methane: typical time t of the catalyst cooling at temperature $\Delta T \approx 100$ °C is only $t \approx C\Delta T/\omega \approx 1$–$10$ s, where C is the heat capacity of the catalyst and ω is the SPL value ensured by the same catalyst. The expected value of t is confirmed by direct measurements (see Fig. 12.6). Note that the value of the cooling time is much shorter than that for nuclear

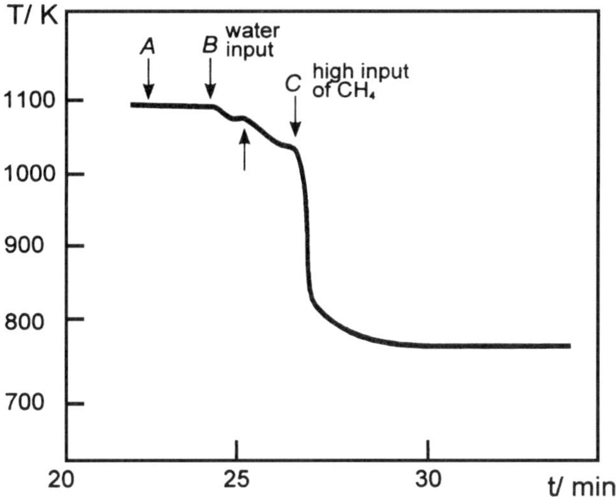

Figure 12.6. Experimental data for adjustment of the catalyst temperature in an emergency situation for the reaction $CH_4 + H_2O \rightleftarrows 3H_2 + CO$. (From [17].)

reactors with chemically inert cooling media, which increases the operating safety of nuclear reactors.

Under continuous consideration also for nuclear energetics are the more complicated thermochemical cycles such as those for storing the evolved nuclear energy via indirect dissociation of water into H_2 and O_2. These cycles are considered extensively for large-scale production of hydrogen from water and thus for creating a technological basis for the so-called nuclear–hydrogen energetics and technology of the future (Chapter 5). A large variety of multistep processes combining both homogeneous–heterogeneous and even electrochemical reactions were discussed and developed as the chemical basis for these thermochemical systems for water cleavage (see [8,9,12,19,20]). Catalysts play an important role in these cycles, because many steps of the thermochemical cycles that are being developed for the above purpose involve reactions between gaseous compounds and highly dispersed heterogeneous regenerative contacts (see [21,22]).

4 Thermochemical conversion of solar energy to fuel energy

In the last two decades, serious efforts have been made to develop catalytic processes for solar energy conversion. One pathway in this research deals with quantum processes, which produce high-quality fuels such as hydrogen via electronic excitation of molecules or other kinds of photosensitive materials by the action of any type of solar irradiation, including highly scattered irradiation (see Chapter 13). However, much more advanced developments today are in the area of non-quantum thermocatalytic conversion of concentrated solar energy.

Thermochemical conversion of solar energy is based on the same ideas that have been proposed for energy conversion at nuclear power plants (see Table 12.1 & Fig. 12.7). In a first step, the solar light should be highly concentrated by a parabolic mirror, for example. This concentrated solar energy is then converted to heat inside a specially designed catalytic reactor where an endothermic catalytic reaction, e.g. methane reforming with steam or carbon dioxide to syn-gas (reactions 4 and 5 in Table 12.1), takes place. The converted solar energy, accumulated in the form of an energy-enriched mixture of chemical compounds (syn-gas), is transported to a storage vessel and then can be used for the production of high-potential heat (Fig. 12.7).

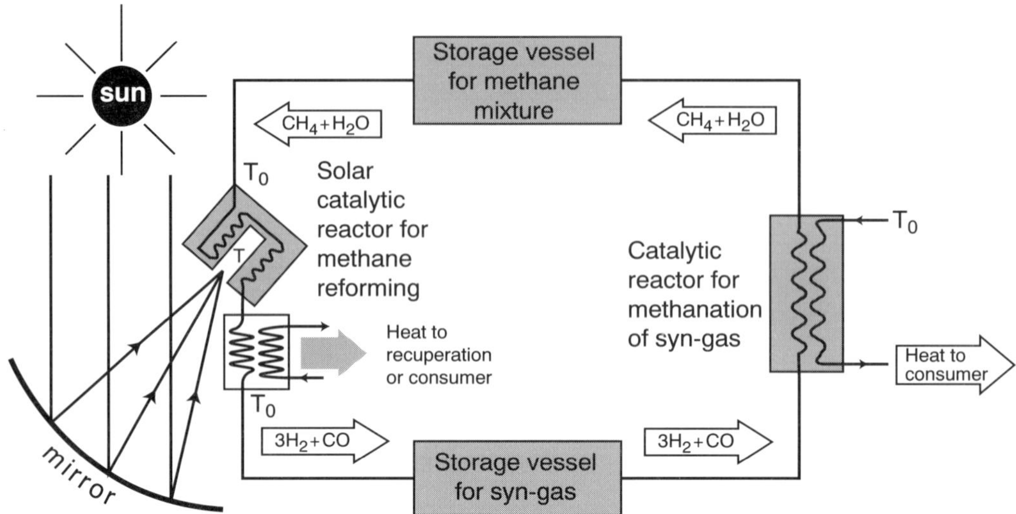

Figure 12.7. Principal scheme of thermocatalytic conversion of solar energy based on the idea of the EVA–ADAM cycle (T, temperature inside the cavity of the solar catalytic reactor; T_0 temperature of the environment). (From [2,3,23,24].)

A most interesting feature of the implementation of thermochemical methods to the conversion of solar energy is their high energetic efficiency, much higher than that for quantum solar energy converters (cf. Figs 12.8 & 13.2 from Chapter 13). Indeed, the overall thermodynamically allowed efficiency η_G of solar energy conversion into Gibbs energy of a chemical fuel via thermochemical reactions occurring at temperature T can be expressed [23] as:

$$\eta_G = \left\{\begin{matrix}\text{Efficiency of} \\ \text{conversion of solar} \\ \text{energy to heat with} \\ \text{temperature } T\end{matrix}\right\} \times \left\{\begin{matrix}\text{Efficiency of} \\ \text{heat-to-chemical} \\ \text{energy } \Delta G \\ \text{conversion}\end{matrix}\right\} \approx \left(1 - \frac{T}{T_{\text{solar}}}\right) \times \left(1 - \frac{T_0}{T}\right)$$

Here $T_{\text{solar}} \approx 5800$ K is the temperature of the surface of the Sun and T_0 is the temperature of the ambient. It is seen from Fig. 12.8, which was calculated with the help of the above formula, that η_G is a maximum of *c.* 60% at $T \approx 1000$–1100 K, which can be easily achieved using simple concentrators of solar light (e.g. mirrors, etc.). Note that for conversion of the concentrated light energy into enthalpy ΔH rather than into the Gibbs energy, the thermodynamic restrictions are not so severe and η_H can be much larger.

Research and development in the direction of thermochemical solar energy conversion is already at the stage of pilot plant design [16,22–29]. In the mid-1980s, prototypes of energy-converting installations of a reasonably useful power (more than 2 kW) and of a very high (> 40%) efficiency of solar-to-chemical energy conversion were tested [24]. Further development of design and other improvements of such solar energy-converting systems [16,25–29] are under way.

Fortunately, the endothermic reactions proposed for thermochemical solar energy conversion are well known in the chemical industry. Therefore, the main task here is to develop solar catalytic reactors of special design and other specialized equipment for such an operation. Typical solar catalytic reactors for steam reforming of methane are cylinders with a cavity in which the solar light is concentrated (see Fig. 12.9). The light provides strong (up to 1000–1300 K) heating of both the cavity and the catalyst placed around it. Syn-gas produced in such reactors can be utilized

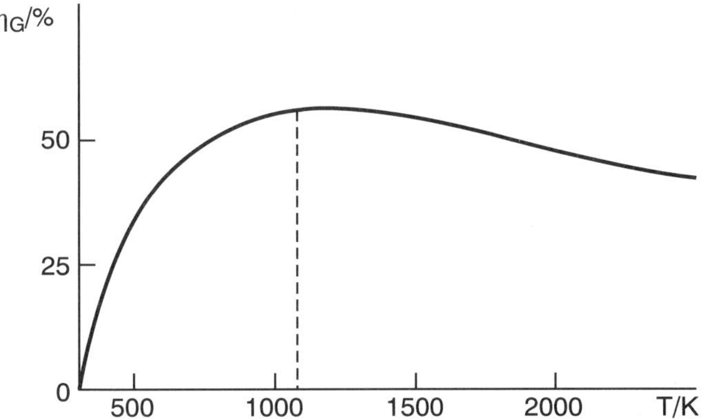

Figure 12.8. Dependence of the thermodynamically allowed efficiency η_G of thermochemical conversion of solar energy into Gibbs energy of chemical fuels on the operating temperature T of the energy converter. (From [23].)

Figure 12.9. 'Field' tests of the thermocatalytic device SCR-5 (designed by the Boreskov Institute of Catalysis, Novosibirsk) in 1984–1985 for solar-to-chemical energy conversion on the basis of steam reforming of methane with a useful power of *c*. 2 kW and an efficiency η_H of 43%. The solar catalytic reactor is located in the focus of the paraboloid mirror of 5 m diameter that concentrates the solar light. A small vertical cylinder to the right (indicated by the arrow) is the catalytic methanator of syn-gas for the release of stored energy. (From [2,24].)

when desired to produce heat of high potential (*c*. 1000 K) during the backward process of methanation. The demands on the catalyst applied for thermocatalytic solar energy conversion are similar to those for thermocatalytic conversion of nuclear energy. However, an extra demand should be added: the catalyst should have the ability to sustain enormous thermal shocks inside the cavity of the solar catalytic reactor due to possible rapid variation of light fluxes.

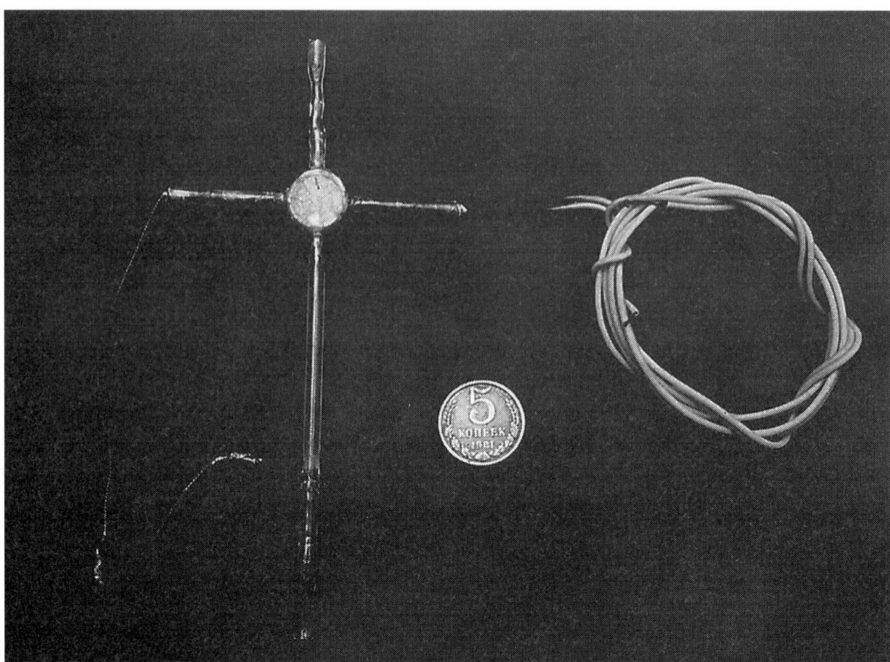

Figure 12.10. Laboratory-scale thermochemical reactor with light-transparent walls to demonstrate a useful power of direct light-to-chemical energy conversion of > 100 W. The light-to-stored enthalpy efficiency η_H of this reactor was $> 60\%$. (From [2,16,29].) The coin indicates the actual size of the reactor.

Tests of small pilot plants containing both solar catalytic reactors for accumulating energy in the form of syn-gas and catalytic reactors for release of the accumulated energy during methanation have been carried out (see Fig. 12.9 and [2,24]). These tests demonstrated the possibility of converting the concentrated solar energy to chemical energy (enthalpy) of syn-gas with good efficiency at the useful power of 2 kW for the devices. Recently, an efficiency η_H of *c.* 70% was achieved for this process [2,16,29], when arranging the light-transparent walls of solar catalytic reactors to allow direct heating of the catalyst bed with solar light (see Fig. 12.10).

In future, in rural areas, it may be reasonable to use the thermocatalytic conversion of solar energy to improve the quality of certain fuels: e.g. bio-gas, which is a mixture of methane with carbon dioxide produced via anaerobic fermentation of biomasses of different origin, can be upgraded easily in calorific value by methane reforming (see Chapter 10).

Of some, though indirect, relation to the solar energy industry are catalytic methods of biomass processing, e.g. biomass gasification, with the help of concentrated solar light to form energy-saturated fuels. These approaches of solar energy conversion are now also under development.

5 'Zero-emission turbines'

Thermochemical cycles provide very efficient accumulation of different kinds of energy in the form of chemical fuels. Unfortunately, wide application of these cycles is usually postponed by the disadvantage of conventional schemes: the accumulated energy can be evolved in the form of low- or moderate-potential heat only, whereas the typical customer needs to receive mechanical energy or electricity.

Table 12.3. Examples of typical reversible 'monofuels' and enthalpies of their conversion.

'Charged' fuel mixture	'Discharged' (spent) fuel mixture	ΔH°_{298} (kJ mol-1)
$3H_2 + CO$	$CH_4 + H_2O$	–206
$3H_2 + N_2$	$2NH_3$	–92
$3H_2 + C_6H_6$	C_6H_{12}	–207

An interesting application of catalytic technologies to the thermochemical methods of nuclear or solar energy conversion could be the creation of 'zero-emission turbines' (ZETs), which use chemically reversible 'monofuels'. As 'monofuels' one can consider mixtures such as H_2 and CO, etc., which do not need separation or additional mixing of their components during enrichment of the mixture with energy (its 'charging') or utilization of the stored energy ('discharging' the monofuel) — see Chapter 6. The main idea of a ZET is to use a hot, recyclable reaction mixture at the outlet of the energy-releasing catalytic reactor directly as the working fluid of a turbine. It is of interest to make some remarks on the peculiarity of this idea.

The usual combustion of organic fuels is accompanied by an increase in the total number of moles of reaction products compared with that of the initial reagents. This feature is widely used in devices for transforming the chemical energy of a fuel into mechanical energy. In this situation, the mechanical work is immediately performed in, for example, a gas turbine where a mixture of hot products of deep oxidation of the fuel ($CO_2 + H_2O$) is expanding. Because reactions of organic fuel combustion are irreversible, no simple regeneration and recycling of the initial fuel mixture exists after its use. Moreover, conventional combustion of fuels produces a lot of waste products (e.g. oxides of carbon and nitrogen) that are harmful to the environment.

A principal peculiarity of all reversible thermochemical reactions in the gas phase is a decrease in the number of moles of the converting mixture at the exothermic stage, which is strongly related also to the necessity to diminish the entropy of the mixture at this stage. This is the case, for instance, when a mixture of hydrogen and oxygen is used, as well as when the fuel is produced in the form of a non-separated reagent mixture via reversible chemical reactions (see Table 12.3).

A complete flow diagram for direct mechanical energy production using these reversible monofuels in a closed-loop cycle, such as that shown in Fig. 12.11, has been used in [30] for calculating its thermodynamic efficiency. Incomplete conversion of the fuel mixture when reaching a chemical equilibrium in the reactors was taken into account. Note that the scheme under consideration avoids any emission from the turbines and thus can be regarded as the real basis for a ZET.

Clearly, ZET allows the combination of total ecological safety of the power plants with a highly efficient conversion of energy of the chemical fuel into mechanical energy without intermediate heat exchangers (see Figs 12.5 & 12.11). For example, in the case of application of the mentioned reversible syn-gas methanation reaction, the efficiency of chemical-to-mechanical (electrical) energy conversion in a ZET can achieve the value of 50% at a reasonable (no greater than 20-fold) compression of syn-gas at the ZET inlet [30].

The suggested scheme of a ZET can be characterized by the thermal efficiency $\eta = W/Q$ of the mechanical energy production (here, $W = Q - Q_1$ is the useful work made by the ZET and Q is the heat evolved by 'discharging' the fuel), by the energy capacity E of a reversible monofuel ($E = W/M$, where M is the total fuel mass in the cycle) and by some other important parameters of the cycles as functions of the maximum allowed (due to some construction material or catalyst

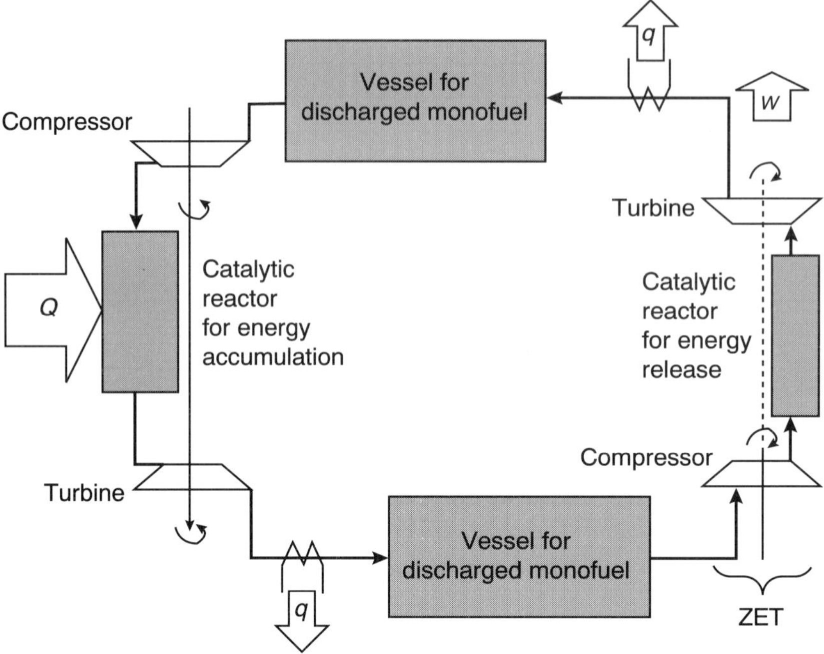

Figure 12.11. Scheme of direct chemical-to-mechanical energy conversion in a closed-loop system for the 'zero-emission turbine' (ZET) for use of reversible fuel mixtures: Q, primary heat energy flux consumed for its accumulation and further release as work W (or electricity as well); q, heat fluxes equilibrating the reaction mixture temperature with that of the ambient.

property restrictions) temperature T of the fuel in the catalytic reactors and compression ratios ε of the gas mixtures before the reactors.

For all monofuels presented in Table 12.3, the efficiency η of the suggested scheme appears to increase monotonically (Fig. 12.12) with a rise of both the temperature T and compression ratio ε in the catalytic reactor for the chemical-to-heat energy transformation in a manner similar to that of a conventional gas turbine. It turns out that, at the moderately high temperatures (< 1100 K) and pressures P_0 (εP_0 < 4 MPa) of the fuel mixture in the scheme, the thermodynamic efficiency of the loop can exceed 40–50%, thus being of considerable practical interest.

The above values of η appear to be slightly lower than those for the ideal Brighton gas turbine cycle at the same compression ratio ε and maximum temperature T. However, the E value in the ZET scheme can significantly exceed that for the Brighton cycle (Fig. 12.12). The high energy capacity might reduce the work expense for feeding the fuel and enhance the efficiency in the real heat engine. Moreover, the temperature of the fuel mixture after the ZET can be high enough to be used for an additional improvement of the overall scheme efficiency.

Note that the suggested application of hydrogen-containing chemically active working media opens new perspectives in the field of heat engine applications in thermochemical energetics. For example, when using the reactions of Tables 12.1 and 12.3, the exhausted fuel mixture can be regenerated simply at a reasonable temperature above T^* via a reverse endothermic reaction and then reused in a closed technological cycle. Because such promising sources of energy as nuclear fuel, concentrated solar light, etc. can be used to accomplish the mentioned endothermic reaction, the suggested scheme makes it possible to design a closed thermochemical loop equipped with a ZET to transform efficiently these kinds of energy into mechanical work and further into

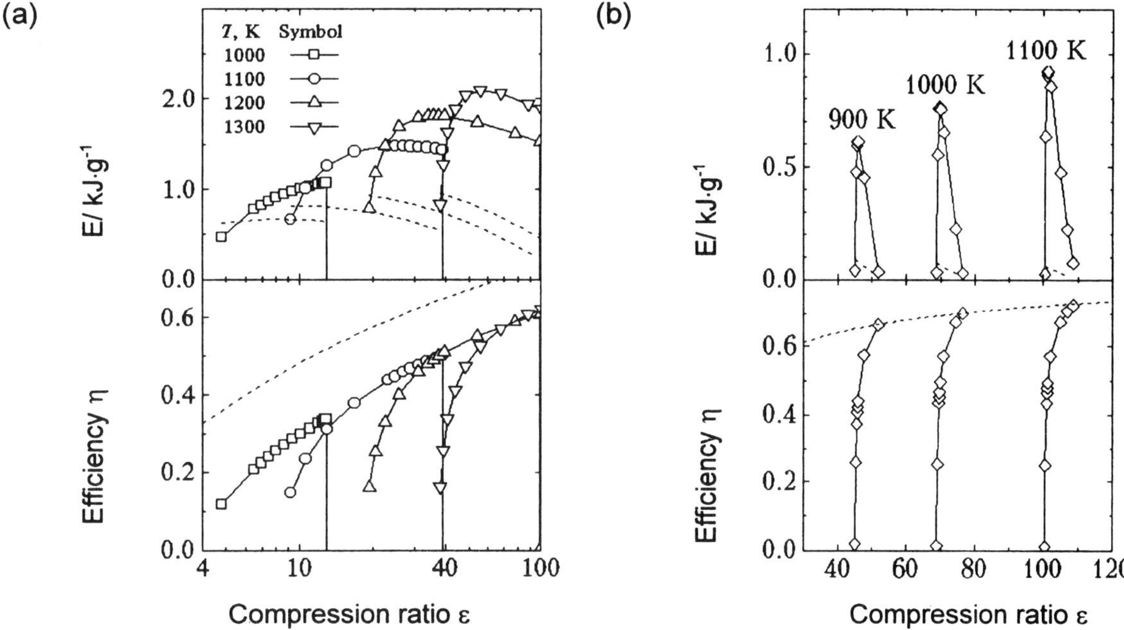

Figure 12.12. Efficiency η of chemical-to-mechanical energy conversion in a ZET and energy capacity E for two different kinds of 'monofuels' versus compression ratio ε for reactions $3H_2 + CO \rightleftarrows CH_4 + H_2O$ (a) and $3H_2 + N_2 \rightleftarrows 2NH_3$ (b). T is the maximum allowed temperature in the catalytic reactor for chemical-to-heat energy transformations. The dotted curves correspond to the ideal Brighton gas turbine cycle parameters used for a comparison. (From [30].)

electricity. This perspective seems to be of importance for the future because of the inevitable depletion of natural resources of fossil organic fuels.

Because in the diagram under consideration the working fluid is heated by the energy released just inside its volume during the chemical reaction, no heat exchangers are required for the generation of mechanical energy or electricity. As a consequence, the mass of the mechanical energy generator can be highly diminished and the difference between the real and the thermodynamic efficiency might be expected to be much less than in the case of conventional gas turbines. This is another point in favor of practical implementation of the ZET scheme in comparison with 'conventional' thermochemical energy-transforming cycles that use various heat exchangers.

In the case of thermochemical cycles based on hydrogen-containing reversible monofuels, the most promising practice seems to be to avoid separation of the fuel mixture into its components after enrichment with energy. Among the other processes, the reaction of syn-gas methanation could provide the highest efficiency at sufficient degrees of conversion of the fuel mixture.

It is of interest to mention that the closed-loop ZET-equipped thermochemical cycles with the use of hydrogen-containing monofuels may appear to be more efficient, compact and reliable for storing and releasing energy than the often-discussed systems based on water electrolysis (for accumulation of energy) with subsequent production of electricity using highly efficient 'hydrogen + oxygen' fuel cells (see Chapters 5 and 8). Indeed, in the case of solar energy storage via photovoltaic devices, the upper energy-converting efficiency of the whole 'solar energy → useful electricity' path is expected to be no more than 10% (see Chapter 13). On the contrary, because the direct thermochemical conversion of solar energy into the energy of chemical fuels can be accomplished with an efficiency of > 40% (see above), the expected overall efficiency of the cycle 'solar energy → syn-gas (accumulation of energy) → ZET-produced electricity' can be up to 20%. Moreover, no sophisticated, low-power-density equipment is needed in the latter case.

Similar estimates for a 'nuclear energy → electricity (by gas turbines) → electrolysis of water → electricity (by fuel cells)' cycle give *c.* 17% of the overall efficiency, whereas direct conversion of nuclear energy into chemical energy of a syn-gas via the ICAR process in combination with a ZET (Fig. 12.5) could provide an overall efficiency η of > 25%, because the energy accumulating efficiency of the ICAR process can be > 50%. The same comments on the necessary energy-converting equipment are valid for this case too.

6 Thermochemical technologies in the conversion and use of middle- and low-potential heat energy

Chemical and especially catalytic technologies based on thermochemical energy conversion can be used efficiently also for middle- and low-potential heat (500–800 and < 500 K), yielding various industrial heat wastes, geothermal water, etc. Mostly, the schemes of such conversion are based on middle- and low-temperature reversible reactions of catalytic dehydrogenation–hydrogenation (see Table 12.1 & Fig. 12.13), as well as of catalytic or non-catalytic dehydration–hydration (see Table 12.4 and [31]).

The main developments in this field are focused on using the so-called 'chemical heat pumps'. Chemical heat pumps allow an increase in the heat temperature during transfer from the heat source to the customer (see Fig. 12.13) and reveal some potential advantages compared to conventional heat pumps that are based on chemically inactive 'working fluid'. The main advantages are the possibility of finding suitable substrates for desirable levels of temperature and of storing the converted heat for any time in the form of chemical energy.

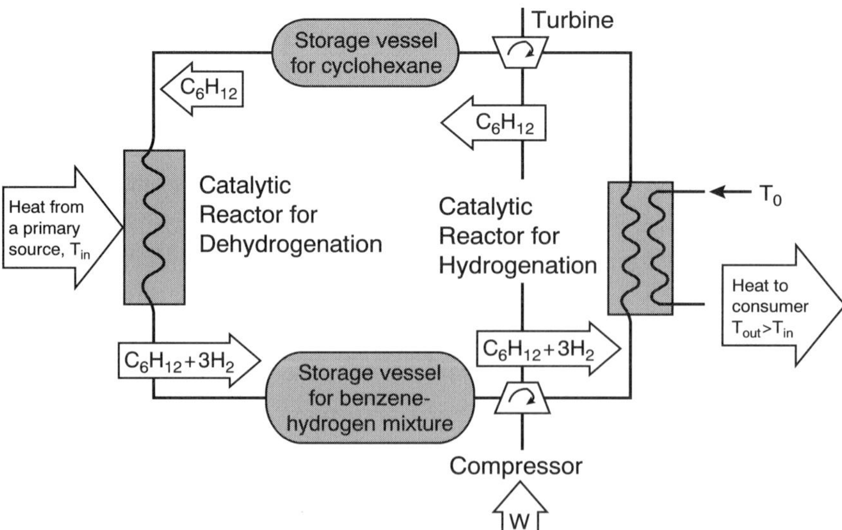

Figure 12.13. An example of the principal scheme of a chemical heat pump based on the reversible catalytic process of cyclohexane dehydrogenation–benzene hydrogenation. The temperature T_{out} at the outlet of the heat pump is higher than the temperature T_{in} at the inlet due to consumption of mechanical energy W provided by the compressor, which increases the pressure in the catalytic reactor and thus shifts the conditions of chemical equilibrium to a higher temperature. (From [32,33].)

Table 12.4. Enthalpy and characteristic temperature of reversible dehydration for some simple crystalline hydrates in bulk.

No.	Formula of crystalline hydrate and number of water molecules capable of release	Average enthalpy of dehydration (kJ mol-1 H_2O)	Specific heat of dehydration* (kJ kg-1)	Reference temperature of decomposition of crystalline hydrates (°C)
1	$NiSO_4 \cdot 7H_2O$ (−1)	50.6	180	31.5
2	$Na_2SO_4 \cdot 10H_2O$ (−10)	51.4	1595	32.4
3	$Na_2SO_3 \cdot 7H_2O$ (−7)	51.7	1436	87
4	$CaCl_2 \cdot 6H_2O$ (−4)	53.4	1061	30
5	$SrCl_2 \cdot 6H_2O$ (−4)	56.7	797	60
6	$MgCl_2 \cdot 6H_2O$ (−6)	70.0	1948	120–280
7	$CuCl_2 \cdot 2H_2O$ (−2)	59.61	1420	200
8	$CoSO_4 \cdot 7H_2O$ (−7)	55.6	1385	300
9	$MgSO_4 \cdot 7H_2O$ (−7)	58.5	1661	280
10	$Na_2B_4O_7 \cdot 10H_2O$ (−5)	54.2	710	25–50

From [31].
* Calculated per unit mass of the hydrated state and for the number of releasing water molecules denoted in parentheses; the final state of water is considered to be gaseous.

Work considering the possibility of thermocatalytic conversion of middle- and low-potential heat concerns various secondary sources of energy, that are now utilized insufficiently. The work of chemistry application in this area is still quite scanty; however, the development of chemical heat pumps that permit the conversion of initial middle-potential heat to more powerful high-potential heat by using mechanical energy seems to be quite promising. Interest in catalytic processes for this application is due to the fact that traditional physical processes, e.g. 'evaporation–condensation', are not always convenient enough to be used in the heat pumps operating within the middle-temperature region. At the same time, application of reversible catalytic reactions, such as cyclohexane dehydrogenation–benzene hydrogenation or methylcyclohexane dehydrogenation–toluene hydrogenation (reactions 6 and 7 in Table 12.1), makes it possible to transform the heat potential within a definite temperature range and keep the accumulated energy in the chemical form for a certain time until it can be used (see Fig. 12.13 and [32–35]). There are also studies on heat pumps based on catalytic reactions that proceed at much lower temperatures.

At present, the work on catalytic devices for conversion of middle- and low-potential heat energy into chemical energy is still mainly within the stage of research, but if consumers should become more interested in the utilization of waste heat, intensive development could rapidly be undertaken. For example, the reaction of selective dehydrogenation of isopropyl alcohol to acetone, which occurs at temperatures of *c*. 370 K, was proposed for the conversion of low-potential heat in catalytic heat pumps (see reaction 12 in Table 12.1 and [36,37]). However, to use heat of a temperature below 400 K, more often thermochemical processes are proposed in which weaker chemical bonds are ruptured: e.g. the processes of reversible degasification, dehydration being the most widely known example of this type. These processes are not truly catalytic, but some such systems are based on the use of materials that are produced commercially as catalysts or adsorbents. For example, experiments have been reported aimed at the development of heat pumps that are based on reversible dehydration or deamination of zeolites [37,38] or dehydrogenation of dispersed metal hydrides [39] as working processes.

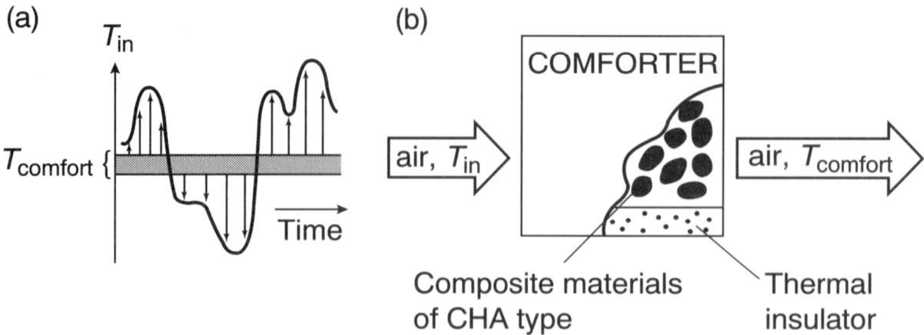

Figure 12.14. Accumulation (recuperation) of low-potential ambient heat with a composite material of a 'chemical heat accumulator' (CHA) type based on crystalline hydrates or other hygroscopic materials (a) and the implementation of this idea in a 'comforter' — a device filled with the CHA granules that are blown around with air flows to be cooled or heated (b). Equilibration of air temperature is accomplished via sorption–desorption processes of moisture contained in the air.

An ability to accumulate and store the low- and middle-potential heat is provided also by new materials with a large absorption capacity. A wide variety of compounds such as crystalline hydrates, with large heat-storing capacities, are well known (see Table 12.4). Of special interest are also materials such as zeolites and the so-called chemisorbents. When using very simple substrates as adsorbates (e.g. water, carbon dioxide, oxygen, etc.), these materials are able to accumulate a large amount of heat even in the ambient temperature interval. This provides a powerful tool for the recovery of heat wastes and for efficient utilization of the natural day-to-night or seasonal temperature gradients, as well as for the design of non-traditional freon-less air-conditioning systems, etc. The idea of using crystalline hydrates or chemisorbents is very simple and is clarified in Fig. 12.14. However, in order to obtain the materials for easy and convenient exploitation in such devices, the crystalline hydrates should be impregnated in a matrix with open pores, otherwise these hydrates melt on adsorbing water and therefore lose their mass- and heat-exchanging ability.

Based on the last idea, recently new composite materials with the mentioned properties — known as chemical heat accumulators (CHA) or selective water sorbents — have been designed especially to promote storage of low-potential heat of a temperature below 370 K [31]. These materials consist of a porous matrix with the open pores filled with hygroscopic substances such as crystalline hydrates, etc., capable of reversible hydration–dehydration processes in a 'comfort' temperature region (10–30 °C). The simplest example of such a material is silica filled with calcium chloride:

$$[CaCl_2{\cdot}nH_2O/SiO_2] \rightleftarrows [CaCl_2{\cdot}(n-m)H_2O/SiO_2] + mH_2O - Q$$

Dehumidification of the bed with such material by dry or hot air leads to cooling of the blown air, whereas passing cool or humid air through the dehumidified bed leads to adsorption of water vapor and therefore liberation of sufficient heat of water adsorption.

Although the accumulation capacity of the physical phase transition (latent heat of melting, etc.) does not exceed 200 kJ dm^{-3}, the corresponding value for the 'chemical heat accumulators' (based on hydration–dehydration reactions) can be an order of magnitude higher. As a result, the combination of the properties of crystalline hydrates in storing the low-potential heat with the improved mass and heat transfer in the granulated CHA beds allowed unique materials to be obtained with heat storing capacities up to 2000 kJ kg^{-1} [31]. Recent studies show that crystalline hydrates inside the fine porous matrices behave as a supersaturated liquid solution rather than as a solid nanocrystal [40].

Note that in countries with a rigid continental climate, the day-to-night and seasonal temperature gradients seem to create a non-expected energy source with enormous potential that is able to substitute large expenditures of electricity or fuel to stabilize the temperature in buildings, etc.

7 Conclusions

The above presentation did not disclose all the abilities of thermochemical processes based on reversible chemical reactions for energy conversion. However, it is seen how numerous and large the areas of application of these thermochemical energy conversion technologies could be. The main problems in the development of suitable thermochemical technologies consist of: (i) finding a highly reversible chemical process suitable for a given temperature region; (ii) finding highly active catalysts for this reaction in order to ensure the necessary SPL of the process; and (iii) the design of special catalytic reactors determined to optimize the energetic efficiency of the above energy transformation.

A very interesting feature of thermochemical processes is the possibility of designing closed-loop cycles (for chemically active 'working bodies' of the energy cycles), which are extremely ecologically benign and seem to be an important argument for fast development of these technologies for energy industries.

Thermochemical conversion of different types of primary energy is now under intensive study and, due to its simplicity, sometimes approaches the level of a pre-commercial use, having, however, only little (if ever) practical application as yet. The reason for these technologies being at the level of mostly pilot scale tests originates from their lower commercial efficiency in comparison with conventional technologies based on non-reversible consumption of available and still cheap organic energy carriers.

One can expect a sufficient rise of interest in these technologies even in the observable near future, because they allow, first of all, the construction of devices or installations with very high efficiency and SPL of conversion of various kinds of primary energy into energy of chemical fuels, which enables long-term storage of the accumulated energy. Also, typically, thermochemical technologies allow the conversion and utilization of sources of heat with moderate (temperatures below 1000 K) or even low potential, which is another evident advantage of these technologies in comparison with those used conventionally in the energy industries.

8 References

1 Bland RB, Ewing FJ. Conversion of heats of chemical reactions to sensible energy. *US Patent 3,225,538*. Filed March 25, 1960, Issued December 28, 1965.

2 Parmon VN. *Catal. Today* 1997; **35**: 153.

3 Parmon VN, Ismagilov ZR, Kerzhentsev MA. In Thomas JM, Zamaraev KI (eds) *Perspectives in Catalysis. Chemistry for the 21st Century Monograph*, Oxford: Blackwell–IUPAC, 1992; 337.

4 Belousov IG. In Legasov VA (ed.) *Atomno-Vodorodnaya Energetika i Tekhnologiya (Atomic-Hydrogen Energetics and Technology)*, Issue 3. Moscow: Atomizdat, 1980; 172 (in Russian).

5 Aristov YuI, Parmon VN. In Zamaraev KI, Parmon VN (eds) *Photocatalytic Conversion of Solar Energy*. Novosibirsk: Nauka, 1991; 315 (in Russian).

6 Shnyp VP, Novikov GI, Poplavskii VV, Glybin VP, Vilivetskii VG. *Vestsi Akad. Navuk BSSR Ser. Khim.* 1989; **1**: 24 (in Russian).

7 Ache von HJ. *Angew. Chem.* 1989; **101**: 1.

8 Legasov VA (ed.) *Atomic-Hydrogen Energetics and Technology*, Issues 8. Moscow: Energoatomizdat, 1988 (in Russian).

9 Bolcich JC, Veziroglu TN (eds) *Hydrogen Energy Progress* XII. Proceedings of the 12th World Hydrogen Energy Conference, Buenos Aires, Argentina, 21–26 June 1998. Buenos Aires: Corradi Impresions, 1998.
10 Hohlein B, Menzer R, Range J. *Appl. Catal.* 1981; **1**: 125.
11 Carty RH, Congev WP. *Int. J. Hydrogen Energy* 1979; **4**: 489.
12 Broggi A, Joels R, Mertel G, Morbeello M. *Int. J. Hydrogen Energy* 1981; **6**: 25.
13 Panteleimonova AA. *Int. J. Hydrogen Energy* 1995; **20**: 429.
14 Petros'yants AM. *Nuclear Energetics*. Moscow: Nauka, 1981 (in Russian).
15 Nesterenko VB, Nichipor GV. *Kinetics and Mechanism of Thermal and Radiation–Thermal Processes in Heat Carrier* N_2O_4*–*NO_2. Minsk: Nauka i Technika, 1989 (in Russian).
16 Tanashev YuYu, Fedoseev VI, Aristov YuI. *Catal. Today* 1997; **39**: 251.
17 Aristov YuI, Tanashev YuYu, Prokopiev SI, Gordeeva LG, Parmon VN. *Int. J. Hydrogen Energy* 1993; **18**: 45.
18 Gordeeva LG, Aristov YuI, Moroz EM, *et al. J. Nucl. Mater.* 1995; **218**: 202.
19 Gamburg DYu, Dubrovkin NF (eds) *Handbook on Hydrogen. Properties, Production, Storage, Transportation, Application*. Moscow: Khimija, 1989 (in Russian).
20 Parmon VN. In Zamaraev KI, Parmon VN (eds) *Photocatalytic Conversion of Solar Energy*. Novosibirsk: Nauka, 1985; 7 (in Russian).
21 Kamejama H, Tomino Y, Sato T, Amir R. *Int. J. Hydrogen Energy* 1989; **14**: 323.
22 Tagawa H, Endo T. *Int. J. Hydrogen Energy* 1989; **14**: 11.
23 Parmon VN, Zamaraev KI. In Serpone N, Pelizzetti E (eds) *Photocatalysis. Fundamentals and Application*. New York: John Wiley, 1989; 565.
24 Anikeev VI, Parmon VN, Kirillov VA, Zamaraev KI. *Int. J. Hydrogen Energy* 1990; **15**: 275.
25 Yogev A, Kribus A, Epstein M, Kogan A. *Int. J. Hydrogen Energy* 1998; **23**: 239.
26 Anikeev VI, Kirillov VA, *Sol. Energy Mater. Solar Cells* 1991; **24**: 1.
27 Becker M, Funken KH (eds) *Solarchemische Technik*, Bands I and II. Berlin: Springer, 1989.
28 Schpilrain EE (ed.) *Proceedings of 7th International Symposium on Solar Thermal Concentrating Technologies, September 26–30, 1994*. Moscow: IVTAN, 1994.
29 Aristov YuI, Fedoseev VI, Parmon VN. *Int. J. Hydrogen Energy* 1997; **22**: 869.
30 Prokopiev SI, Aristov YuI, Parmon VN. *Int J. Hydrogen Energy* 1997; **22**: 415.
31 Levitskij EA, Tokarev MM, Aristov YuI, Parmon VN. *Sol. Energy Mater. Solar Cells* 1996; **44**: 219.
32 Aristov YuI, Parmon VN. *Int. J. Energy Res.* 1993; **17**: 293.
33 Murata K, Yamamoto K, Kameyama H. *Int. J. Hydrogen Energy* 1996; **21**: 201.
34 Beckmann G, Gilli PV. *Thermal Energy State*. Wien: Springer Verlag, 1989.
35 Apinner B (ed.) *Recent Prog. Gen. Proced.* 1988; **2**: 1.
36 Saito Y. *Seisan-Kenkyu* 1986; **38**: 459 (in Japanese).
37 Ito E, Yamashita M, Hagiwara SL, Saito Y. *Chem. Lett.* 1991; 351.
38 Restuccia G, Recupero V, Cacciola G, Rothmeyer M. *Energy* 1988; **13**: 333.
39 Bjurstrom H, Suda S. *Int. J. Hydrogen Energy* 1989; **14**: 19.
40 Aristov YuI, Tokarev MM. *React. Kinet. Catal. Lett.* 1996; **59**: 325, 335.

13 Quantum Chemical Processes for the Conversion of Energy

H. TRIBUTSCH[1] and V.N. PARMON[2]

[1] *Freie Universität Berlin and Hahn-Meitner Institut, Dept. Solare Energetik, 14109 Berlin, Germany*

[2] *Boreskov Institute of Catalysis, Siberian Branch of the Russian Academy of Sciences, Prospekt Akademika Lavrentieva 5, Novosibirsk 630090, Russia*

1 Introduction

Quantum chemical processes for the conversion of energy have developed successfully during the evolution of life and are now present throughout the world in the form of efficiently photosynthesizing bacteria and plants.

In molecular systems, photons of solar light are able to excite electrons, thereby temporarily increasing their free energy. Originally, it was believed that by tailoring appropriate molecular photochemical mechanisms it would be possible to convert light into chemical energy even via simple homogeneous photochemical processes, but it is now established that this cannot occur with reasonable efficiencies because the products generated would rapidly react back to dissipate the temporarily converted energy into heat. Heterogeneous or microheterogeneous structures (colloids, microcrystals, membranes, solid/liquid interfaces) are needed to provide vectorial transport of separated charge carriers in order to reduce the probability for back reactions. Natural photosynthesizing systems are convincing evidence for the need to have heterogeneous structures for quantum chemical energy conversion. In all known cases, the conversion of light into electrochemical and chemical energy is bound to heterogeneous structures (membranes, vesicles) through which vectorial processes are generated and where oxidized species are separated from reduced species.

During recent decades many molecules, assemblies of molecules and organic materials have been tested for quantum chemical energy conversion. General experience has been that molecular systems used in photosynthetic structures cannot easily be considered for artificial energy conversion systems. The main reason for this is that the molecules involved in biological systems (electron-transfer mediators, proteins, catalytic enzymes) are typically not stable for long periods of time. They are functioning in biological systems because they are operated far from equilibrium under conditions of self-organization; this means that they are self-regenerated, repaired and replaced. When introduced into artificial systems they can only survive for relatively short periods of time before irreversibly degrading. Nevertheless, many important basic research studies have been made on molecular systems extracted from natural quantum energy-converting structures. These studies have allowed us to deduce some basic principles that can be applied for energy conversion and to learn about the structure of the electron transfer assemblies required for charge separation and energy conversion.

In order to obtain reasonable efficiencies and reasonable stability over long periods of time, inorganic semiconducting materials (in the form of colloids, particles, layers and single crystals) have been employed for artificial quantum energy-converting systems. The semiconductors involved are materials characterized by energy gaps in the range 0.8–2.5 eV. Hole–electron recombination across these energy gaps occurs typically only in time periods of 10^{-9}–10^{-5} s which is sufficiently long for charge separation and energy conversion. In addition, these materials are

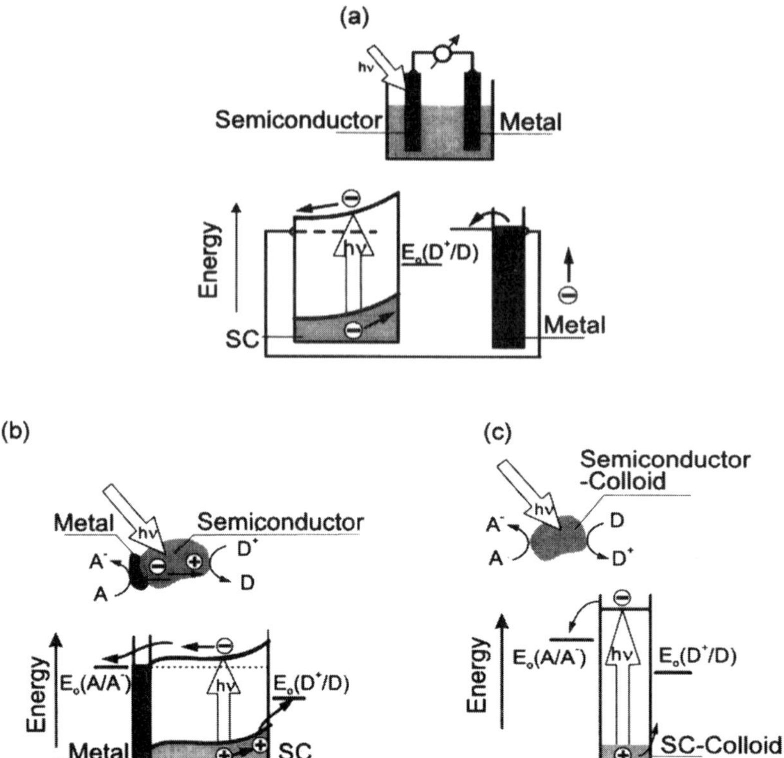

Figure 13.1. Drawing of a photoelectrochemical cell with a semiconductor (SC) anode (a) as well as a photochemical diode (b) (semiconductor grain with metal particle), and a colloid (c), with energy schemes explaining the photoelectrochemical process.

electrically conducting, which allows the separation of charges and the build-up of electrical potentials. Typical arrangements used for quantum chemical energy conversion are shown in Fig. 13.1.

In the energy scheme of Fig. 13.1 it can be seen that a colloid may have an extended forbidden energy region due to quantum size phenomena. A microscopic particle is made asymmetric by attachment of a localized metal deposit, so that the system can function as a photochemical diode. In addition, the most commonly used photoelectrochemical geometry is shown in which the photoactive material and a metal counter-electrode form part of a photoelectrochemical cell in which ion conduction closes the electrical circuit. A photoelectrochemical cell is, in principle, the equivalent of the membrane arrangement used in natural photosynthesis, where proteins capable of electron transfer across the membrane serve as electronic conductors and the two membrane surfaces act as anode and cathode, respectively.

2 Principal restrictions on quantum energy conversion efficiency

The thermodynamic limitations for conversion of light into chemical and electrical energy are well investigated. Basically, a distinction must be made between pure quantum energy conversion and combined quantum–thermal energy conversion. As Fig. 13.2 shows, *c*. 31% energy conversion can be obtained with a 'one-sun' light intensity. This efficiency, however, decreases with increasing temperature of a light-absorbing system. Quantum energy conversion is therefore more efficient at a lower absorber temperature.

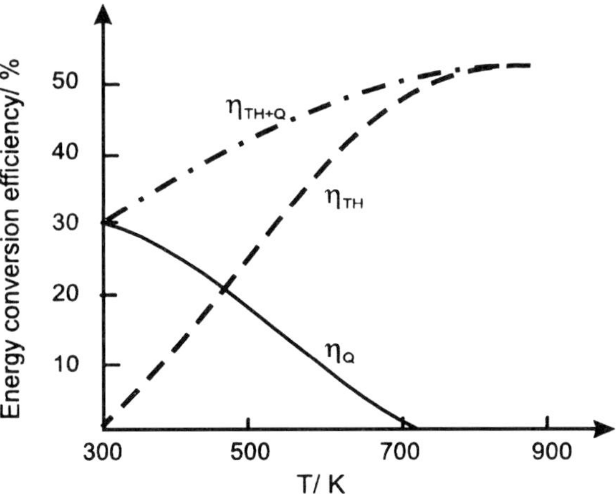

Figure 13.2. Maximum thermodynamic energy conversion efficiencies for the quantum energy conversion and combined quantum–thermal energy conversion dependent on the temperature of the absorber; η_Q and η_{TH} denote the values of the quantum and thermal conversion efficiencies [1,2].

If solar energy is converted by a thermal machine, high absorber temperatures must be reached because of the requirements of the Carnot cycle. For this reason, the efficiency for thermal energy conversion increases gradually to reach a value of *c*. 54% (again for 'one-sun' intensity) at high temperature [1–3]. For the combined quantum–thermal energy conversion, a superposition of both efficiencies defines the limit of the total energy conversion efficiency. It is important to note that the efficiencies for energy conversion increase with increasing solar energy concentrations.

In the case of pure quantum energy conversion, however, it is necessary to avoid the energy-converting absorbers being heated too much. For these converters, the energy conversion efficiencies are, of course, dependent on the energy gap (i.e. 'red wavelength thresholds' for initiation of photochemical activity) across which the excitation of electrons occurs. Figure 13.3 shows the theoretically expected maximum energy conversion efficiency as a function of the energy gap, indicating that the efficiency decreases towards large and small energy gaps: in the first case, too few photons from the solar spectrum are absorbed; in the second case, the absorbed photons produce excitation of electrons to high energy levels from where they rapidly reach the edge of the conduction band by releasing a large portion of thermal energy. When several absorbers are combined to increase the total absorption of solar energy, much higher efficiency limits can be obtained. In an ideal case of infinite absorbers with varying energy gaps, a total solar energy conversion efficiency of 68% has been estimated [1,2].

Figure 13.3 shows the solar energy conversion efficiency reached for different materials depending on their energy gap. It can be seen that solid-state photovoltaic systems like silicon and gallium arsenide have reached efficiencies relatively close to the thermodynamic limit of energy conversion efficiency. A series of electrochemical cells have reached energy conversion efficiencies of more than 10%; however, they are not yet interesting for technical applications due to their limited stability against corrosion and degradation. A special case of an electrochemical solar cell is the dye-sensitized solar cell in which photons excite a dye adsorbed onto the interface of a large-gap oxide, which leads to electron injection and energy conversion. Here, the absorption edge of the photon-absorbing dye limits the energy conversion efficiency. A 10% energy conversion efficiency has been obtained with the dye-sensitized cells, which have very large oxide surfaces for dye absorption and efficient capture of photons.

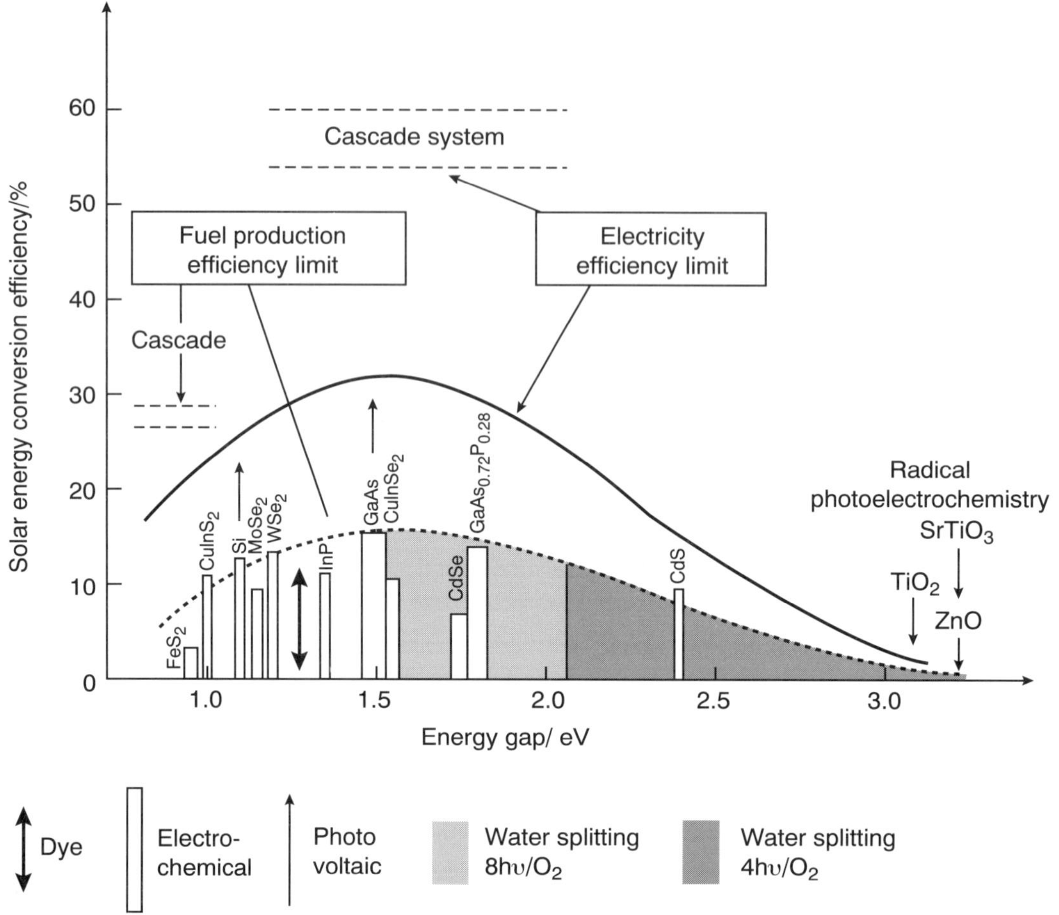

Figure 13.3. Energy conversion efficiency as a function of the energy gap of the quantum absorber.

Table 13.1. Some processes suggested for artificial systems that mimic plant photosynthesis for quantum (photochemical or photocatalytic) conversion of solar energy.

No.	Process	n*	ΔG°_{298} (kJ mol^{-1})	ΔH°_{298} (kJ mol^{-1})	ΔG°_{298} per electron (eV)	λ_o† (nm)
1	$H_2O \rightarrow H_2 + \frac{1}{2}O_2$	2	237.0	285.5	1.23	1008
2	$2H_2O + CO_2 \rightarrow CH_4 + 2O_2$	8	817.2	889.5	1.06	1176
3	$2H_2O + CO_2 \rightarrow CH_3OH + \frac{3}{2}O_2$	6	701.8	725.6	1.21	1025
4	$3H_2O + 2CO_2 \rightarrow C_2H_5OH + 3O_2$	12	1330.5	1407.8	1.15	1077
5	$\frac{3}{2}H_2O + \frac{1}{2}N_2 \rightarrow NH_3 + \frac{3}{4}O_2$	3	339.0	382.1	1.17	1059
6	$2H_2O + N_2 \rightarrow N_2H_4 + O_2$	4	757.8	620.3	1.97	629

* n, number of electrons to be transferred.

† λ_o, 'red' threshold for the denoted n-quantum process.

A large variety of possible quantum processes have been considered as candidates for photochemical or photocatalytic conversion of solar energy into chemical fuels (see [3] and Table 13.1). The calculated values for the excitation threshold, however, do not consider that for solar light typically 0.8–1.0 eV per photon is additionally lost due to various loss processes ranging from insufficient spectral matching and wavelength-dependent absorption efficiency of pigments to overpotentials in heterogeneous (membrane) processes.

Energy conversion efficiencies for the generation of chemical fuels are lower than for the generation of electricity, the reason being that energy is typically lost during intermediate chemical transformations leading to energy storage. It is also important for calculating the efficiency for the generation of chemical fuels that consideration of the typical electron turnovers needed for the generation of desired chemical energy carriers is given. Water molecules can, for example, be split into molecular hydrogen and oxygen by eight photons (with respect to one molecule of O_2 evolved), as occurs in green plant photosynthesis. Here, electrons are excited twice in series in order to be transferred from the redox potential of water to the redox potential of the species to be reduced on the opposite side of the membrane (nicotinamide adenine dinucleotide phosphate, $NADP^+$). Water molecules can, however, also be split into molecular products by using only four photons to transfer four electrons [1,2]. This limits the use of visible light due to expected losses of energy in the intermediate transformations. A minimum energy corresponding to *c*. 2.1 eV would be required for such a process (compare this with the corresponding thermodynamic limits in Table 13.1). Thus, if nature applied such a mechanism, the leaves of trees would have to be yellow for maximum energy absorption. Depending on whether fuel-generating systems are designed for four or eight quantum processes in the case of water splitting, different energy conversion efficiencies are obtained. As Fig. 13.3 shows, an eight-photon system for oxygen evolution, as applied in photosynthesis, allows a much higher energy conversion efficiency of *c*. 16%. If a water-splitting system was designed to involve two absorbers in series, in order to capture more light, a maximum theoretical light efficiency of 28% would be obtained. This means that it is probable for a technical system with *c*. 20% energy conversion efficiency to be achieved in the future.

It is interesting to compare such an estimated technical energy conversion efficiency with the energy conversion efficiency typically obtained for the production of biomass. A sugar-cane field in Brazil with three harvests per year can yield an average total energy conversion efficiency for the generation of biomass (chemical fuel) of *c*. 2.0%. This is relatively low, because the plant is not efficiently synthesizing chemicals throughout the year and there are periods when the field is not active at all. Nevertheless, it has to be said that very efficient so-called 'C4-plants' (like the mentioned sugar-cane) reach the maximum energy conversion efficiencies for the synthesis of chemicals (biomass) of 3–4%. This is, however, the case during the active growing season. The efficiency for primary energy conversion should be higher because part of the energy converted is used for the living processes of the plant itself. In conclusion, it can be said that high energy conversion efficiencies for the generation of fuels via quantum chemical processes can be expected. If properly designed, the artificial systems could become much more efficient than natural photosynth esizing systems, which are optimized according to very different boundary conditions and not specialized for energy production alone.

3 Quantum chemical processes and photoproduction of hydrogen

Here, we briefly discuss some achievements and modern trends in research on the conversion of solar energy via the most popular quantum chemical process — hydrogen photoproduction. We shall also give a short physicochemical assessment of the various methods of hydrogen photoproduction being developed, especially its production in quantities essential for use in the small- or large-scale energetics or hydrogen-based technologies. In so doing, we will not discuss methods of hydrogen photoproduction that are based on light-stimulated exothermic (or, more correctly, exergonic) chemical processes. In such processes, due to their thermodynamic allowance, hydrogen production can occur spontaneously, with the light only accelerating these reactions. Also, the proposed definition of the term 'photoproduction' makes it possible to exclude from discussion the well-known method of obtaining hydrogen by conventional electrolysis of water or some other substrates at the expense of electricity produced by solid semiconductor solar cells or other solar generators of electricity, although this method is popular enough and the solar-generated electricity is in the process of being produced on a semi-industrial scale (see [4] as well as earlier issues of this set of books).

Thus, the methods of direct photoproduction of hydrogen may be classified into four broad groups: photobiotechnological, photoelectrochemical, photochemical and thermochemical. The first three methods are quantum whereas the fourth is non-quantum and is discussed in Chapter 12.

The most desirable raw material for the production of hydrogen is indeed water. For this case, the overall reaction is expressed as a four-electron redox process (see also Table 13.1):

$$2H_2O \rightarrow 2H_2 + O_2$$

The main steps here are the two-electron reduction of water by the light-generated reducing agents:

$$2H_2O + 2e^- \rightarrow H_2\uparrow + 2OH^-$$

as well as a four-electron oxidation of water by the light-generated oxidants:

$$4OH^- - 4e^- \rightarrow O_2 + 2H_2O$$

However, for some particular needs one can also consider as raw materials for hydrogen photoproduction various electron-donating compounds in the form of dissolved or water-suspended organic substances, sulfide ions, etc.

In principle, for a possible customer of developing technologies of solar hydrogen photoproduction, the basic parameter for their assessment and comparison should be the surface area needed for the production of a desired amount of hydrogen (resulting from the available flux of solar light and the energetic efficiency of solar energy utilization), as well as the specific productivity of the hydrogen generator and a large set of economic characteristics. Unfortunately, for the majority of systems producing hydrogen under direct action of solar light, the level of development reached still does not allow one to make serious assessments of this kind. Therefore, at the moment we can usually only operate with highly abstract and easily obtained estimates of the maximum productivity of the systems for the photoproduction of hydrogen.

Thus, for the quantum methods of hydrogen production where one quantum of light may initiate no more than one endothermic primary elementary chemical reaction, the limiting productivity of the system corresponds to the number of photoactive light quanta falling on the surface of a photoproducing device (a reactor with microorganisms, photoelectrolyzer, photochemical reactor, etc.). It is not difficult to calculate that when using light quanta more energetic than, for example, those of red light with $\lambda = 700$ nm (which is the 'red threshold' for most photosynthesizing organisms), on sunny summer days one may obtain from 1 m^2 of the device up to 85 dm^3 of hydrogen per hour if only two light quanta are needed to evolve one hydrogen molecule (the peak flux of solar energy on a sunny day at noon is considered to be *c.* 1 kW m^{-2}). For medium latitudes (e.g. the region of Berlin, London, Moscow, Novosibirsk), this corresponds to estimates of *c.* 0.6 and 70 m^3 m^{-2} of hydrogen per sunny summer day and per year, respectively.

In fact, the figures given above usually have to be reduced by a factor of at least two or even four (the latter for the photosynthesis of green plants). The maximum available productivity of an ideal terrestrial quantum generator of hydrogen with respect to its 'red threshold' is presented in

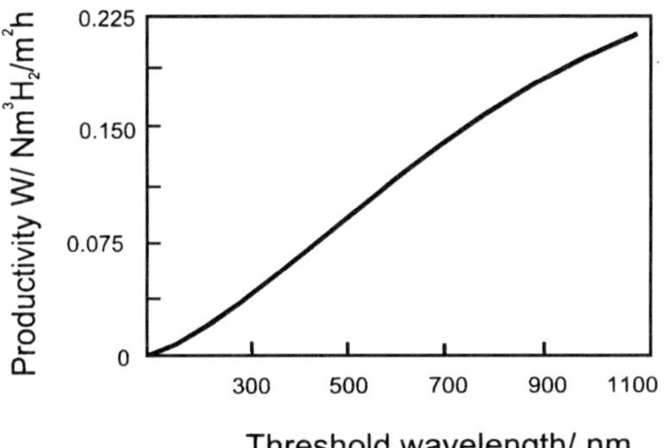

Figure 13.4. Dependence of maximum available productivity W (in Nm^3 H_2 m^{-2} h^{-1}) of an ideal terrestrial quantum generator of hydrogen on the wavelength of its 'red threshold' at the solar light intensity 1 kW m^{-2}, which is typical for a sunny summer at noon and close to the total flux of solar light out of the Earth's atmosphere (1.3 kW m^{-2}). The hydrogen generator is assumed to have a quantum efficiency of 0.5 (i.e. it utilizes two light quanta for the generation of one H_2 molecule) for the photoactive part of the spectrum. For the calculations, the solar spectrum was approximated with that of a black body at the temperature 5800 K. Light of wavelength $\lambda < 300$ nm is considered to not reach the Earth's surface due to its total absorption by the atmosphere.

Fig. 13.4. Still more precise estimates that take into account the particular thermodynamics of the processes being accomplished, as well as some particular properties of light-absorbing substances or objects, may also reduce these values.

For non-quantum thermochemical processes, the limiting productivity of a hydrogen-generating device is not correlated with the number of quanta incident on the device but depends on the density of the energy flux, on the particular thermodynamics of the chemical process used and on the shape of the chemical reactor used simultaneously as a receiver of solar radiation (see [3] and Chapter 12). Estimates and available experimental data with the pilot installations show that the productivity of these systems with respect to the amount of hydrogen produced may considerably exceed the limiting productivity of quantum systems. Unfortunately, thermochemical systems can operate only under the action of highly concentrated light, which is usually possible only on bright sunny days in the absence of high light scattering, which is a frequent problem at northern latitudes. This compares to the quantum systems, which operate under mild conditions and are capable of using even weak scattered sunlight fluxes.

3.1 *Photobiotechnological methods of hydrogen production*

All photobiotechnological methods of hydrogen production are based on the application of living photosynthesizing organisms or of their functioning fragments (e.g. chloroplasts).

The principal limit of productivity of biotechnological systems is restricted by peculiarities of photosynthesizing systems of living organisms, which are able to use only light of wavelength $\lambda < 900$ nm.

3.1.1 BIOPHOTOLYSIS OF WATER

There is no doubt that the most attractive photobiotechnological method of hydrogen production could be the application of strains of plants (e.g. green micro-algae) that are capable of performing direct biophotolysis of water via the reaction:

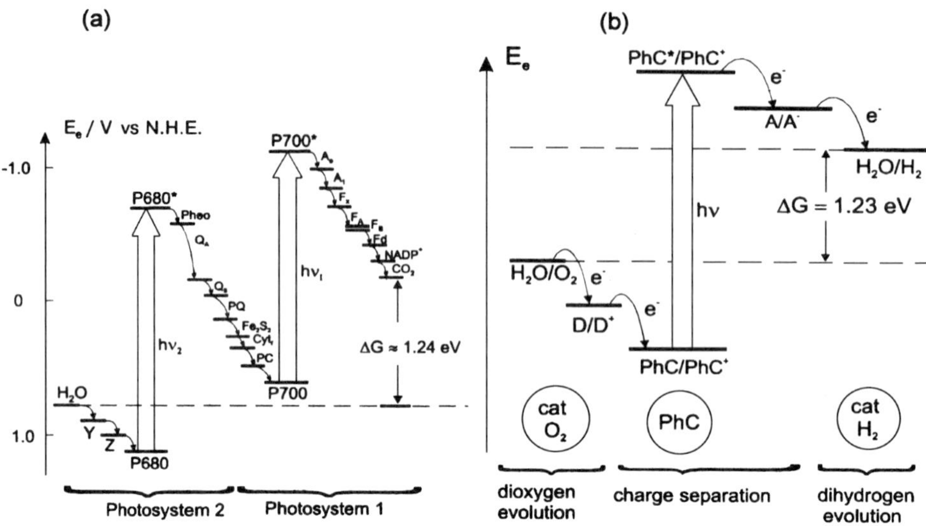

Figure 13.5. A simplified energy diagram and scheme of electron transfer in plant photosynthesis (a) and during water cleavage in an artificial molecular photocatalytic system (b); E_e is the electrochemical potential of reacting electrons; ΔG is the accumulated Gibbs free energy; PhC is a photocatalyst capable of charge separation under the action of light quanta.

$$H_2O \rightarrow H_2 + \tfrac{1}{2}O_2 \quad (1)$$

immediately in their cells, instead of the more complicated process of traditional plant photosynthesis that results in the formation of carbohydrates from carbon dioxide and water:

$$CO_2 + H_2O \rightarrow (CH_2O) + O_2 \quad (2)$$

In fact, the electron transfer chain of green plants is created by nature especially as a biotechnological adaptation of the more simple process of decomposition of water into hydrogen and oxygen (see [5–7]). Because this takes place as a result of the particular structure of the photosynthesizing apparatus of plants, composed of two photosystems ('Photosystem 1' and 'Photosystem 2'), it may be hoped that the energy efficiency of the direct biophotolysis of water could be doubled in comparison with the efficiency of the process of traditional plant photosynthesis (see Fig. 13.5), with the final efficiency of solar energy conversion achieving a factor of *c.* 20–30% when using a cascade system. Indeed, in principle, 'Photosystem 2' alone is already sufficient for providing the evolution of hydrogen and oxygen; this follows from a comparison of the thermodynamic parameters of 'Photosystem 2' and those of water cleavage. The most attractive feature of systems based on living hydrogen-producing organisms is that they are generally capable of self-maintenance and self-reproduction.

Many photosynthetic microorganisms are known to be capable of producing hydrogen under the action of light in anaerobic media in the absence of carbon dioxide [7–9]. Nevertheless, attempts to create efficient strains of such hydrogen-producing microorganisms or to select conditions to obtain a reasonable output of hydrogen at the expense of the process taking place in a single organism at steady-state illumination have not yet achieved reasonable success.

Attempts to create systems based on free or immobilized chloroplasts of plants that are able to provide photolysis of water have not been very successful either. Moreover, even in the known cases of hydrogen photoproduction by plants or their chloroplasts, hydrogen evolution arises very often from *secondary* processes, which means that the hydrogen formed appears to be a side-

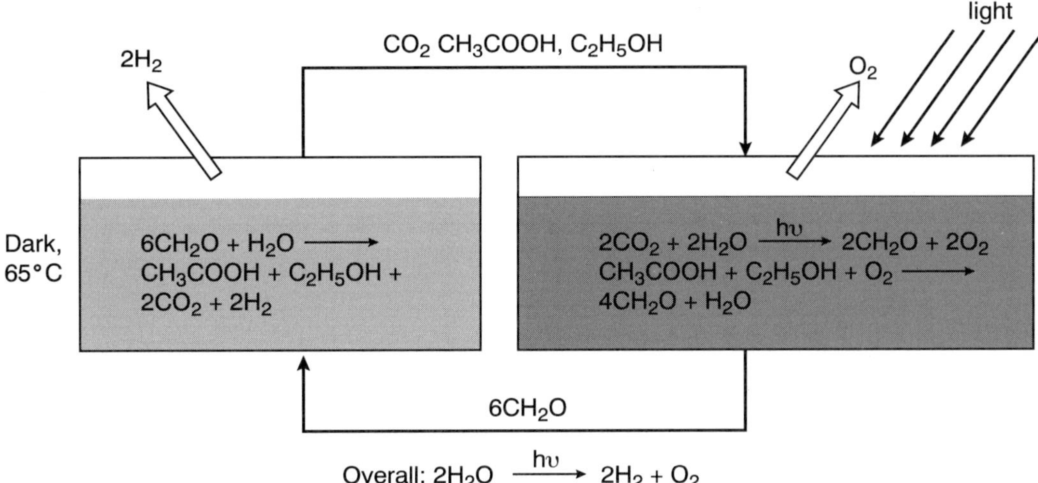

Figure 13.6. A flow diagram for the photoproduction of hydrogen in communities of microorganisms coupled by products of their metabolism. As a particular example, communities of *Chlorella* or *Synechocystis* with a thermophilic hydrogenase-active microorganism were used. (From [8,10].)

product of the organism's living processes (metabolism) and processing of the carbohydrates and other compounds, i.e. substances that are already accumulated.

It is difficult to foresee whether it is possible to find or to create plants producing hydrogen efficiently. In fact, there are some doubts about such a possibility because, among other things, the hydrogen-evolving enzymes ('hydrogenases') are well known to be inhibited by even traces of oxygen. These considerations point to the high possibility of a 'short circuit' of the electron-transfer chain of the steady-state water photodecomposition process in living plants.

Much more success was achieved in accomplishing biophotolysis of water by declining to perform the whole process inside the single cell of a plant, and using indirect biophotolysis of water in communities of microorganisms coupled by the products of their metabolism.

For example, a group of micro-algae (of the *Chlorella* or *Synechocystis* type) has been tested together with thermophilic hydrogenase-active bacteria (see [8,10]). Experiments were conducted in a flow system with two bioreactors, one of which was transparent and illuminated. In the illuminated bioreactor, the suspension of microorganisms was fed with the products of dark metabolism (in this particular case these are CO_2, acetic acid and ethyl alcohol accumulated in the second bioreactor), with the micro-algae providing process (2) of traditional photosynthesis, with oxygen evolution and the simultaneous consumption of metabolic CO_2, acetic acid and alcohol for the formation of carbohydrates. In the second bioreactor, these carbohydrates were processed by thermophilic bacteria with hydrogenase activity into hydrogen, and again to CO_2, acetic acid and alcohol (see Fig. 13.6). As only the gaseous products (namely, molecular hydrogen and oxygen) of the metabolism of the community of living organisms were released from the bioreactors, the process as a whole is described by process (1) of water biophotolysis. The total productivity of such a process in steady-state conditions was significant, reaching 0.4 cm^3 H_2 cm^{-3} suspension h^{-1} in a system illuminated with light of a high intensity. One may hope that future research in this direction could make it possible to create more effective systems of practical interest.

Impressive results were obtained with attempts to chemically modify micro-algae chloroplasts [11,12]. Such modifications were made by introducing into the chloroplasts exogenous catalysts of finely dispersed platinum that interrupt the electron-transfer chain of usual photosynthesis and direct the light generated, highly energetic electrons to the process of reduction of water to hydrogen, instead of reduction of carbon-containing compounds. It was reported that such chemically modified chloroplasts show a significant activity of stoichiometric hydrogen and oxygen photoproduction and have the ability to function for some days.

The reported calculated energetic efficiency of solar energy conversion by such systems appeared to be up to *c.* 10%, which is only two to three times less than the limiting theoretical efficiency of the photosynthesizing devices. Unfortunately, chemically modified organisms are certainly not capable of self-reproduction and for this reason the systems are practically intermediate between living and dead. The practical perspective of these systems is still uncertain.

3.1.2 PHOTOBIOTECHNOLOGICAL HYDROGEN PHOTOPRODUCTION FROM SUBSTRATES CONTAINING ELECTRON-DONATING COMPOUNDS

The cornerstone of biophotolysis of water is indeed oxidation of water to molecular oxygen. This is a very complicated chemical process and in order to perform it green plants have a special 'Photosystem 2' with an attached manganese-containing catalytic center for oxygen evolution. At the same time photobiotechnological production of hydrogen from substrates containing various electron-donating compounds, both organic and inorganic (e.g. carbohydrates, amines, sulfide ions, hydrogen sulfide, etc.), is performed in the process of bacterial photosynthesis, which is more effective than photosynthesis by plants. The bacterial photosynthesizing systems are much simpler in comparison with those of plants due to the absence of 'Photosystem 2'.

Thus, photosynthetic bacteria cultivated in an electron-donating medium make it possible to create systems for hydrogen photoproduction that are reasonably efficient. A rate of hydrogen photoproduction of $\geq 1\ cm^3\ h^{-1}\ cm^{-3}$ of the reactor system has been reported [13,14]. Photosynthetic bacteria are known that coexist in culture with other microorganisms, having a high dark hydrogenase activity. This could create a basis for developing practically interesting photobiotechnological methods for hydrogen production with the use of various electron-donating substrates, especially those that are available in large quantities, such as organic wastes, natural sulfide-containing waters, etc. Of most interest should be systems in which hydrogen-producing microorganisms are fixed or immobilized on heterogeneous supports, making it possible to create a flow-through of expendable electron-donating substrates without any disturbance of the living processes of the microorganisms themselves.

The highest theoretical productivity of these systems (which also have to use no less than two light quanta to generate one H_2 molecule), according to some estimates, could be only slightly less than that shown in Fig. 13.4. Until now, no serious restrictions on the application of a somewhat higher than 'one-sun' light intensity were reported. Consequently, light concentrators might be used in photobiotechnological devices to increase the specific productivity of the working systems, but care will need to be taken to prevent photoinhibition effects, which can be a problem with low light-saturating microorganisms.

3.2 *Hydrogen production by photoelectrolysis of water*

Photoelectrochemical systems can be applied for both regenerative solar energy conversion and solar fuel production. In the first case, the electron transfer between anode and cathode is simply maintained by suitable electron-transfer agents (Fe^{2+}/Fe^{3+}, I^-/I_2). In a fuel-producing photoelectrochemical cell, the photoinduced current has to generate chemical fuels at one of the two solid

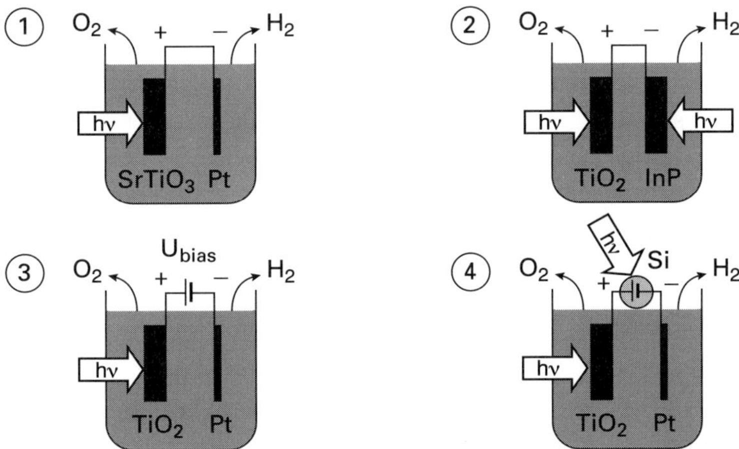

Figure 13.7. Schemes of operation of the main types of photoelectrochemical devices for water cleavage: (1) one photoelectrode (photoanode, made of TiO_2) without an external bias potential (energetic efficiency of solar energy conversion η up to 0.7%); (2) two photoelectrodes without an external bias potential (η up to 10%?); (3) one photoelectrode with an external bias potential (η up to 2–3%); (4) one photoelectrode with an external bias potential generated by a solid solar cell (η up to 7%).

electrolyte interfaces involved. This can, for example, be oxygen evolution from water at the anode and hydrogen evolution at the cathode. Many regenerative electrochemical solar cells have been designed, even though their long-term stability is not yet proven, but relatively few fuel-producing systems have resulted. The most prominent example is TiO_2, an oxide with an energy gap of *c*. 3 eV that mediates water decomposition into oxygen and hydrogen upon illumination with near-ultraviolet light and the application of a small potential. This process, which was discovered in 1972 [15], has triggered a lot of studies on water splitting; however, it turned out that TiO_2, because of the high energy gap involved, can oxidize water species in a one-electron-transfer step during illumination. This means that free or adsorbed OH radicals can be produced, which require an oxidation ability equal to the redox potential of *c*. $E_o = +2.8$ V versus normal hydrogen electrode (NHE). Even though this interesting photocatalytic property of TiO_2 has been applied for radical chemistry-mediated decomposition of chemical pollutants in water, this system did not turn out to be a helpful model system for learning to split water with visible light.

According to thermodynamics, in order to split water into molecular hydrogen and oxygen in a photoelectrochemical way, light quanta with a 'red threshold' of $\lambda < 1000$ nm (> 1.23 eV) are needed (see Table 13.1). However, the real photoelectrochemical systems are far from this ideality, and their efficiency is dependent mainly on the scheme of operation of the system (Fig. 13.3). Figure 13.7 shows the main types of photoelectrochemical systems tested for the production of hydrogen from water. In order to use visible light, it is not sufficient to generate radical species, but necessary to oxidize water via the thermodynamic redox potential of 1.23 V versus NHE, which requires a multi-electron transfer step with well-controlled redox potentials of intermediate species.

Another strategy for the photoelectrochemical production of fuels is the use of classical semiconductors (which typically corrode) covered with highly catalytic inorganic or organic layers. Many experiments dealing with water oxidation, hydrogen evolution or CO_2 fixation have been performed in this way. It was shown that catalytic reactions can indeed be stimulated by photoreactions; however, general experience was that the thinly deposited layers turned out not to be stable in the long term.

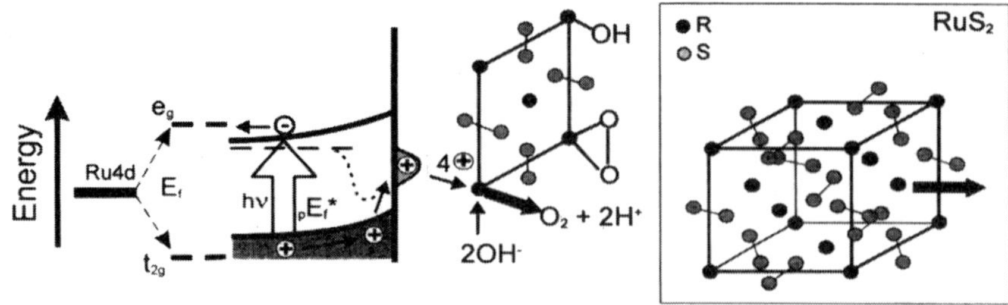

Figure 13.8. Scheme of operation of an RuS_2-based semiconductor photoanode with the valence band derived from d-states of the metal. The crystalline structure of RuS_2 is shown on the right.

Another strategy that has been developed over the last two decades is the selection of semiconductor materials that have energy bands derived from transition metal d-states (Fig. 13.8). Upon illumination, photogenerated charge carriers can thus be transferred via metal-centered interfacial reactions. Photoinduced interfacial-coordination chemical mechanisms thus become possible. If only water species are available for reaction, they will coordinate to the transition metal centers on the electrode surface during the oxidation process. If the transition metal centers involved can reach a sufficiently high oxidation state, such as with ruthenium, photoinduced oxygen evolution is possible. This has been demonstrated to occur with high quantum efficiency (70%) with the low-gap semiconductor RuS_2 (E_g = 1.3 eV). Because the energy gap of this material is much lower than the minimum energy gap (E_g = 2.1 eV) required for water splitting without supporting potential, only photoassisted water oxidation is thus possible. Nevertheless, this strategy has shown that it is necessary to produce photoinduced interfacial-coordination chemical mechanisms to accomplish difficult catalytic reactions such as the multi-electron oxidation of water.

At present, the most stable photoelectrolytic systems for simultaneous production of hydrogen and oxygen are still those with photoanodes made of semiconducting oxides, generally of titanium dioxide, strontium titanate and some others [4,6,16–20]. The main disadvantage of these photoanodes is the large (*c.* 3 eV) width of the semiconductor band gap, which requires energetic enough light quanta, usually ultraviolet (e.g. 3 eV, which corresponds to $\lambda \approx 400$ nm), to carry out the photoprocess. As a consequence, the energetic efficiency of the solar energy conversion by these systems is not high (usually no larger than 0.7% — see Fig. 13.7 and [4,6,15–20]) and the expected productivity of H_2 generation is also not very high (see Fig. 13.4). Nevertheless, the systems for photoelectrolysis of water on the basis of titanium dioxide and some other oxide photoanodes are expected to be easily and cheaply manufactured, allowing real possibilities for their practical application. There are examples of pilot tests of large-demonstration photoelectrolysis devices using TiO_2 photoanodes (with the bias potential generated by a second solid-silicon semiconductor solar cell of efficiency *c.* 14% — see Fig. 13.7, Scheme 4) that showed the possibility of obtaining up to 7 dm^3 of hydrogen per light-day from a device with a light-exposed working area of *c.* 2–3 m^2 (Fig. 13.9) [21].

Systems for water photoelectrolysis with non-oxide photoanodes, as well as photocathodes made of narrow-band-gap semiconductors, are generally able to provide a much greater solar energy conversion efficiency. Unfortunately, the chemical stability of such photoelectrodes (mostly due to their photocorrosion) needs significant improvement.

Considerable advances have been made in the development of systems having both photoanode and photocathode (usually with a narrow band gap — see Fig. 13.7 and references in [4] and earlier issues of this set of books). These systems do not need an external bias potential for opera-

Figure 13.9. A roof-top pilot device for photoelectrochemical generation of H_2 via water cleavage, erected and tested in Erevan (Armenia) at the end of the 1980s. (From [21].) The device follows Scheme 4 of Fig. 13.7. The top of the panel is covered by Fresnel lenses, concentrating solar light on modular water electrolyzers with TiO_2 photoanodes. The arrow shows the silicon solar cells, which provide a bias cathode potential.

tion but are also prone to photocorrosion, unless special precautions are taken in the electrolyte composition.

A quite promising but still unexplored field in the construction of photoelectrochemical devices for water splitting appears to be 'stack systems' or so-called 'septum cells' [22,23] (Fig. 13.10). Such devices have two or more light-exposed semiconductor photoelectrodes, as well as the systems mentioned above and shown in Fig. 13.7. However, in contrast to the systems shown in Fig. 13.7, the 'septum cells' have several compartments with different electrolytes separated by electron-conducting (usually metal) membranes, covered from one side with a thin layer of a semiconductor and from the other side with a layer of a suitable redox catalyst, e.g. that for reducing or oxidizing water. It is clear that when the device under discussion has two photosensitive electrodes, it resembles the 'Z-scheme' of green plant photosynthesis, with separate well-tuned electron chains for giant 'reaction centers' of artificial 'Photosystem 1' and 'Photosystem 2' based on semiconductor layers. Being easily manufactured and adjusted to particular conditions of illumination, etc., such systems could be of a practical interest for the future.

Sensitization of the wide-gap semiconductor photoelectrodes to longer wavelength light to increase the overall efficiency of the photoelectrochemical systems is also under way. The most promising results seem to stem from the use of very stable metallocomplex compounds of polypyridine or tetrapyrrole types (see [24]).

The knowledge that dyes may sensitize the photoactivity of inorganic semiconductors accompanied the development of photography [25]. Since the demonstration that, with chlorophyll, this

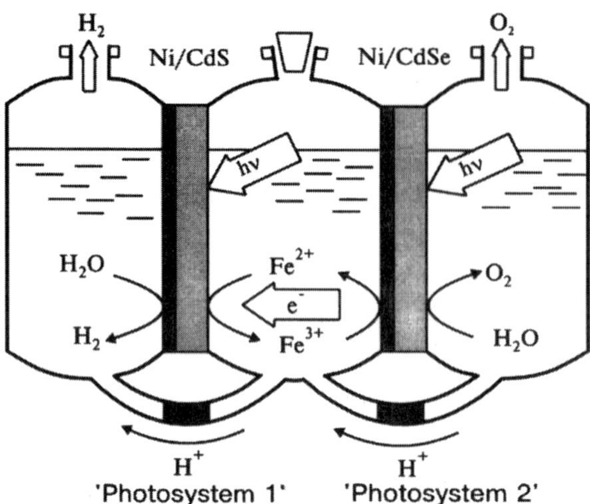

Figure 13.10. Diagram of operation of a possible 'septum cell' for water splitting. Such cells can be considered as macroscopic biomimetic analogs of photosynthesizing devices of green plants operating on the basis of the 'Z-scheme' (see Fig. 13.5a).

phenomenon may serve as a model system for photosynthesis and may be used to convert light into electrical energy, many efforts have been made to develop this mechanism [26–28]. On the basis of very rough TiO_2 surfaces based on nanostructured materials and with Ru complexes as the sensitizer, a solar energy conversion efficiency of 10% was reached [24]. Although very optimistic hopes were raised with respect to the technical applicability of such cells for solar energy conversion, it still has to be demonstrated that they remain stable for many years. Stability problems concern the sealing of the cell against evaporation of the electrolyte, decomposition of the electrolyte and the stability of the ruthenium complex itself.

3.3 *Hydrogen production by photoelectrolysis of electron-donating compounds*

In the studies of the photoelectrolysis of water, research in this field is known as 'hydrogen production with sacrificial electron donors'. It is evident that in electrolytic cells with photoelectrodes it is possible to carry out not only electrolysis of water but also processes similar to that of hydrogen production by bacterial photosynthesis. In the case of electrolysis of a mixture with electron-donating compounds, hydrogen is evolved on the cathode, as in typical water electrolysis, whereas on the anode, instead of water oxidation to oxygen, a more simple process of oxidation of an organic or inorganic electron-donating compound takes place. Usually, for the latter process a lower oxidation potential is needed on the anode than for water oxidation. This indicates the applicability of narrow-band-gap semiconductors, which convert solar energy more efficiently than wide-band-gap semiconductors. However, the problem of electrode photocorrosion control seems to be no less serious than in the case of photoelectrolysis of pure water, due to the fact that the majority of electron-donating compounds of practical interest (i.e. low cost and abundant) are usually available in the form of water solutions or suspensions. There are, for example, industrial and agricultural wastes with relatively high contents of organic compounds, suspensions of biomass or coal, etc. Indeed, in these cases one should compare the possibility of using these substrates to produce hydrogen in a 'conventional' thermochemical way, with the photoelectrolytic one, which could appear to be more efficient for customers.

Among the photoelectrochemical systems with easily corroding photoelectrodes of narrow-band-gap semiconductors, the exceptions are systems with narrow-band-gap chalcogenides (usually the sulfide type), which allow metal-centered electron transfer from their photoanodes [6,16]. Under the influence of visible light, these photoelectrodes provide highly efficient photoelectrolysis of hydrogen sulfide:

$$H_2S \rightarrow H_2 + S \quad (3)$$

Sulfide ions present in the solution serve to 'heal' the photocorroding chalcogenides.

Serious research and development in the field of hydrogen production by photoelectrolysis of electron-donating substrates might produce results that are interesting for practical applications.

3.4 *Photochemical and photocatalytic hydrogen production*

In photochemical processes a chemical reaction is initiated by light quanta, which are absorbed either by the reactive substances themselves or by special substances called photosensitizers (photocatalysts). For this reason, photochemical methods of hydrogen production include, in general, both the direct photolysis of water or other hydrogen-containing compounds (at the expense of light absorption by molecules of water or these compounds themselves) and indirect photolysis, i.e. photosensitized photolysis (or photocatalytic processes) of various hydrogen-containing or electron-donating compounds.

The theoretical limitations on the productivity of such systems in hydrogen production are nearly the same as for photoelectrochemical systems [3].

3.4.1 HYDROGEN PRODUCTION BY DIRECT PHOTOLYSIS OF WATER OR OTHER HYDROGEN-CONTAINING COMPOUNDS

It is well known that an efficient direct photolysis of water is indeed possible (see reviews [20,29]). However, unfortunately, this requires very intense ('vacuum') UV light, which is absent from the solar spectrum at ground level. Consequently, this method of hydrogen production is of no practical interest for terrestrial solar-hydrogen and is not discussed below. The same is true for systems with 'physical' photosensitizers, which absorb light and transfer the energy of their electronic excitation to water molecules.

The possibility of hydrogen production by the direct photolysis of hydrogen-containing compounds, which differ from water, is also well documented. The best-known example is reaction (3), i.e. direct photolysis of hydrogen sulfide, which occurs very efficiently both in the gas phase and in water solutions (see review [30]).

However, when analyzing the widely available hydrogen-containing compounds, which for this reason could be of practical interest for hydrogen energetics, one finds that all these compounds (at least those known to the authors of this chapter) are able, like water, to absorb only short-wavelength UV light. For this reason, direct photolysis processes using these compounds also have no prospects for terrestrial hydrogen energy.

It is worth mentioning here also a non-quantum system of direct thermolysis of water obtained directly by highly concentrated light capable of providing temperatures over 1300 K in special light-transparent chemical reactors (see Fig. 13.11 and [31–33]). The recorded energy efficiencies of such a system for the storage of solar energy appeared to be not very high and restricted to

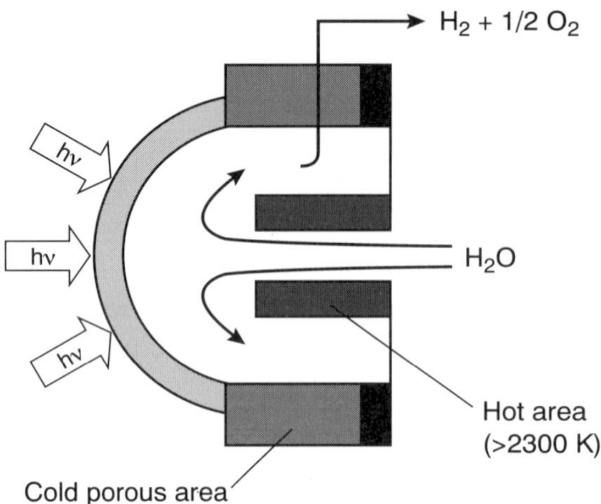

Figure 13.11. Sketch of a particular design of the reactor tested for direct thermolysis of water with the use of highly concentrated light. (From [31,32].)

several per cent. The main problem is to perform very fast cooling ('hardening') of the H_2- and O_2-containing mixture, which creates obstacles for scaling up the devices.

3.4.2 HYDROGEN PRODUCTION BY PHOTOCATALYTIC CLEAVAGE OF WATER OR BY PHOTOCATALYTIC PROCESSES WITH ELECTRON-DONATING COMPOUNDS

At present photochemical processes of hydrogen production are universally recognized to be of practical interest only when they are photocatalytic and accomplished under the influence of visible light (see [6,16,20,29,30,34]). In these processes, the photoactive quanta of visible light are absorbed by special brightly-colored photosensitizers, the electron-excited state of which participates in chemical reactions with the reagent molecules and is regenerated after each cycle (hence, these photosensitizers may be called photocatalysts, PhC) [35].

According to the nature of the photosensitive substance, molecular and semiconductor photocatalysts and photocatalytic systems are distinguished.

3.5 *Molecular photocatalytic systems*

The initial energy-storing reaction in these systems is usually the generation of the energy-saturated pair 'oxidizer D^+ – reducing agent A^-' under the influence of light on the photocatalyst molecules:

$$A + D \xrightarrow[\text{PhC}]{h\nu} A^- + D^+ \tag{4}$$

The next chemical transformations, leading in particular to hydrogen production, may be carried out without light quanta (see Fig. 13.5b and [6,16,20,29,30,34]). Thus hydrogen production is usually followed by the reduction of protons (or molecules of water) with photogenerated particles A^- over a special catalyst:

$$H^+(\text{or } H_2O) + A^- \xrightarrow[\text{cat}_{H_2}]{} \tfrac{1}{2}H_2(\text{or } + OH^-) + A$$

The photogenerated oxidizer D^+ is used to close the chain of catalytic transformation by oxidizing either water (to oxygen) or some other electron-donating substrate.

At present, essentially no-one believes that an efficient photocatalytic cleavage of water to hydrogen and oxygen in simple homogeneous photocatalytic systems is possible because of a strong possibility of a 'short circuit' of the electron-transfer chains in the systems designed for simultaneous production of hydrogen and oxygen. The basic reason for this 'short circuit' in homogeneous systems is the complexity and thus slowness of the processes of catalytic reduction and oxidation of water in comparison with the rapid recombination of the primary photogenerated and energy-saturated pair A^- and D^+

$$A^- + D^+ \rightarrow A + D \quad (5)$$

which is the reverse of reaction (4).

Thus, organized media (micelles, vesicles, micro-emulsions, colloids, cavities of zeolites, etc.) are required for molecular photocatalytic systems in order to achieve the vectorial processes needed for separation of the reactants and products to an extent that back reactions can be slowed down. Many microheterogeneous systems have been investigated and it was convincingly shown that reactions can be achieved that differ from those observed in purely homogeneous systems [6,20,34]. The difference is that charge carriers may be temporarily separated so that the light-induced interfacial mechanisms can proceed.

Much effort was made to demonstrate the decomposition of water into hydrogen and oxygen using microheterogeneous systems. In many experiments, ruthenium(II) trisbipyridyl ions were used to absorb light, but different donors and acceptors were employed and very different substrates to which the Ru(II) complex was adsorbed or complexed were used. A very controversial process was based on the four-component system, including ruthenium trisbipyridyl, methyl viologen, platinum and RuO_2 [20]. Although platinum was used for hydrogen evolution, RuO_2, which is known to be a very good catalyst for oxygen evolution from water, was intended for oxygen evolution. For many years it has been claimed that the system functions but it could never be reproduced by other groups. A major problem appears to be that the microscopic electron circuits cannot easily be established and maintained so that photon energy is efficiently converted into chemical energy.

Considerable ingenuity was used in the construction of molecular systems for photocatalysis. The main problems are still the long-term stability of such systems and the lack of efficiency. Not all projected applications of microheterogeneous photocatalytic systems appear to be practical. Joint generation of hydrogen and oxygen does not seem to be a reasonable basis for a solar-hydrogen technology because a separation of the gases would require additional energy and the oxygen–hydrogen mixture would also produce safety problems. On the other hand, light-induced generation of fertilizers, e.g. in a water tank, may be an interesting opportunity but would still have to be demonstrated.

The present consensus is that an obligatory condition for the construction of efficient molecular photocatalytic systems for water cleavage should be a rigorous spatial organization, thereby preventing the occurrence of reaction (5). At present, the most convenient and proven method to do this appears to be a structural organization of the molecular system in suspensions of lipid vesicles, which are microscopic spherical species with a bilayer membrane with a surface that divides two non-miscible water phases (see Fig. 13.12 and [6,16,34,36–39]). Over the last decade the amount of research in this field has sharply increased. Currently, vesicular systems have been developed that allow the evolution of both hydrogen and oxygen simultaneously in the same volume [36,38]; unfortunately, in these systems, the cycle of light-induced water cleavage was not closed, because hydrogen and oxygen were being evolved in separate subsystems that are not

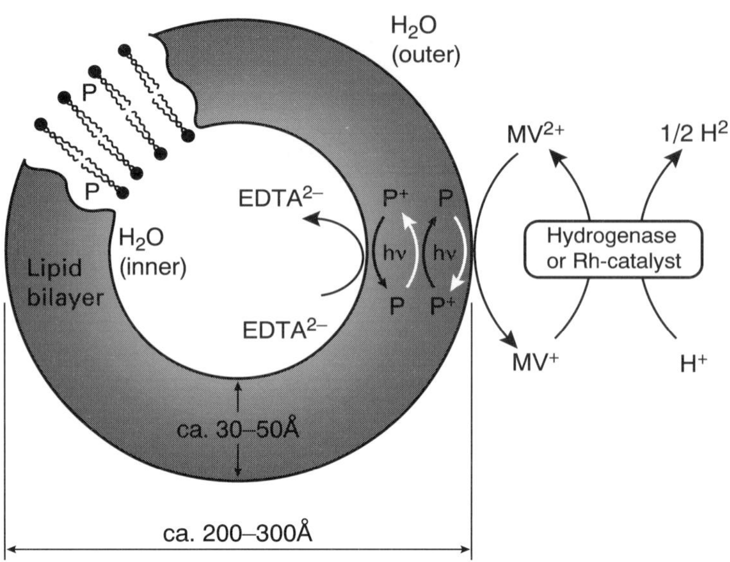

Figure 13.12. An example of photocatalytic evolution of hydrogen integrated with vectorial electron phototransfer across the membrane of a lipid vesicle. The vesicle contains the photocatalyst molecules — zinc(II)tetraphenylporphine (indicated as P) inside its membrane, water-soluble methyl viologen (MV^{2+}) as another electron relay, a catalyst for H_2 evolution (hydrogenase or finely dispersed Rh) in the outer water phase and ethylenediaminetetraacetate ($EDTA^{2-}$) as an irreversible electron donor in the inner water phase. (From [39].)

connected by an electron-transfer chain. Nevertheless, these experiments are of importance for planning a strategy of R&D on structurally organized molecular systems for the photocatalytic cleavage of water.

For instance, it has been shown that in systems with lipid vesicles, an effective mutual inhibition of the processes of hydrogen and oxygen evolution takes place. This is mainly due to the great permeability of the lipid bilayer membranes to oxygen molecules. Thus, searching for novel membranes that are non-permeable to oxygen (and hydrogen) and of molecular thickness is important.

Another possible but still not investigated approach may exist in the construction of vesicular systems in which the steps of hydrogen and oxygen evolution will be divided in time and/or in space, as occurs for example in systems with hydrogen-producing communities of living microorganisms.

Unlike molecular photocatalytic systems for water cleavage, efficient molecular systems for photocatalytic hydrogen production from solution or suspensions of electron-donating compounds are well known (usually called 'sacrificial systems'; see [6]). From the viewpoint of hydrogen energy the main disadvantage of known systems is that up to now there have been practically no attempts to immobilize such photocatalytic system components on a heterogeneous support to allow easy separation of these components from the flow of solution with the electron-donating substrate. Construction of immobilized 'integral' molecular photocatalytic systems, containing all the elements necessary for hydrogen evolution, is equivalent to the construction of functional, and possibly much more stable, analogs of reaction centers of photosynthetic bacteria or 'Photosystem 1' of plants. Successful solution of similar problems for semiconductor photocatalytic systems may guarantee construction of such a system if a serious 'customer' were available.

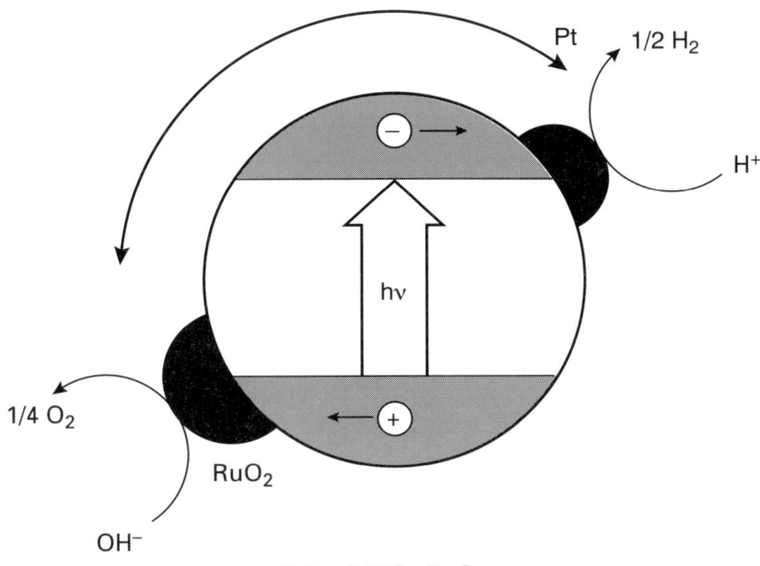

Figure 13.13. The simplest one-particle dispersed semiconductor photocatalysts widely tested for water cleavage. (From [41].) The quantum yield and thus the energy efficiency of such systems are not sufficient, probably due to a 'short circuiting' of H_2- and O_2-evolving reaction centers (shown by the two-headed arrow).

3.6 *Semiconductor photocatalytic systems*

Unlike molecular photocatalytic systems, in semiconductor systems the primary photoseparation of charges is performed just inside the semiconductor particle (e.g. TiO_2, CdS, etc.), the primary energy-saturated pair of particles being the electron, 'promoted' to the semiconductor conduction band, and the 'hole', rather than the A^- and D^+ molecules (see Figs 13.1 & 13.13 and [6,16,34,40]). Semiconductor photocatalytic systems have many similarities with photoelectrochemical systems but they differ due to the absence of a macroscopic electric circuit, as well as in the macroscopic and space-separated cathode and anode.

Photocatalytic cleavage of water to hydrogen and oxygen on dispersed semiconductors has been known for about 25 years (see [41] and reviews in [6,16,20,34]). However, no system capable of photocatalytic cleavage of water with a quantum yield over *c.* 1% under steady-state conditions has been devised. Developed systems of such a type are capable of operating mostly under the influence of near-UV light. Failure to create simple effective systems for photocatalytic cleavage of water with dispersed semiconductors is probably also a consequence of the inevitable 'short circuit' of electron-transfer chains evolving hydrogen and oxygen in the same volume with spatially non-separated reaction centers (see Fig. 13.13).

At the same time, considerable progress was achieved in developing dispersed semiconductor photocatalysts for the production of hydrogen from solutions and suspensions of electron-donating compounds. It was also demonstrated that for hydrogen production a large number of available substances, such as alcohols and carbohydrate solutions, suspensions of biomass, coal, etc., can be used (see [6,16,20,34,40]). Also of importance for hydrogen energy is the fact that photocatalytic systems were developed that efficiently produce hydrogen under the action of the entire spectrum of visible light, thus allowing efficient use of solar energy. At present, the most developed system

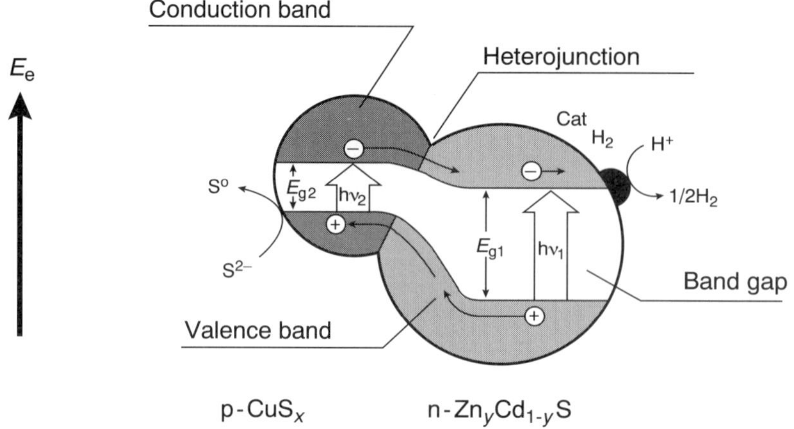

Figure 13.14. Energy diagram and scheme of photoseparation of charges and catalytic processes in a suspended particle of a semiconductor with a microheterojunction $CuS_x/Zn_yCd_{1-y}S$, which is presently one of the most efficient photocatalysts for H_2S cleavage in water solutions. E_e is the electrochemical potential of an electron; E_{g1} and E_{g2} are the widths of forbidden gaps of the semiconductor phases; and $h\nu_1$ and $h\nu_1$ are the quanta of exciting light corresponding to them. (From [3,34].)

of this type seems to be a dispersion of particles with 'microheterojunctions' formed at the point of contact of two different semiconductor phases (see Fig. 13.14 and [3,6,34,42,43]). Using this system allowed the production of hydrogen from water solutions of sulfide ions or hydrogen sulfide according to reaction (3), with a quantum yield of 50% under illumination by green and more energy-rich light and with a quantum yield of > 10% under less energy-rich light, including the red and the near-infrared.

Dispersed semiconductor photocatalysts can easily be heterogenized on a polymeric or ceramic support to create photocatalytic installations of practical interest for hydrogen generation in sunlight. Tests of a pilot device of *c.* 0.25 m^2 area based on this type of system were capable of producing up to several dm^3 of hydrogen on a clear sunny day [43]. Undoubtedly, serious demand for such a system could develop the technology to the level of commercial application.

Research on dispersed semiconductors for splitting of water into hydrogen and oxygen anticipates a breakthrough in constructing a spatially well-organized system on the basis of polymeric or some other dielectric membranes incorporating semiconductor particles that penetrate the membrane (Fig. 13.15). By generating an asymmetry in the membrane-separated solutions (e.g. by modifying asymmetrically the semiconductor particles or by varying the composition of the solutions) one could obtain a very efficient and reliable hydrogen-producing device. Until now experiments in this field were restricted mostly to the production of hydrogen from HBr or other electron-donating substrates (see [44,45]). However, no principal restrictions are foreseen in improving these systems in order to satisfy thermodynamic requirements for water cleavage.

A more elegant way to combine the advantages of photocatalysts based on dispersed semiconductors with those of membrane-structured systems seems to be the inclusion of semiconductor nanoparticles into microscopic vesicular systems with bilayer lipid membranes (Fig. 13.16). It is anticipated that semiconductor nanoparticles in such systems can serve the role of very efficient and stable integral photoreaction centers mimicking completely the spatially well-organized reaction centers based on chlorophylls in photosynthesizing organisms. The first results on such systems confirm their potential [45–49], so serious research is now obviously needed. The main problem still to be overcome in designing such systems is the elaboration of ways to embed highly

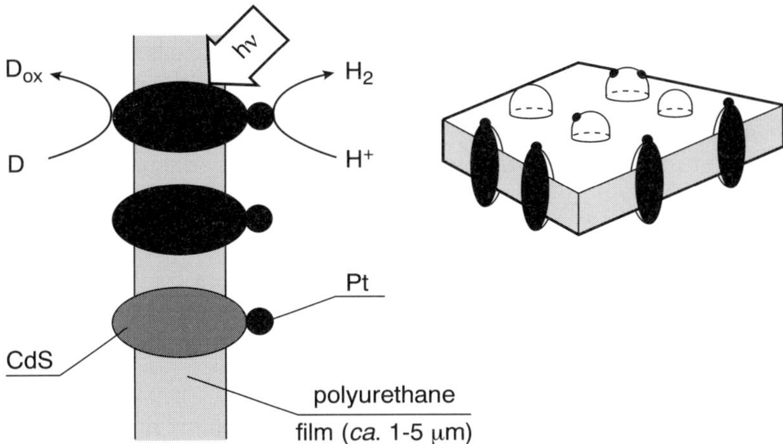

Figure 13.15. A diagrammatic view of the macroscopic 'monograin membrane' suggested in [45] as an efficient tool for photocatalytic production of hydrogen at the expense of oxidation of some electron-donating compounds. The thickness of the polymeric membranes in the experiments was a few micrometers.

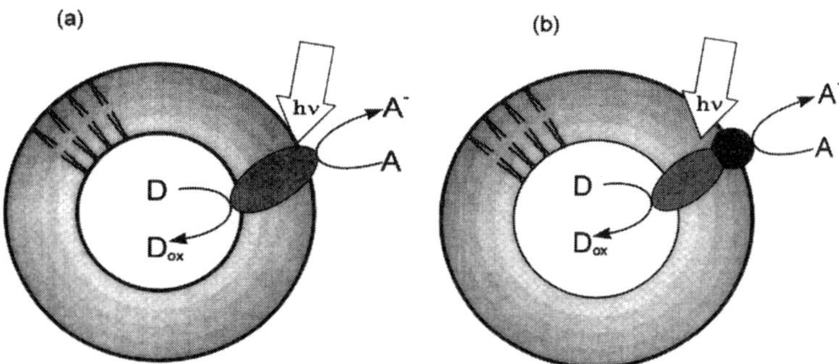

Figure 13.16. A diagrammatic view of hypothetical biomimetic devices for photocatalytic charge separation by using semiconductor nanoparticles penetrating the bilayer membrane of a lipid vesicle. Such systems combine the ideas of systems shown in Figs 13.14 and 13.15 and are now under elaboration: (a) conventional version with a simple nanoparticle of CdS, CdSe, etc.; (b) an improved version with a microheterojunction between two semiconductor nanoparticles of a different nature.

hydrophilic inorganic nanoparticles into the body of the lipid bilayer. The biggest challenge will be the long-term stability of the photocatalyzing substrates, because it is well known that most polymer layers degrade in intensive sunlight.

4 Conclusions

The theoretical limits of thermodynamics and physical laws allow a high energy conversion efficiency of conversion of light into chemical energy, 20–30% being a realistic long-term aim for solar fuel production with cascade photochemical systems. This provides a significant challenge for the present and the next generation of scientists, especially as such techniques of energy conversion do not, in principle, harm the environment because they are renewable. While many basic mechanisms have already been explored it is clear that mankind still has a long way to go towards the exploitation of such energy conversion mechanisms. The most challenging problems are the long-term stability of photoactive systems and the difficulty of developing photoactive materials that are efficient light absorbers, act as semiconducting compounds and are highly catalytically active. Nature solves the problem by using relatively unstable but highly reactive compounds, which are stabilized through self-organization. In this way, high catalytic reactivity can be coupled with long-term or medium-term stability. Because technically the problem of self-organization has not yet been dealt with, stable catalysts have to be selected, the supply of which is only limited.

Semiconductor electrochemical systems have a better medium-term possibility of succeeding, firstly because they are more efficient, and secondly because they have a better long-term stability than the molecular catalytic systems tested up to now. Because only a few materials are thermodynamically stable under illumination in contact with aqueous electrolytes, research should be aimed at kinetic stability and possibly at stabilization through intensive dynamic reactions that can result in self-organization of the reactive system. This latter possibility, which has not yet been explored, appears to be the basis for stabilization of labile membranes within biological systems. Stabilization works in such a way that a local entropy decrease is accomplished through autocatalytic processes at the expense of overall entropy production.

Because natural photosynthetic systems demonstrate that quantum chemical energy conversion is the basis for sustaining life and because physical laws promise much higher efficiencies than those generally exploited by nature, basic research should be continued systematically in order to lay the foundation for a technology based on quantum chemical energy conversion.

5 References

1 Bolton JR, Haugh AF, Ross RT. In *Photochemical Conversion and Storage of Solar Energy*. Connolly JS (ed.). New York: Academic Press, 1981; 297–339.
2 Ross RT, Hsiao T-L. *J. Appl. Phys.* 1977; **48**: 4784.
3 Parmon VN, Zamaraev KI. In Serpone N, Pelizzetti E (eds) *Photocatalysis. Fundamentals and Applications*. New York: Wiley, 1989; 565.
4 Veziroglu TN (ed.) *Hydrogen Energy Progress XI*. Frankfurt: Shön & Wetzel, 1996.
5 Govindjee (ed.) *Photosynthesis*. New York: Academic Press, 1982.
6 Serpone N, Pelizzetti E (eds) *Photocatalysis. Fundamentals and Applications*. New York: Wiley, 1989.
7 Weaver PF, Lien S, Seibert M. *Sol. Energy* 1980; **24**: 3. (David: Please, add more new reference!?)
8 Varfolomeev SD. *Energy Conversion by Biocatalytic Systems*. Moscow: Moscow State University, 1981 (in Russian).
9 Rao KK, Hall DO. *J. Mar. Biotechnol.* 1996; **4**: 10.
10 Varfolomeev SD, Medman DY, Pinchukova EE, Toaj CD. *Dokl. Akad. Nauk SSSR* 1985; **284**: 1275.
11 Greenbaum E. In Norris JR, Meisel D (eds) *Photochemical Energy Conversion*. New York: Elsevier, 1989; 184.
12 Greenbaum E. *J. Phys. Chem.* 1990; **94**: 6151.
13 Kondrat'eva EN, Gogotov IN. *Molecular Hydrogen in Metabolism of Microorganisms*. Moscow: Nauka, 1981 (in Russian).
14 Salih FM. *Int. J. Hydrogen Energy* 1989; **14**: 661.
15 Fujishima A, Honda K. *Nature* 1972; **238**: 37.
16 Norris JR, Meisel D (eds) *Photochemical Energy Conversion*. New York: Elsevier, 1989.
17 Pleskov YV. *Solar Energy Conversion. A Photoelectrochemical Approach*. Berlin: Springer, 1990.
18 Arutyunyan VM. In Zamaraev KI, Parmon VN (eds) *Photocatalytic Conversion of Solar Energy*. Novosibirsk: Nauka, 1985; 74 and 1991; 228 (in Russian).
19 Pelizzetti E, Schiavello M (eds) *Photochemical Conversion and Storage of Solar Energy*. Dordrecht: Kluwer, 1991.
20 Grätzel M (ed.) *Energy Resources through Photochemistry and Catalysis*. New York: Academic Press, 1983.

21 Vardapetyan TA, Vartanyan AV. In Zamaraev KI, Parmon VN (eds) *Photocatalytic Conversion of Solar Energy*. Novosibirsk: Nauka, 1991; 294 (in Russian).
22 Tien HT, Chen JW. *Int. J. Hydrogen Energy* 1990; **15**: 563; *Sol. Energy* 1992; **48**: 199.
23 Bard AJ, Mallouk TE. *J. Phys. Chem.* 1993; **97**: 7127.
24 O'Ragan, Grätzel M. *Nature* 1991; **353**: 373.
25 Meier H. *Spectral Sensitization*. London: The Focal Press, 1967.
26 Tributsch H, Calvin M. *Photochem. Photobiol.* 1971; **14**: 95.
27 Tributsch H. *Photochem. Photobiol.* 1972; **16**: 261.
28 Tsubomura H, Matsumura M, Nomura Y, Amamya T. *Nature* 1976; **261**: 402.
29 Parmon VN. In Zamaraev KI, Parmon VN (eds) *Photocatalytic Conversion of Solar Energy*, Part 2. Novosibirsk: Nauka, 1985; 6 (in Russian).
30 Parmon VN. *Zh. Obshch. Khim.* 1992; **62**: 1703 (in Russian).
31 Lede J, Lapicque F, Villermaux J. *Int. J. Hydrogen Energy* 1983; **7**: 939; 1982; **8**: 675; 1987; **12**: 3.
32 Yalcin S. *Int. J. Hydrogen Energy* 1989; **14**: 551.
33 Kogan A. *Int. J. Hydrogen Energy* 1998; **23**: 89.
34 Zamaraev KI. In Hightower JW, Delgass WN, Iglesia E, Bell AT (eds) *Studies in Surface Science and Catalysis*, Vol. 101, Part A. Amsterdam: Elsevier, 1996; 35.
35 Parmon VN. *Catal. Today* 1997; **39**: 137.
36 Shafirovich VY, Shilov AE. In Norris JR, Meisel D (eds) *Photochemical Energy Conversion*. New York: Elsevier, 1989; 173.
37 Parmon VN, Zamaraev KI. In Norris JR, Meisel D (eds) *Photochemical Energy Conversion*. New York: Elsevier, 1989; 316.
38 Shafirovich VY, Lymar SV, Parmon VN, Shilov AE, Zamaraev KI. In Zamaraev KI, Parmon VN (eds) *Photocatalytic Conversion of Solar Energy*. Novosibirsk: Nauka, 1991; 18 (in Russian).
39 Lymar SV, Parmon VN, Zamaraev KI. *Topics in Current Chemistry*, Vol. 159. Springer: Berlin, 1991; 1.
40 Gruzdkov YuA, Savinov EN, Parmon VN. In Zamaraev KI, Parmon VN (eds) *Photocatalytic Conversion of Solar Energy*. Novosibirsk: Nauka, 1991; 138 (in Russian).
41 Bulatov AV, Khidekel ML. *Izv. Akad. Nauk SSSR Ser. Khim.* 1976; 1902 (in Russian).
42 Gruzdkov YA, Savinov EN, Parmon VN. *Int. J. Hydrogen Energy* 1989; **14**: 1.
43 Gruzdkov YA, Savinov EN, Makarshin LL, Parmon VN. In Zamaraev KI, Parmon VN (eds) *Photocatalytic Conversion of Solar Energy*. Novosibirsk: Nauka, 1991; 186 (in Russian).
44 Meissner D, Memming R, Kastening B. *Chem. Phys. Lett.* 1983; **96**: 34.
45 Fendler JH. *Chem. Rev.* 1987; **87**: 877.
46 Tian Y, Wu Ch, Fendler JH. *J. Phys. Chem.* 1994; **98**: 4913.
47 Khramov MI, Parmon VN. *J. Photochem. Photobiol. A* 1993; **71**: 279.
48 Igumenova TI, Vasil'tsova OV, Parmon VN. *J. Photochem. Photobiol. A* 1996; **94**: 205.
49 Vassil'tsova OV, Parmon VN. *J. Photochem Photobiol. A* 1996; (in press).

14 General Conclusions

V.N. PARMON

Boreskov Institute of Catalysis, Siberian Branch of the Russian Academy of Sciences, Prospekt Akademika Lavrentieva 5, Novosibirsk 630090, Russia

This book gives only a cursory glance at the possible part that chemistry may play in resolving one of the most important problems of our future — the problem of a sustainable energy base for society. However, we have tried to demonstrate that the role of chemistry in the energy industry of the future should be numerous and decisive.

What will be the particular path taken by chemists in this area? A path that is actually demanded by our lifestyle.

In the near future, many very important features of the strategy of large-scale energy will be dictated mainly by resolving a crucial question: do large anthropogenic emissions of CO_2 actually result in an irreversible 'greenhousing' of our planet. The answer to this question will generate attempts either to continue mainly with the use of the presently available carbon-containing fossil fuels, such as low-quality coals, etc., or to change dramatically the nature of the main 'integrating' chemical energy carriers.

If CO_2 turns out not to be very detrimental for the planet there will be no problem, at least for the coming century, in continuing to use the conventional and most convenient hydrocarbon energy carriers with only a stepwise and smooth shift in the basis of the large-scale energy industry, as well as transport, etc., to more clean but also more expensive energy carriers such as hydrogen or raw materials such as renewable biomass. The most important role of chemistry in such a scenario will be, first of all, to ensure the environmental purity of the energy and to provide more efficient utilization of the available chemical energy carriers. Also, new chemical technologies will be needed to produce the fuels from new available raw materials such as plant biomass, etc.

If the enhanced emissions of CO_2 appears to be decisive in the undesirable change of the Earth's climate, then the role of chemistry will be even more important because a much faster shift to the newest technologies based on inexhaustible and CO_2-pollution-free energy resources will be needed. In this case a wide implementation of non-traditional approaches of chemistry to energy, e.g. thermochemical cycles, utilization or transformation of energy released by fission or fusion nuclear power plants, and numerous solar energy conversion devices could be expected.

Even the brief overview presented in this book shows the great potential for chemical technologies, particularly catalysis application, in the various branches of the energy industry of the future. In many cases catalytic processes used for converting one type of energy into another via heating are well developed; some of them have been used in the chemical industry for a long time. In particular, these reactions include complete oxidation, reforming of light hydrocarbons and selective dehydrogenation of organic compounds.

When direct conversion of noble types of energy is needed, e.g. of light, radiation and electric energy, then new non-traditional catalytic processes should be developed. Although such developments are only at the initial stages, some quite encouraging results have been obtained already. This opens up interesting and promising fields of joint activity for researchers in the areas of chemistry, catalysis and energy production. An additional stimulus to this will be the gradual, though perhaps inevitable, transition to the 'non-traditional' energy industry, as well as to the necessity for the energy industry of the future to be safer ecologically than the energy industry of today.

For example, advances made so far in the thermocatalytic conversion of solar energy suggest that this method may, perhaps, become important for energy in the 21st century. Whether or not this actually happens will depend primarily on economic and ecological factors:

1 At present, all existing methods of energy production via conversion of solar energy are far more expensive than traditional methods such as combustion of fossil fuels at electricity generating plants and nuclear fission. In this situation the shift towards a wide application of solar energy is expected to occur gradually, following the trends of solar energy becoming cheaper with every new scientific and technological breakthrough and traditional fossil energy resources becoming more expensive due to their gradual exhaustion.

2 Much will depend on how dangerous the global consequences prove to be of the Earth's pollution by man-made extra heat, chemicals, etc. associated with traditional fossil types of energy. It should be noted that nuclear fusion energy, which sooner or later is expected to be developed, is also likely to pollute the Earth with extra heat. If such pollution proves to be intolerable, the development of solar energy, including that based on the primary growth of biomass via natural or improved photosynthesis, which produces no extra heating of the Earth and in other respects seems to be ecologically more sound than any other kind of energy, may obtain high priority.

The decisive factors for creating the real driving forces of dramatic changes in the structure of the large-scale world energy will no doubt, be:

1 the rate of exhaustion of traditional resources of fossil organic energy carriers;

2 the reality of the 'greenhouse' effect due to CO_2 emissions;

3 the reality of overheating of the Earth due to anthropogenic emission of heat.

In any situation, chemistry will be able to create the conditions for a normal-style life for billions of the Earth's population. However, in order to ensure this, the joint and continuing efforts of both chemists and other scientific and social communities of the world are needed.

The history of civilization provides a challenge for chemists that the chemists have to win.

Index

Page numbers in *italics* refer to tables, those in **bold** refer to tables.